TURBULENCE IN THE OCEAN

ENVIRONMENTAL FLUID MECHANICS

A. S. MONIN AND R. V. OZMIDOV

P. P. Shirshov Institute of Oceanology,
Academy of Sciences, Moscow, U.S.S.R.

Turbulence in the Ocean

Translation edited by H. Tennekes

Royal Netherlands Meteorological Institute, de Bilt

and Free University, Amsterdam

D. Reidel Publishing Company

A MEMBER OF THE KLUWER ACADEMIC PUBLISHERS GROUP 

Dordrecht / Boston / Lancaster

Library of Congress Cataloging in Publication Data

Monin, A. S. (Andreĭ Sergeevich), 1921–
Turbulence in the ocean.

(Environmental fluid mechanics)
Translation of: Okeanskaîa turbulentnost′.
Bibliography: p.
Includes index.
1. Turbulence. 2. Hydrodynamics. 3. Ocean currents. I. Ozmidov, R. V. (Rostislav Vsevolodovich) II. Tennekes, H. (Hendrik) III. Title. IV. Series.
GC203.M6613 1985 551.47′01 85-8215
ISBN 90-277-1735-4

Published by D. Reidel Publishing Company,
P.O. Box 17, 3300 AA Dordrecht, Holland

Sold and distributed in the U.S.A. and Canada
by Kluwer Academic Publishers
190 Old Derby Street, Hingham, MA 02043, U.S.A.

In all other countries, sold and distributed
by Kluwer Academic Publishers Group,
P.O. Box 322, 3300 AH Dordrecht, Holland

Originally published in Russian by Gidrometeoizdat under the title
ОКЕАНСКАЯ ТУРБУЛЕНТНОСТЬ

Translated by L. J. Usina and G. Z. Ribina

Printed in The Netherlands

Table of Contents

DIASPORA

Though your state defies description,
the shining armies of my will
have gone out to count you:

unintelligible mumbling
finally will coincide
with our encounter
deep inside language.

Then your body emerges from my sums
because I call back
all molecules
from their dispersion.
All.

Gerrit Achterberg, 1905–1962. Translated from the Dutch into English by H. Tennekes. Taken from Gerrit Achterberg, *Stof* (Matter), The Hague, 1946, translated into French as *Matière*, Montpellier, 1952; from the French into Arabian as *Hay̅ulā*, Damascus, 1962. See also *Verzamelde gedichten* (Collected Poems), Querido, Amsterdam, 1963.

Preface to the English Edition

Four years have elapsed since the preparation of the original Russian version of this book. This is a long time when dealing with such actively expanding fields of oceanography as research into small-scale structures and the investigation of hydrophysical processes. Over this period new quick-response devices have been developed and successfully used for measurements taken in various ocean areas. Improvements in high-frequency meters used to measure hydrophysical parameters has enabled workers to obtain more accurate absolute values of the fluctuations measured by such devices.

In view of this scientific progress, some of the ideas presented in this book now require additional explanation. Great care should be used in dealing with the absolute fluctuation values of hydrophysical fields, since the methods used for the determination of the accuracy of the high-frequency measuring devices have been imperfect in the past. Nevertheless, it would appear that the results of the investigations summarized in this book have not lost their importance, and that the established laws governing small-scale processes in the ocean are of a sufficiently universal nature and, as such, have not been shattered with the qualitative and quantitative advances in devices used for measurements taken in oceans. The authors feel that their work is of interest to English-speaking readers.

The appearance of the English translation of the book is, to a very large extent, due to the tremendous amount of editing work brilliantly done by Prof. H. Tennekes. The high professional qualifications of this editor, his devotion to his chosen branch of knowledge, and his friendly attitude towards foreign colleagues have contributed to a wonderful piece of work and have led to a book which is not lacking in any respect in comparison with the original Russian edition. The authors are happy to take this opportunity to express their sincere gratitude to Prof. H. Tennekes, and to express their hope for useful collaboration with him in the future. They would also like to thank Dr. V. B. Kouznetsov and Mrs. A. V. Ozmidova of the P. P. Shirshov Institute of Oceanology for their help in editing the English translation.

P. P. Shirshov Institute of Oceanology,
Moscow,
U.S.S.R.
February, 1985

ANDREI MONIN
ROSTISLAV OZMIDOV

Editor's Preface

Working on a translation is a trip into a different state of consciousness. It is not unlike writing poetry. Translator and editor both need some sensitivity for the structure of language, for its patterns of rhythm and reference, for its myths and images. As a Dutchman editing the English translation of a Russian monograph, I could not help getting involved in the structure of the assembly languages that human brains in different cultures use in the process of making sentences. I cherish the frequent moments of meditation that I encountered on my trip through this book.

Editorial work aims at effective communication. The editor of a translation has the opportunity to exercise his skills at detecting and removing the stumbling blocks that were missed – or perhaps created – by author and translator. To the reader, a perfect editorial job is invisible. I do not claim that I have come close to this ultimate goal, but what better exercise in consciousness raising could I want?

Very visible in this book are the erudition and experience of its two authors. In Chapter 1, Monin presents a concise and lucid introduction to the theory of turbulence in stratified flows. The chapter includes a fairly detailed account of recent developments in the theory of the chaotic behavior of nonlinear systems. This is the first turbulence book for a geophysical audience in which bifurcations and strange attractors receive as much attention as spectra and dissipation rates.

In more than one respect, Chapter 2 is the centerpiece of this book. It clearly bears Ozmidov's mark; one can almost smell the atmosphere on board one of the Russian research vessels. Ozmidov presents a thorough and exhaustive review of sensors, instruments, experimental methods and data processing techniques. All of the turbulence data obtained on Russian ocean expeditions are discussed, with emphasis on the spectra and dissipation rates of kinetic energy, temperature variance and salinity fluctuations. Ozmidov also provides thoughtful judgements of the influence of local background conditions on the structure of the data obtained. This chapter contains the kind of information that young oceanographers need to absorb fully before they design their first experiment or embark on their first cruise.

Chapter 3 again demonstrates Monin's skillful hand. Like a surgeon he cuts to the heart of the matter, giving ample references to other disciplines within the realm of physics as he proceeds. In this chapter the subject is two-dimensional or geostrophic turbulence. Synoptic eddies in the ocean play a crucial role in the global climate system; it is altogether fitting that this monograph concludes with a chapter that provides a link between small-scale turbulence and the planetary scales of ocean circulation. The attention of turbulence researchers in oceanography and meteorology is evolving toward larger scales; a chapter such as this hopefully accelerates that process.

I have enjoyed the cooperation of Mr. J. van der Lingen, the librarian of the Royal Netherlands Meteorological Institute, who checked the bibliography, of Mrs. M. L. Collet, my secretary, who assisted with proofreading and other chores, and of Professor A. S. Monin, who resolved a number of questions that had me baffled. I trust that this monograph will find a receptive audience throughout the Western community of oceanographers.

De Bilt,
December, 1983

H. TENNEKES

Preface

There is at present a keen interest in ocean turbulence research. This has been kindled, first of all, by the paramount importance of turbulence in the formation of hydrological fields in the ocean. Due to turbulent mixing in the tropics, the ocean accumulates a substantial amount of heat which is then transferred by sea currents towards the temperate and Arctic zones. This indicates that the ocean is a major factor governing climate. If it was not for the intensive turbulent motion of the water, the resources of biogenic material in the upper photosynthesizing zone of the ocean and those of oxygen in the abyssal layers would soon be exhausted. The ocean would then change from being a vast reservoir of organic life into a lifeless desert. Of particular importance are turbulence studies concerned with the urgent problem of ocean pollution control. Because of the presence of turbulence, the ocean can 'process' a certain amount of foreign matter since turbulent diffusion fairly quickly reduces the concentration of contaminants. However, the maximum tolerable amounts of polluting substances and the size or lifetime of zones with high contaminant concentrations can be predicted only from the turbulence properties that prevail in different hydrometeorological situations.

In the first stage of turbulence research in the ocean, which started approximately three-quarters of a century ago, special attention was paid to estimates of the turbulent exchange coefficients. These are certain fitting parameters that ensure agreement between observed and calculated average fields of temperature, salinity and velocity. Studies of the structure of turbulence and of the characteristics of hydrological field fluctuations started much later. It was not until the late thirties that V. B. Shtokman proposed an observational technique that took into account the turbulent character of the flow field at sea. In lieu of single measurements at widely spaced points, he proposed long-term (or repeated) observations at one or several adjacent sites in an oceanic region (or 'polygon'). Processing of the data obtained on a polygon required statistical methods and a model of random fields. The first, comparatively modest, observations made in 1956 in the Black Sea provided interesting information on the large-scale structure of turbulence and the statistical characteristics of the large-scale components of the velocity field in the sea. Further field observations, including the recent large-scale Soviet–American investigations in the 'POLYMODE' project, carried out in the Atlantic Ocean, have furnished a wealth of information and have led to a fundamentally new understanding of the ocean. The universally accepted notion of the ocean as a nearly stationary system with a pattern of steady-scale gyres had to be replaced by a new conception, which admits that most of the energy in the motion of the ocean is contained in vortices having various sizes and lifetimes, rather than in the average circulation.

Studies of small-scale turbulence, which is ultimately responsible for the kinetic energy dissipation in the ocean, were initiated later, since measurements of this turbulence required sensitive, low-inertia devices which were not designed until the mid-fifties. At present, a great variety of devices are used to collect all kinds of data concerning small-scale turbulence fluctuations of the hydrophysical fields in various hydrometeorological conditions. The data obtained have made it possible to state some general features of small-scale turbulence and to demonstrate that it is not universal, in general because of local energy sources that operate at low Reynolds numbers. These generation mechanisms result in the interesting phenomenon of intermittency of turbulent and non-turbulent zones, which is of critical importance for life in the ocean.

Some specific features of ocean turbulence are associated with strong density stratification in the ocean. Therefore, internal waves are of great importance. When overturning, these seem to generate most of the small-scale turbulence in the bulk of the ocean. The resulting turbulent spots collapse in a stratified medium and become pancake-shaped. The turbulence intensity in each of the spots reduces in time, and when the spots vanish they leave traces in the form of steps and internal layers of mixed fluid in the vertical profiles of the hydrological characteristics. These fine-structure features in the profiles are then subject to erosion by molecular effects, the situation becoming still more complex because of new episodes of turbulence generation. A better understanding of the close relationship between these phenomena – small-scale turbulence, internal waves, and the fine structure of the hydrophysical fields – has resulted in a new strategy of measurements in the ocean. Complex new measurement systems that can simultaneously detect high-frequency turbulence fluctuations, and elements of both internal waves and of fine-scale structures, have made it possible to conduct studies of the causes and effects of the above phenomena and to establish the basic patterns of their relationship.

As information on ocean turbulence was being gained, there appeared the first reviews on various aspects of the problem (Benilov, 1969; Bowden, 1964, 1965; Monin, 1970a, 1973a, 1977; Ozmidov, 1961, 1965a, 1967, 1969, 1977, 1978b; Defant, 1954). Some communications concerning ocean turbulence were made at the Second International Congress on Oceanology (Monin *et al.*, 1966), the Joint Oceanographic Assembly in Tokyo (1971), and at the Sixteenth General Assembly of the International Geophysical and Geodesic Association (see Brekhovskikh *et al.*, 1976). The subject was further discussed at the First and Second Symposia on Ocean Turbulence in Vancouver and Liège, respectively (see Monin, 1969; Ozmidov and Fedorov, 1979) and at the First All-Union Symposium on Small-Scale Turbulence Research (Ozmidov, 1975). The first monograph on the subject was written by Ozmidov (1968), who paid particular attention to large-scale turbulence and turbulent diffusion of contaminants in the ocean. These had been fairly well studied by that time. Some aspects of ocean turbulence were also discussed in the monographs by Phillips (1967) and Csanady (1973), in the book *Oceanic Acoustics* edited by Brekhovskikh (1974), in *Variability of the World Ocean* by Monin *et al.* (1974), in *Investigations of Ocean Turbulence* and *Investigations of the Variability of Hydrophysical Fields in the Ocean*, edited by Ozmidov (1973, 1974). A brief but consistent review of theoretical and experimental results of studies on turbulent processes in the ocean can also be found in the chapter by Monin and Ozmidov (1978) in the many-volume publication *Oceanology* (see Volume I: *Hydrophysics of the Ocean*, Chapter IV: 'Turbulence in the Ocean'). The information available on small-scale turbulence was reviewed in detail by Ozmidov (1978) in his paper 'Turbulence in the Upper Ocean'.

However, to obtain contemporary information concerning instrumentation, measurement techniques, data processing methods, and the characteristics of turbulence of various scales in various hydrophysical fields, and also the theoretical aspects, one must consult a large number of publications which often report contradicting results. The aim of the present monograph is to meet the need for a consistent and critical statement of the problem.

The book consists of three chapters. Chapter I is devoted to the theory of turbulence in stratified flows. The notion of turbulence and the basic equations of turbulence dynamics are given, turbulence generation mechanisms are discussed, and the effects of ocean stratification on turbulence are analyzed. Also, current knowledge of the spectral structure of turbulent fields and of the influence of turbulent energy dissipation rate fluctuations on the spectral structure of the various fields are reviewed.

In Chapter 2, the available data on small-scale turbulence in the fields of velocity, temperature and electrical conductivity are summarized, information on the main statistical parameters of turbulence fluctuations (mean-square values, correlations and spectral functions) is presented, and their dependence on the hydrological conditions in the local background is analyzed. Much attention is paid here to the analysis of dissipation rates of turbulent energy and temperature inhomogeneities and to the characteristics of the intermittency of turbulent fields in the ocean. Possible approaches to the climatology of small-scale turbulence are discussed, which can pave the way for predicting its parameters by mean hydrometeorological conditions on a polygon.

Chapter 3 is devoted to large-scale quasi-two-dimensional turbulence. This turbulence is associated with a very interesting phenomenon, namely that of 'negative viscosity', which occurs when the energy of irregular turbulent disturbances can be transferred to a less chaotic mean flow. This phenomenon, first discovered in the atmosphere, is discussed in Section 12. This is followed by a section devoted to the general theory of two-dimensional turbulence. Finally, Section 14 summarizes the available data concerning large-scale oceanic turbulence spectra and, in particular, gives information on the synoptic vortices that were discovered in the ocean recently.

This monograph is based on published papers as well as on reports of expeditions devoted to oceanic turbulence research. Some sections of the book are written on the basis of papers published in cooperation with investigators from the Shirshov Institute of Oceanology of the U.S.S.R. Academy of Sciences (V. S. Beliayev, I. D. Lozovatsky, M. M. Liubimtsev, M. L. Pyzhevich, N. N. Korchashkin, and others). Sections 1–6 and 13 were written by A. S. Monin, the rest of the book was written by R. V. Ozmidov. The authors are very grateful to the workers of the Laboratory of Marine Turbulence of the Institute of Oceanology for their assistance in collecting and preparing the material for this book.

CHAPTER I

Theory of Turbulence in Stratified Flows

1. DEFINITION OF TURBULENCE

Turbulence is a phenomenon observed in a large number of rotational flows in liquids and gases, both in nature and in technical devices. The thermodynamic and hydrodynamic variables of these flows (velocity vector, temperature, pressure, contaminant concentration, density, sound velocity, electrical conductivity, refractive index, etc.) experience chaotic fluctuations, which are induced by numerous vortices of various dimensions. These variables therefore change randomly in space and time. The Fourier components in wave number space that correspond to the spatial distributions of these variables occur over a wide range of frequencies (there are no one-valued dispersion relations). Also, the phase shifts between fluctuations in the variables at fixed points in space change randomly with the frequency of these fluctuations.

Thus, turbulent flow is rotational flow, characterized by a great number of excited degrees of freedom and by random distributions of dispersion relations and phase shifts.

Accordingly, the basic feature of turbulence is the random character of the changes in space and time of the hydrodynamic characteristics. The term turbulence, however, should not be used to refer to all flows of this type; sometimes it is necessary to distinguish turbulent flows in liquids and gases from other random motions that are characterized by a certain degree of order. For instance, the waves induced in liquids by particles shifting from equilibrium result from certain restoring forces: the pressure force in acoustic oscillations of compressed liquids, surface tension in capillary waves on free surfaces, the buoyancy force in internal gravitational waves of stratified liquids, the rotational part of the vertical Coriolis force in a rotating spherical liquid layer which causes meridian particle shifts, etc.

The superposition of a great number of waves of various types, with different wavenumber vectors and random amplitudes and phases, can result in a flow which changes irregularly in space and time. In many cases such a flow can be distinguished from turbulence by the properties of its elementary wave components: by a specific (e.g., longitudinal or transverse) orientation of particle shifts relative to the direction of the wavenumber vector, or by the fact that the phase shift between the fluctuations caused by an elementary wave at a fixed point in space are unambiguously determined by the wavenumber vector (the so-called dispersion relation).

The vorticity of the flow is of paramount importance in turbulence mechanics. It gives rise to a cascade process, in which small vortices are generated by larger ones (as in the case of the hydrodynamic instability of large vortices). This cascade, which occurs over the entire scale spectrum, leads to the transfer of kinetic energy toward smaller scales. Therefore, turbulence is defined here as an ensemble of random fluctuations of thermodynamic characteristics in rotational flows. Thus, it is distinguished from any other random non-rotational or potential flow and hence from any waves in an ideal fluid induced by potential forces. This includes all linear acoustic, and surface waves, as well as all nonlinear potential surface waves.

Some knowledge of the origin of turbulence may prove to be useful for an understanding of its nature. Let us introduce the concept of degrees of freedom of a fluid flow by expanding the flow into elementary components, whose states are characterized by a few parameters. The sum of the energies is equal to the energy of the flow as a whole. Mathematically this amounts to expanding the velocity field throughout the volume occupied by the fluid into an appropriate orthogonal set of functions of points in space. The coefficients of this expansion are generalized flow coordinates. The number of time-dependent coordinates is the number of degrees of freedom of the flow. An instantaneous state of the flow will be determined by a set of values of all generalized coordinates. This corresponds to a point in a certain multi-dimensional space, referred to as a phase space. The process of flow evolution can be presented in the phase space by a certain line, i.e., a phase-space trajectory, consisting of a single point for a stationary flow and forming a closed line (a cycle) for a periodic one.

Let us now consider the process of the generation of turbulence in cases where a stationary flow, $\mathbf{u}_0(\mathbf{x})$, loses its stability with respect to disturbances. The velocity field corresponding to an infinitesimal distrubance can be found as the solution of a set of linearized equations, which has the form

$$\mathbf{u}'(\mathbf{x}, t) = A(t)\mathbf{f}_0(\mathbf{x}); \qquad A(t) = e^{\lambda t}, \quad \lambda = \gamma + i\omega_1. \tag{1.1}$$

In this case, when the Reynolds number

$$\mathrm{Re} = \frac{LU}{\nu} \tag{1.2}$$

is small (L and U are length and velocity scales typical of the laminar flow analyzed, ν is the kinematic coefficient of molecular viscosity, or kinematic viscosity in short), all eigenvalues λ of the linearized equations have negative real parts ($\gamma < 0$), so that any weak disturbance (1.1) is damped in time. The stationary flow then is stable with respect to weak disturbances. However, the real parts of some eigenvalues increase with Re, and there is always a critical value of the Reynolds number $\mathrm{Re}_{1\,\mathrm{cr}}$, at which an eigenvalue $\lambda(\mathrm{Re})$ crosses the imaginary axis in the complex λ-plane for the first time, i.e., $\lambda(\mathrm{Re}_{1\,\mathrm{cr}}) = 0$. The corresponding disturbance (1.1) will neither increase nor decrease in time, i.e., it will be neutral. At $\mathrm{Re} > \mathrm{Re}_{1\,\mathrm{cr}}$ there exist eigenvalues λ with positive real parts, $\gamma > 0$. The disturbances (1.1) then increase in time so that the stationary flow under discussion will be unstable with respect to weak disturbances.

According to the Hopf bifurcation theorem (Hopf, 1942), there is a one-parametric set of closed phase trajectories of a flow at Re values in a certain vicinity of $\mathrm{Re}_{1\,\mathrm{cr}}$. Consider first the case of 'standard bifurcation' when a set of closed phase trajectories is observed at $\mathrm{Re} > \mathrm{Re}_{1\,\mathrm{cr}}$. In this case, these are limit cycles which correspond to flows that are periodic in time. The transition of the unstable weak disturbance (1.1) to a periodic flow was described by Landau (1944) and Landau and Lifshitz (1953). As long as the disturbance (1.1) is weak, its amplitude $A(t)$ obeys the linear equation

$$\frac{d|A|^2}{dt} = 2\gamma|A|^2, \tag{1.3}$$

but at finite values of $|A|$ the right-hand side of (1.3) must contain further terms of its expansion in powers of A and A^* (where the asterisk denotes a complex conjugate). In that case, the high-frequency fluctuations in (1.1) (with frequencies $|\omega_1| \gg \gamma$) can be removed by averaging them over a period τ such that $2\pi/|\omega_1| \ll \tau \ll \gamma^{-1}$. The third-power terms then vanish, while the fourth-power terms, proportional to $|A|^4$, remain. Hence, instead of (1.3) we have the Landau expansion

$$\frac{d|A|^2}{dt} = 2\gamma|A|^2 - \delta|A|^4. \tag{1.4}$$

When $\delta > 0$, the solution of Eq. (1.4) is

$$|A(t)| = \frac{A_0^2 A_\infty^2}{A_0^2 + (A_\infty^2 - A_0^2)e^{-2\gamma t}}; \quad A_\infty = \left(\frac{2\gamma}{\delta}\right)^{1/2}, \tag{1.5}$$

so that, at a small initial value A_0, the amplitude $|A(t)|$ first increases exponentially (as $A_0 e^{\gamma t}$ does, according to linear theory), but then the rate of increase becomes smaller. At $t \to \infty$, the amplitude tends to a finite value A_∞, which is independent of A_0 and proportional to $(\mathrm{Re} - \mathrm{Re}_{1\,\mathrm{cr}})^{1/2}$ at small $\mathrm{Re} - \mathrm{Re}_{1\,\mathrm{cr}}$, since $\gamma \simeq \mathrm{Re} - \mathrm{Re}_{1\,\mathrm{cr}}$ and $\delta \neq 0$ at $\mathrm{Re} \to \mathrm{Re}_{1\,\mathrm{cr}}$. Thus, at small $\mathrm{Re} - \mathrm{Re}_{1\,\mathrm{cr}} > 0$ and increasing t the disturbance (1.1) tends towards a periodic fluctuation $\mathbf{u}_1(\mathbf{x}, t)$ with a given, finite amplitude and an arbitrary phase. Since the phase is determined by the random phase of the initial disturbance, this is, in fact, a degree of freedom for a finite flow. As Re increases, it can acquire another critical value, $\mathrm{Re}_{2\,\mathrm{cr}}$, corresponding to the second bifurcation. The periodic flow $\mathbf{u}_0(\mathbf{x}) + \mathbf{u}_1(\mathbf{x}, t)$ then becomes unstable with respect to disturbances of the type $e^{\lambda t}\mathbf{f}_1(\mathbf{x}, t)$. Here, $\mathbf{f}_1$ is a periodic function of t with period $2\pi/\omega_1$, and the eigenvalue λ has an imaginary part $\pm i\omega_2$. At small $\mathrm{Re} - \mathrm{Re}_{2\,\mathrm{cr}} > 0$, this disturbance increases in time up to a finite limit, at which it becomes a quasi-periodic fluctuation with two periods, $2\pi/\omega_1$ and $2\pi/\omega_2$, and two degrees of freedom (phases).

By Landau's assumption, a further increase in Re results in an ever-growing number of standard bifurcations. As t increases, the phase trajectory approaches a limit cycle corresponding to the quasi-periodic flow $\mathbf{u}[\mathbf{x}, \varphi_1(t), \ldots, \varphi_n(t)]$ with a period of $2\pi/\omega_k$ relative to each of the arguments $\varphi_k(t) = \omega_k t + \alpha_k$. In phase space, this limit cycle would occupy a region that corresponds to all possible sets of initial phases $\alpha_1, \ldots, \alpha_n$, while the phase trajectory spiralling towards it would pass through almost all points of the above region. Indeed, for the moments $t_n = 2\pi n/\omega_1$, $n = 0, 1, 2, \ldots$ at which the phase $\varphi_1(t)$ is equal to α_1, the phase of any other oscillation $\varphi_1(t)$, reduced to the range $(0, 2\pi)$, acquires the values

$$\frac{2\pi n\omega_2}{\omega_1} + \alpha_2, \quad n = 0, 1, 2, \ldots.$$

This set contains numbers infinitely close to any given number within this range, so that the frequencies ω_1 and ω_2 are, generally speaking, incommensurable. Landau defines developed turbulence as ergodic, in this sense quasi-periodic flow $\mathbf{u}[\mathbf{x}, \varphi_1(t), \ldots, \varphi_n(t)]$, with a large number of degrees of freedom. Note, however, that in general the temporal velocity correlation functions do not tend to zero at infinity. A mathematical model of the case discussed here was proposed by Hopf (1948).

A number of laboratory experiments and numerical simulations of turbulence in Couette flow between rotating cylinders, in convection at small Prandtl numbers, in the boundary layer on a flat plane, in the mixing layer between flows with different velocities, in the wake formed by a fluid flowing along a cylinder, and in multilayer models of the atmospheric circulation, has to a certain extent confirmed Landau's assumptions concerning the development of quasi-periodic flows. In these experiments, however, only a small number of successive bifurcations were traced. Thereafter the flow unexpectedly became quite irregular in time (with a continuous frequency spectrum), although the wavenumber spectrum still remained discrete. In most cases, the latter also turned into a continuous one with increasing Re.

For instance, a succession of standard bifurcations was observed by Willis and Deardorff (1970) in experiments on natural convection in fluids heated from below. They used a liquid with a small Prandtl number (air, with Pr = 0.71). The Prandtl number is defined as $\mathrm{Pr} = \nu/\chi$, where χ is the kinematic coefficient of molecular thermal conductivity, thermal diffusivity for short. Here, mention should also be made of experiments by Krishnamurti (1970, 1973) in mercury and air, by Rossby (1969) in mercury, by Ahlers (1974) in classical liquid helium with Pr = 0.86, and by Moller and Riste (1975) in liquid crystals. First of all, in convection the stability of the flow changes. At a certain Rayleigh number, $\mathrm{Ra}_{1\,\mathrm{cr}}$,

$$\mathrm{Ra} = \alpha g H^3 \, \delta T/\nu\chi \tag{1.6}$$

a new kind of stationary flow emerges. In (1.6), α is the thermal expansion coefficient of the medium, H is the layer width, and δT is the temperature difference between its lower and upper boundaries. This flow can have the form of either horizontally periodic two-dimensional rollers (provided the material properties of the medium, α, ν, and χ, change negligibly with temperature throughout the width; see Whitehead (1971), or that of hexahedral Benard cells if the material properties are temperature-dependent (Busse, 1967; Schülter *et al.*, 1965).

Let us consider the stability of two-dimensional rollers. Busse (1972) took their axes to be parallel to the Y-axis and described their stream function ψ and temperature deviation θ in the (x, z) plane by three modes (with amplitudes X, Y and Z). For infinitesimal disturbances he found that ψ and θ are given by:

$$\begin{aligned} k_1/\pi(1 + k_1^2/\pi^2)^{-1}\chi^{-1}\psi &= X\sqrt{2}\sin(k_1 x/H)\sin(\pi z/H) \\ k_1^2/\pi^6(1 + k_1^2/\pi^2)^{-3}(g\alpha H^3/\nu\chi)\theta &= Y\sqrt{2}\cos(k_1 x/H)\sin(\pi z/H) - \\ &\quad - Z\sin(2\pi z/H). \end{aligned} \tag{1.7}$$

The amplitudes of other modes are infinitesimal quantities of higher order, in particular when the Prandtl number is small. The modes in (1.7) pertain to the case when both boundaries are free surfaces, this limitation being, in all probability, non-essential (e.g., Palm *et al.*, 1967). With experiments in air it has been found that, at $\mathrm{Ra}_{1\mathrm{cr}}$ equal to about one-third of $\mathrm{Ra}_{2\,\mathrm{cr}}$, there arise transverse oscillations of the rollers, i.e., waves either standing or travelling along their axes, with nearly coinciding phases and with amplitudes relatively constant everywhere, except in the vicinity of the boundaries. This indicates their low sensitivity to boundary conditions. These oscillations were calculated within the framework of both linear (Busse, 1972) and nonlinear (McLaughlin and Martin, 1975) theories. McLaughlin and Martin computed the eight-mode motion involving non-stationary rollers (1.7) and one harmonic along the Y-axis. They determined $\mathrm{Ra}_{2\,\mathrm{cr}}$ analytically and carried out the Landau expansion (1.4). They proved that in this expansion $\delta > 0$, i.e., this is a case of standard bifurcation. Furthermore, they calculated the thirty-nine-mode motion containing the rollers (1.7) and four harmonics along the Y-axis. It was convenient to express the temperature difference causing convection in the following units

$$\mathrm{R}^* = (k_1^2/\pi^6)(1 + k_1^2/\pi^2)^{-3}\,\mathrm{Ra}.$$

In the calculations it was assumed that $k_2/\pi = 0.1\,k_1/\pi = 0.072$ and $\mathrm{Pr} = 1$; here, $\mathrm{R}^*_{2\,\mathrm{cr}} \simeq 1.25$. The calculations with $\mathrm{R}^* = 1.4$ yielded periodic conditions and those with $\mathrm{R}^* = 1.45$ slightly non-periodic ones. However, at $\mathrm{R}^* = 1.5$ and 1.55 the conditions were periodic again (due to the disappearance of the internal motion, which contributes to the increase of disturbances with high wavenumbers), and at $\mathrm{R}^* = 1.6$ the conditions became sharply non-periodic. With the fourth harmonic excluded, the calculations at $\mathrm{R}^* = 1.6$, 2, and even 20, yielded periodic conditions.

Consider now the case of 'reverse bifurcation', when the one-parametric set of closed phase trajectories predicted by the Hopf bifurcation theory is observed even at $\mathrm{Re} < \mathrm{Re}_{1\,\mathrm{cr}}$. In this case, the coefficient δ of the second term in the Landau expansion (1.4) must be negative. Equation (1.4), however, can be used to investigate the evolution of disturbances in the range $\mathrm{Re} < \mathrm{Re}_{1\,\mathrm{cr}}$ if written in the form

$$\mathrm{d}|A|^2/\mathrm{d}t = -2|\gamma| \cdot |A|^2 + |\delta| \cdot |A|^4. \tag{1.8}$$

At $\mathrm{Re} < \mathrm{Re}_{1\,\mathrm{cr}}$, the limit cycle in phase space is unstable. First, the phase trajectories within this cycle spiral towards a stationary point. In other words, disturbances with small amplitudes

$$|A| < A_1 = (2|\gamma|/|\delta|)^{1/2}$$

decay in time. Second, the phase trajectories outside this limit cycle spiral away from it and go to other regions of phase space. Disturbances with finite amplitudes $|A| > A_1$ thus grow in time so that at $\mathrm{Re}_{1\,\mathrm{cr}} > \mathrm{Re} > \mathrm{Re}_{A\,\mathrm{cr}} = \mathrm{Re}_{1\,\mathrm{cr}} - \alpha^2|A|^2$ the motion is unstable with respect to finite disturbances with amplitudes $|A| > A_1$. When Re approaches $\mathrm{Re}_{1\,\mathrm{cr}}$ from below, i.e. $\mathrm{Re} < \mathrm{Re}_{1\,\mathrm{cr}}$, the limit cycle becomes smaller, and when Re exceeds $\mathrm{Re}_{1\,\mathrm{cr}}$ it vanishes completely. For $\mathrm{Re} > \mathrm{Re}_{1\,\mathrm{cr}}$ the solution of (1.4) (with coefficients $\gamma > 0, \delta < 0$) takes the form

$$|A(t)|^2 = \frac{A_0^2 A_1^2}{(A_0^2 + A_1^2)\,\mathrm{e}^{-2\gamma t} - A_0^2}; \quad A_1 = \left(\frac{2\gamma}{|\delta|}\right)^{1/2}, \tag{1.9}$$

which becomes infinite at the finite time

$$t = 1/2\gamma \ln(1 + A_1^2/A_0^2).$$

It is clear, however, that (1.4) becomes invalid at an earlier time and that it must be modified with the subsequent terms of the Landau expansion.

One of the most fully studied kinds of viscous fluid flow with reverse bifurcation is perhaps the plane-parallel flow in a channel (Monin and Yaglom, 1971; Stuart, 1971). For the case of laminar flow (the so-called plane Poiseuille flow with a parabolic velocity profile) linear theory predicts instability ($\gamma > 0$) for Reynolds numbers and dimensionless longitudinal disturbance wavenumbers k within the range that is marked by a solid line in Figure 1.1. Note that as Re increases, both branches of the neutral curve asymptotically approach the X-axis, $k = 0$. On this curve, the lowest critical Re, based on the maximum velocity and the half-width of the channel, is about 5800. At the same time, experimental data from Davis and White (1928) and recent results by Tillman, cited by Stuart (1971) show that turbulence in plane Poiseuille flow begins at much lower values of Re, somewhere between 1000 and 2500. This leads to the assumption that there must exist reverse bifurcation and instability with respect to finite-amplitude disturbances.

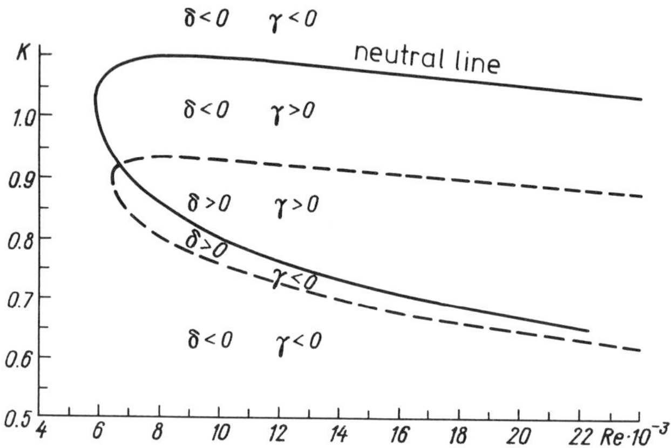

Fig. 1.1. Instability ranges of plane Poiseuille flow. The solid line surrounds the range of instability for infinitely small disturbances ($\gamma > 0$), the dashed line shows the range of $\delta > 0$ calculated by Pekeris and Shkoller (1967).

A number of theoretical calculations did anticipate the negative sign of δ in the Landau expansion (1.4) and instability with respect to finite disturbances at $\mathrm{Re} > \mathrm{Re}_{\mathrm{cr\,min}} \simeq$ 2500–2900. This was in good agreement with experiment. For instance, δ-values at various Re and k were computed by Pekeris and Shkoller (1967) and Reynolds and Potter (1967). Pekeris and Shkoller's neutral curve $\delta(k, \mathrm{Re}) = 0$, shown in Figure 1.1 by a dashed line, and the neutral curve $\gamma(k, \mathrm{Re}) = 0$ of linear stability theory, divide the plane (k, Re) into four areas with different combinations of the signs of γ and δ.

One more interesting example of a system with reverse bifurcation is the idealized three-mode roller convection in a liquid layer with a high Prandtl number, which obeys (1.7). If the interactions with all the other modes are neglected, the hydrodynamic equations yield in the Boussinesq approximation the following equations for the dimensionless amplitudes X, Y, Z of these three modes:

$$X' = -\sigma X + \sigma Y; \qquad Y' = rX - Y - XZ; \qquad Z' = -bZ + XY. \tag{1.10}$$

Here primes designate the derivative with respect to dimensionless time

$$\pi^2 H^{-2}(1 + k_1^2/\pi^2)\chi t.$$

The coefficients σ, b, and r are defined by

$$\sigma = \mathrm{Pr}; \quad b = 4(1 + k_1^2/\pi^2)^{-1}; \quad r = \mathrm{Ra}/\mathrm{Ra}_{1\,\mathrm{cr}},$$

where

$$\mathrm{Ra}_{1\,\mathrm{cr}} = \pi^4 (k_1/\pi)^{-2}(1 + k_1^2/\pi^2)^3.$$

The smallest $\mathrm{Ra}_{1\mathrm{cr}}$, obtained at $k_1 = \pi/\sqrt{2}$, equals $27\pi^4/4 \simeq 657.5$ as calculated by Rayleigh. The resulting value of $b = 8/3$ will be used below in the analysis of (1.10). More general equations for finite-mode two-dimensional convection were simplified by Lorenz (1960) and integrated numerically by Saltzman (1962). In most cases, all the unknown functions appeared to tend towards zero with time, while the functions X, Y, Z varied non-periodically. This fact most probably inspired Lorenz (1963) to investigate (1.10).

The phase space of the system (1.10) is the three-dimensional space (X, Y, Z). When $Z_1 = Z - r - \sigma$ is substituted for Z, (1.10) become a system of the hydrodynamic type (Dolzhansky *et al.*, 1974) and thus, on attaining fairly high values, the sum $X^2 + Y^2 + Z_1^2$ must decrease with time. Consequently, all the phase trajectories remain in a certain limited region for long times. Moreover, their divergence $\partial X'/\partial X + \partial Y'/\partial Y + \partial Z'/\partial Z$ has a constant negative value, $-(\sigma + b + 1)$, so that each small phase volume decreases in time and all the trajectories tend towards a certain subspace of zero volume. At $r < 1$, the system (1.10) has one stationary point $0 = (0,0,0)$, this point being stable. At $r > 1$, this point becomes unstable and there appear two other stationary points: $C = (\sqrt{b(r-1)}, \sqrt{b(r-1)}, r-1)$ and $C' = (-\sqrt{b(r-1)}, -\sqrt{b(r-1)}, r-1)$, which are equitable since the system (1.10) does not vary under a $(X, Y, Z) \rightarrow (-X, -Y, Z)$ transformation. For $\sigma < b + 1$ the points C and C' are stable, for $\sigma > b + 1$ they are stable if $1 < r < r_{\mathrm{cr}} = \sigma(\sigma + b + 3)(\sigma - b - 1)^{-1}$ and unstable if $r > r_{\mathrm{cr}}$, the last case being of particular interest. Note, however, that at high Prandtl $(\sigma > b + 1)$ and Rayleigh $(r > r_{\mathrm{cr}})$ numbers, the three-mode system (1.10) stops corresponding to any real convection. Experiments by Willis and Deardorff (1970) with silicon oil $(\sigma = 57)$ and by Krishnamurti (1970, 1973) with water $(\sigma = 6.7)$ and other liquids with large Prandtl numbers showed the non-stationary state to manifest itself, not in the motion in the rollers (1.7), but in the convection arising from the thermal boundary layer at the lower limit of the fluid.

According to linear theory, at $r = r_{\mathrm{cr}}$ neutral infinitesimal disturbance can arise in the stationary motion of the rollers (1.7) described by the phase point C. Their frequencies ω_{cr} obey the formula $\omega_{\mathrm{cr}}^2 = 2b\sigma(\sigma + 1)(\sigma - b - 1)^{-1}$, while, for the X-coordinate, the neutral disturbances can be written as $\delta X = A \cos \omega_{\mathrm{cr}} t$. For r much lower than r_{cr}, these disturbances require a small non-linear correction. The main term of this correction, of the order of A^2, contains a non-periodic additive and a harmonic of frequency $2\omega_{\mathrm{cr}}$. Also, A becomes a slow function of time. McLaughlin and Martin (1975) derived the Landau equation (1.4) for the squared disturbance amplitude $|A|^2$ accurate to zero order relative to $\sqrt{r-1} - \sqrt{r_{\mathrm{cr}} - 1}$. With an accuracy of order σ^{-1}, they determined its coefficients as $\gamma = (b/2\sqrt{\sigma})(\sqrt{r-1} - \sqrt{r_{\mathrm{cr}} - 1})$, $\delta = -37/72$, which indicated the presence of reverse bifurcations.

Lorenz (1963) integrated (1.10) numerically for $b = 8/3$, $\sigma = 10$ and $r_{cr} = 470/19 \simeq 24.74$, assuming r to have a slightly supercritical value, $r = 28$. Sooner or later, each trajectory reaches the vicinity of one of the stationary points, either C or C', makes some diverging loops around it and, then, when at a sufficient distance away, passes over to the vicinity of the other point, etc. The succession of these transitions is quite irregular and depends substantially on the initial conditions. Figure 1.2 depicts an example of such trajectories, kindly supplied by M. I. Rabinovich who integrated (1.10) using an analog computer. Lorenz assumed the above trajectories to cover a two-dimensional infinite-leaved surface, the intersection of which with any straight line being a Cantor set of points (i.e., a nowhere-dense closed set without isolated points, having the power of a continuum).

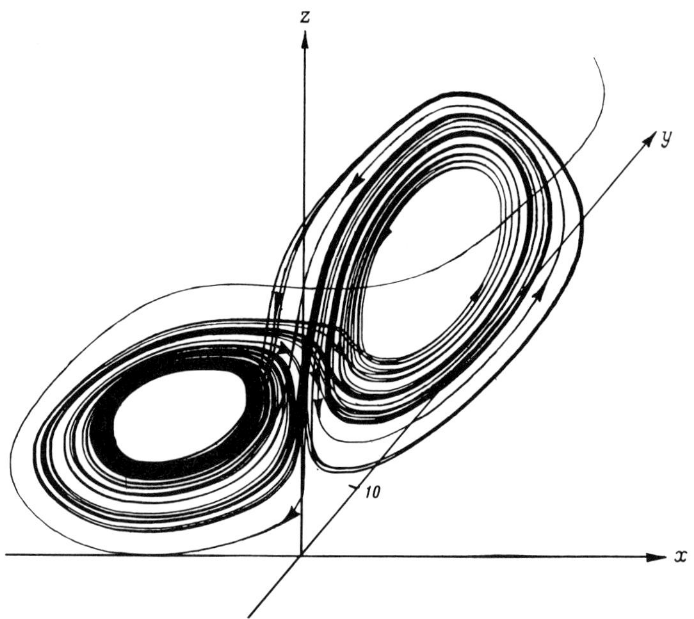

Fig. 1.2. Trajectory of the system calculated by Rabinovitch with an analog computer.

A set of points in phase space, to which the phase trajectories of a dynamic system approach (are attracted to) asymptotically, is called an attractor. Ten years ago, it was conventionally assumed that only stationary points, and either closed or quasi-periodic orbits can serve as attractors for phase trajectories of dynamic systems. Irregularity (stochasticity) in the behavior of such systems could be induced either by the introduction of randomness into their initial data or by random external effects, or, lastly, by an extremely complex limiting trajectory corresponding to the excitation of a great number of degrees of freedom.

However, during the last ten years, mathematicians have discovered the so-called strange attractors, which differ from stationary points and closed or quasi-periodic orbits. As far as one can judge from numerical calculations, the Lorenz infinite-leaved surface presents an example of a strange attractor. To formulate more general results, let us introduce the notion of a non-traveling phase point so that a certain phase trajectory crosses its surroundings at least twice. The simplest particular cases are: stationary points, corresponding to stationary solutions of the dynamic equations, and periodic points, belonging to closed trajectories and corresponding to solutions which are periodic in time. A stationary point x of the f transformation of the phase space M within a fixed time t is termed hyperbolic if the limited linear operator Df used as the f transformation differential at this point is hyperbolic, i.e., its spectrum does not cross a unit circumference. The periodic point of the f transformation is called hyperbolic if it is a hyperbolic stationary point of a definite degree of the f transformation. A set of points p of the phase space for which the iteration sequence $f^m(p)$ of the f transformation converges to x at $m \to \infty$ is referred to as a stable variety. The stable variety with respect to f^{-1} is called an unstable variety of the point x with respect to f. Dynamic systems with a set of non-traveling points consisting of a finite number of stationary points and closed trajectories are called Morse–Smale systems. In this case, all the periodic points are hyperbolic and the stable and unstable sets that correspond to any of these two points are transverse. Consider the situation when all generalized coordinates of a dynamic system ω are quasi-periodic time functions with m fixed incommensurable periods. In this case, the phase trajectories of the system ω appear to belong to a certain m-dimensional torus T^m in phase space, and the system itself can be presented as a constant vector field on this torus.

Ruelle and Takens (1971), see also Ruelle (1975, 1976) proved that for $m \geqslant 3$ in each variety of the dynamic systems obtained from ω by small disturbances (i.e., in the direct vicinity of ω, the notion 'direct vicinity' defined precisely) there exists an open subvariety of dynamic systems other than the Morse–Smale ones. Namely, at $m = 3$ on $T^3 = T^2 \times T^1$ there is a variety of systems inducing transformations of the two-dimensional torus T^2, which have sets of non-traveling points containing the Cantor set, while at $m \geqslant 4$ there are systems containing strange attractors in their phase space.

In particular, at $m = 4$ in the direct vicinity of ω there exists an open subvariety of dynamic systems ω' with strange attractors of the following type. Let Σ be a three-dimensional subvariety in T^4 transversely intersected by the phase trajectories of the system ω'. Let us define the Σ self-mapping $P(x)$ (the so-called reflection of the Poincaré series) as the point of intersection of Σ with the phase trajectory outgoing from the point x of this subvariety. In this case, we can consider the systems ω' in which $P(x)$ maps the interior U of the two-dimensional Σ-immersed torus onto itself so that $P(U)$ is the interior of the U-immersed torus with one loop shown in Figure 1.3. The circumference S, which is the U cross-section, is thus transformed into two circles $P(S)$ within S. The next iteration of $P^2(S)$ yields two more circles within each of the $P(S)$ circles, etc. Intersecting all the $P^n(S)$ iterations gives the Cantor set of points on S so that intersecting all the $P^n(U)$ iterations results in the Cantor set of lines, the so-called one-dimensional Williams solenoid. Here, the dynamic system ω' has in its four-dimensional phase space a strange attractor which is a local Cantor set of two-dimensional surfaces.

The Ruelle and Takens theorem proves that the appearance of strange attractors in the phase spaces of dynamic systems, which is due to several standard bifurcations (four or even three, provided stability changes are neglected), is a typical phenomenon in the sense used in this theorem. Whether liquid and gas flows possess such typical properties is still a problem which has to be solved both analytically, taking into account concrete

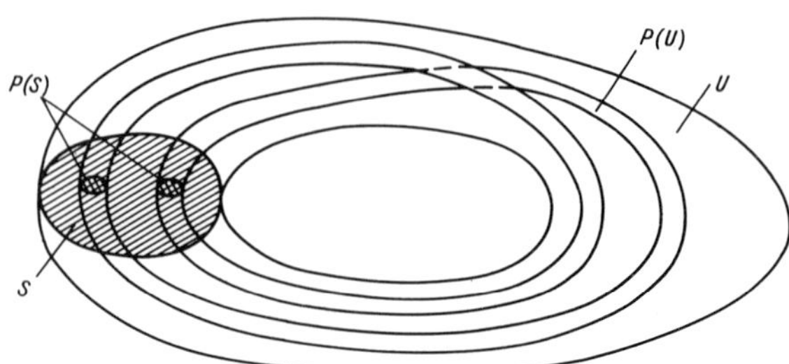

Fig. 1.3. Self-mapping of the series $P(U)$ of the interior U of a two-dimensional torus resulting in a strange attractor.

types of hydrodynamic equations, and experimentally, studying subsequent bifurcations associated with the instability of laminar flows.

The complex structure of strange attractors that do not completely fill phase space (since they have a smaller scale) and contain Cantor discontinua in some of their cross-sections results in a non-periodic behavior of systems developing at such attractors. The attractors are pseudo-random in the sense that the time functions describing these systems have continuous frequency spectra and correlation functions decaying at infinity. Most dynamic systems thus give rise to pseudo-random time functions without requiring the introduction of a random element into the initial data and without needing to take into account random external effects and excitations of a great number of degrees of freedom. At present, however, it is still not clear whether any liquid or gas flow possesses the above properties.

The liquid or gas flow developing at a strange attractor (if any) cannot yet be called turbulent: the definition of turbulence includes that the thermodynamic characteristics should change irregularly in space, i.e., should be described by a great number of space modes or, empirically, should have a continuous space spectrum.

When turbulence occurs in a strong mean flow, the small-scale part of the space spectrum along stream lines appears to be similar to the corresponding part of the frequency spectrum of the fluctuations at fixed points (the so-called 'frozen turbulence' hypothesis of Taylor). Therefore, a flow with continuous frequency spectrum and a discrete space spectrum (i.e., containing only a few modes) is not yet a turbulent one. From this point of view, the roller convection developing at the Lorenz attractor in liquids with a high Prandtl number possesses quite a regular and simple three-mode spatial structure (1.7) and hence is not turbulent. Thus, a turbulent attractor must be multi-dimensional, and the idea of turbulence as a system with a very great number of excited degrees of freedom remains valid. In this case, the question of the time evolution of turbulent space spectra or, in other words, of the succession of bifurcations that increases the scale of the turbulent attractor, remains a valid one, too.

In this book, we are going to study turbulence properties in conditions typical of the ocean. These conditions are characterized by very stable stratification. Stratification is defined as density variations of the medium in the vertical direction, i.e., in the direction of gravity. Indeed, vertical density changes of some 0.001 g cm^{-3} per 10 m, i.e., 10^{-6} g cm^{-4}, are typical of surfaces of discontinuity (pycnoclines) in the ocean. By comparison, a strong temperature inversion in the lower atmosphere, of the order of magnitude of $1°C\ m^{-1}$, corresponds to a vertical density change of 4×10^{-8} g cm^{-4}. Moreover, the free ocean surface, which is also a surface of density discontinuity, can serve as an example of the large stable stratification of the medium.

Vertical turbulent mixing of a medium with very stable stratification demands great energy to counteract the Archimedean forces. Investigations in the ocean have shown that long-term turbulence then develops only in thin layers, which are made quasi-homogeneous by mixing and are separated by surfaces of density micro-discontinuity. In other words, when subjected to turbulent mixing, a very strongly stratified medium acquires a vertical microstructure consisting of thin layers.

The presence of a free surface, an increase in density with depth (especially rapid in a pycnocline) and the existence of surfaces of density micro-discontinuity between the micro-structure layers all create conditions favorable for the development of both surface and internal gravity waves. Indeed, surface and internal waves observed practically at all times and everywhere are typical of conditions in the ocean. From a physical perspective, there arise multiple interactions between microstructure, gravitational waves and turbulence. At the same time, small-scale fluctuation measurements run into the problem of distinguishing between turbulence and waves.

Owing to the viscosity of water, waves arising in reality on the ocean surface have a small vorticity. In the boundary layer at the free surface, which is rather thin for a clean surface and much thicker in the presence of an incompressible surface film, this vorticity is small to first order relative to the wave height, and is of the second order of smallness below the surface layer (see §3.4 in Phillips, 1967). The fluctuations created by a random field of these waves in the upper ocean differ from turbulence by the low vorticity and by the dispersion and phase relations of the elementary wave components. The wave-induced fluctuations are coherent with the surface waves and can, at least approximately, be filtered out from the total fluctuation measurements if the surface waves are detected. The mathematics of this procedure has been developed by Benilov and Filiushkin (1970). See also Benilov (1973) and §3.8 in Monin and Kamenkovich (1974).

The problem of distinguishing between internal waves and turbulence in a stratified ocean proves to be even more complicated. First, it is impossible to detect internal waves separate from total fluctuations. Second, internal waves are not potential flows; they are substantially turbulent in their vertical plane of propagation. At low frequencies, comparable with the inertial frequency (i.e., with the so-called Coriolis parameter $f = 2\Omega \sin \varphi$, where Ω is the angular velocity of the Earth's rotation, and φ is the geographical latitude), they are turbulent in the horizontal plane as well. Third, non-linear effects in internal wave dynamics can turn out to be essential. Thus, internal waves can become turbulent, as will be shown below.

Despite these difficulties, the problem of distinguishing between a random wave field and turbulence seems to be solvable in the case of linear internal waves. Even a random field of linear internal waves possesses a certain regularity, which is not typical of turbulence. Indeed, an arbitrary field $\zeta(\mathbf{x}, z, t)$ of vertical shear in linear internal waves at depth z, statistically homogeneous along the **X**-axis and stationary in time t, can be presented as

$$\zeta(\mathbf{x}, z, t) = \int e^{(i\mathbf{k}\cdot\mathbf{x} - \omega t)}\, \zeta(k, z, \omega)\, dZ(\mathbf{k}, \omega), \tag{1.11}$$

where $\mathbf{k}$ is a horizontal wave vector, ω is a frequency within the range $f \leqslant \omega \leqslant N$ (f is the inertial frequency and N is the so-called Brunt–Väisälä frequency, determined by $N^2 = g\,\mathrm{d}\rho_*/\rho_*\,\mathrm{d}z$, where g is the acceleration due to gravity and ρ_* is the potential density of the medium).

In (1.11), the spectral functions $\zeta(k, z, \omega)$ are regular (not random) functions of z. They satisfy the following linear equation and the boundary condition at the ocean surface ($z = 0$) or at the bottom ($z = H$):

$$\frac{\mathrm{d}}{\mathrm{d}z}\rho_*\frac{\mathrm{d}\zeta}{\mathrm{d}z} + k^2\rho_*\frac{N^2 - \omega^2}{\omega^2 - f^2}\zeta = 0, \quad z = 0;$$
$$(\omega^2 - f^2)\frac{\mathrm{d}\zeta}{\mathrm{d}z} + gk^2\zeta = 0, \quad z = H, \quad \zeta = 0. \tag{1.12}$$

Substituting the above $\zeta(k, z, \omega)$ into the second boundary condition, at a fixed horizontal wavenumber k, we obtain an equation for ω with a countable number of roots. The roots are the eigenfunctions $\omega = \omega_n(k)$, $n = 1, 2, \ldots$, of the various modes. Hence, the random spectral measure $Z(\mathbf{k}, \omega)$ in (1.11) is localized on the dispersion surfaces $\omega = \omega_n(k)$:

$$\mathrm{d}Z(\mathbf{k}, \omega) = \sum_n \delta[\omega - \omega_n(k)]\,\mathrm{d}\omega\,\mathrm{d}Z_n(\mathbf{k}). \tag{1.13}$$

In linear internal wave fields, the velocity components u, v, w and the pressure and density fluctuations, p and ρ, are expressed by equations of the type (1.11), with $\zeta(k, z, \omega)$ replaced by

$$u = \frac{k_1\omega - ik_2 f}{k^2}\frac{\mathrm{d}\zeta}{\mathrm{d}z}; \qquad v = \frac{k_2\omega + ik_1 f}{k^2}\frac{\mathrm{d}\zeta}{\mathrm{d}z}; \qquad w = -i\omega\zeta;$$
$$p = \frac{\omega^2 - f^2}{k^2}\rho_0\frac{\mathrm{d}\zeta}{\mathrm{d}z}; \qquad \rho = \frac{\omega^2 - f^2}{k^2}\frac{\rho_0}{c_0^2}\frac{\mathrm{d}\zeta}{\mathrm{d}z} - \frac{\rho_0 N^2}{g}\zeta, \tag{1.14}$$

respectively. Here k_1, k_2 are the Cartesian components of the wave vector, while $\rho_0(z)$ and $c_0(z)$ are the undisturbed density and the velocity of sound.

In order to determine the nature of observed fluctuations, with the function $\rho_0(z)$ approximated by a suitable analytical expression, it is necessary to derive from (1.12) the eigenfrequencies $\omega_n(k)$ and the corresponding eigenfunctions $\zeta_n(k, r) = \zeta[k, z, \omega_n(k)]$. These can be taken orthonormal to the total energy of the waves. They then obey the condition $\zeta_m \circ \zeta_n = \delta_{mn}$, the functional scalar product being determined by

$$\zeta_m \circ \zeta_n = \frac{\omega_m\omega_n}{(\omega_m^2 - f^2)^{1/2}(\omega_n^2 - f^2)^{1/2}}\left[g(\rho_0\zeta_m\zeta_n)_{z=0} + \int_0^H (N^2 - f^2)\zeta_m\zeta_n\rho_*\,\mathrm{d}z\right]. \tag{1.15}$$

In this example the expansion of the measured field $\zeta(\mathbf{x}, z, t)$ in the functions $\exp(i\mathbf{k}\cdot\mathbf{x})\zeta_n(k, z)$ makes it possible to verify the dispersion relations of linear internal waves, while the expansion of the measured fields in the functions $\exp\{i[\mathbf{k}\cdot\mathbf{x} - \omega_n(\mathbf{k})t]\}$ allows one to check relations (1.14).

The phase difference of any two hydrodynamic characteristics a and b at a fixed point in space can be estimated by their mutual correlation function $B_{ab}(\tau) = \overline{a(t+\tau)b(t)}$. Here, the bar over the symbols denotes a statistical mean value, i.e., a mathematical expectation. The Fourier transform (mutual spectrum) of $B_{ab}(\tau)$ is denoted by $C_{ab}(\omega) - iQ_{ab}(\omega)$, where C_{ab} is the co-spectrum and Q_{ab} the quadrature spectrum. The phase shift spectrum is given by $\varphi_{ab}(\omega) = \arctan[Q_{ab}(\omega)/C_{ab}(\omega)]$. For linear internal waves this spectrum must coincide with that calculated from (1.14). For instance, at $\omega \gg f$ the oscillations u, v, ζ, p, ρ, and T are shifted by $\pi/2$ relative to oscillations at frequency ω. At the same time, in the case of turbulence no regularity in phase shifts can be expected.

In the case of slightly nonlinear internal waves, the dispersion relations and phase shifts appear to be somewhat blurred around the values predicted by linear theory. In the nonlinear case, interactions occur between internal waves with different three-dimensional wave vectors κ_1, κ_2. If these interactions are resonant, i.e., if the resulting wave with wave vector $\kappa = \kappa_1 \pm \kappa_2$ has the frequency $\omega(\kappa) = \omega(\kappa_1) \pm \omega(\kappa_2)$, the typical time of the interaction is $\tau \simeq (\kappa_1\omega_1 \cdot \kappa_2\omega_2)^{-1/2} \gg N^{-1}$. If the interactions are non-resonant, this results in so-called induced modes, which are internal waves with wave vectors $\kappa = \kappa_1 \pm \kappa_2$ and frequencies $\omega = \omega(\kappa_1) \pm \omega(\kappa_2)$ that violate the dispersion relation (i.e., $\omega \neq \omega(\kappa)$). For $\tau \gg N^{-1}$ the amplitudes of the induced modes are small, while for $\tau \simeq N^{-1}$ they prove to be comparable with the amplitudes of the initial waves. In the latter case, the interactions of such induced modes with one another and with free internal waves give rise to rotational motion with a spectrum that obeys no definite dispersion relation, i.e., to turbulence, if the amplitudes are large. We follow Miropolsky and Filiushkin (1971) and assume that the interaction time can be estimated by $\tau = [k^3E(k)]^{-1/2}$, where $E(k)$ is the spectral density of the kinetic energy of the fluctuations per unit mass. For $\tau \gg N^{-1}$, $E(k)$ can be taken to be the spectrum of interacting internal waves; for $\tau \lesssim N^{-1}$, the turbulence spectrum can be used.

2. EQUATIONS OF TURBULENT FLOW

Let us consider the hydrodynamic equations in the Boussinesq approximation. The equations of motion then contain the Archimedean force (buoyancy force) and the continuity equation can be written as the non-divergence of the velocity field (see Kamenkovich, 1973, §2, Chapter IV). The equations are:

$$\partial\rho_0 u_i/\partial t = -\partial/\partial x_\alpha(\rho_0 u_i u_\alpha + p\delta_{i\alpha} - \sigma_{i\alpha}) + g\rho\delta_{i3} + 2\rho_0 e_{i\alpha\beta}\omega_\beta u_\alpha; \tag{2.1}$$

$$\partial u_\alpha/\partial x_\alpha = 0. \tag{2.2}$$

Cartesian coordinates $\mathbf{x} = (x_1, x_2, x_3)$ are used. The axis $x_1 = x$ is eastward, $x_2 = y$ is northward and $x_3 = z$ is downward. The depth z is measured from the equilibrium level of the ocean surface. Also, t is time, u_i are Cartesian velocity components ($u_1 = u$, $u_2 = v$ in the horizontal plane and $u_3 = w$ in the vertical plane). Furthermore, ρ_0 is the equilibrium density, which depends solely on the depth z. The corresponding hydrostatic equilibrium pressure is designated as p_0, so that $\partial p_0/\partial z = g\rho_0$, where g is the acceleration due to gravity. The components of the angular velocity vector of the Earth's rotation are represented by ω_i ($\omega_1 = 0$, $\omega_2 = \omega \cos \varphi$ and $\omega_3 = -\omega \sin \varphi = -f/2$, where φ is the geographical latitude and $f = 2\omega \sin \varphi$ is the Coriolis parameter). Finally, $\sigma_{ij} = \rho_0 \nu(\partial u_i/\partial x_j + \partial u_j/\partial x_i)$ is the viscous stress tensor, ν is the kinematic viscosity (varying from 1.826×10^{-2} cm^2 s^{-1} at 0°C to 1.049×10^{-2} cm^2 s^{-1} at 20°C for sea water with a salinity of 35‰ at atmospheric pressure), δ_{ij} is the unit tensor, e_{ijk} is the unit antisymmetric tensor, and Greek indices imply summation from 1 to 3.

The system (2.1)–(2.2) contains four equations with five unknowns: u_1, u_2, u_3, p and ρ. We supplement it with an equation of state expressing the dependence of sea water density on pressure, temperature and salinity, and with equations for heat transfer and salt diffusion. We employ the equation of state in the linear form,

$$\frac{\rho - \rho_0}{\rho_0} \simeq -\alpha(T - T_0) + \beta(S - S_0), \tag{2.3}$$

where T_0 and S_0 are the equilibrium temperature and salinity, which depend solely on z and are related to p_0 and ρ_0 by the equation of state. The coefficients $-\alpha$ and β are logarithmic derivatives of the equilibrium density with respect to temperature and salinity. Typical values of α and β are approximately 2×10^{-4} °C^{-1} and 1, respectively. In the right-hand side of (2.3) we have neglected a small term of order $p - p_0/(\rho_0 c^2)$ (where c is the speed of sound), which describes the contribution of pressure deviations.

We use the equations of heat transfer and salt diffusion in the following conventionally simplified form:

$$\frac{\partial(c_p \rho_0 T)}{\partial t} = -\frac{\partial}{\partial x_\alpha}\left(c_p \rho_0 T u_\alpha - c_p \rho_0 \chi \frac{\partial T}{\partial x_\alpha}\right) + $$

$$+ c_p \rho_0 (g \rho_0 \Gamma w); \tag{2.4}$$

$$\frac{\partial(\rho_0 S)}{\partial t} = -\frac{\partial}{\partial x_\alpha}\left(\rho_0 S u_\alpha - \rho_0 D \frac{\partial S}{\partial x_\alpha}\right). \tag{2.5}$$

Here χ and D are the kinematic diffusivities for heat and salt. The thermal diffusivity χ changes from 1.39×10^{-3} to 1.49×10^{-3} cm^2 s^{-1} at atmospheric pressure and 35‰ salinity when the temperature changes from 0° to 20°C, while D varies from 0.68×10^{-5} to 1.29×10^{-5} cm^2 s^{-1} so that the Prandtl number $\mathrm{Pr} = \nu/\chi$ ranges from 13.1 to 7.0 and the ratio χ/D from 204.4 to 115.5.

It is convenient to introduce the symbol Γ, defined by $\Gamma = (\partial T/\partial p)_{\eta, s} = \alpha T/c_p \rho$, for the adiabatic temperature gradient. This varies from 0.035 to 0.181°C/100 bar at $T = 0$°C and $S = 35$‰ when the pressure changes from 1 to 1000 bar.

On a free surface, $z = \zeta(x, y, t)$, the following conditions must hold. First, the kinematic boundary condition $w = \partial\zeta/\partial t + u(\partial\zeta/\partial x) + v(\partial\zeta/\partial y)$, but also dynamic boundary conditions: continuity of the velocity and shear stress vectors, the pressure jump condition $p - p_a = -\gamma \rho_0 K_0$, where p_a is the atmospheric pressure, γ is a surface tension coefficient, and $K_0 = \mathrm{div}[(1 + |\nabla\zeta|^2)^{-1/2} \nabla\zeta]$ is the Gaussian surface curvature. Furthermore, the heat and salt fluxes at the surface must be continuous and equal to $q_{\mathrm{turb}} + q_{\mathrm{rad}} - LE + c_p(T_p - T_w)P$ and $\beta^{-1}(E - P)S$, respectively. Here, E and P are evaporation and precipitation rates, L is the latent heat of evaporation, while T_p and T_w are the temperatures of the precipitation and of the water surface, respectively. For small and gently sloping waves these conditions can be linearized and written for an equilibrium surface level $z = 0$ where, in particular, $p \simeq p_a - g\rho_0\zeta - \gamma\rho_0\Delta\zeta$. At the bottom of the ocean, $z = H(x, y)$, both the velocity (u, v, w) and the salt flux must become zero, while the heat flux must equal the local geothermic flux.

In practice it is impossible to employ (2.1)–(2.5) to compute the evolution of particular cases of turbulent flow, because the hydrodynamic fields u_1, u_2, u_3, $u_4 = p$, $u_5 = \rho$, $u_6 = T$ and $u_7 = S$ are extremely irregular. This fact, however, does not impede the solution of practical problems. Often no more is needed than information on the *total* effects of the fluctuations over certain time and space intervals, e.g., data on turbulent friction and heat and mass transfer. Therefore, instead of trying to describe particular cases, it is expedient to consider $u_1, \ldots, u_7$ as *random fields* that obey (2.1)–(2.5), and to confine ourselves to a *statistical description*. To make a complete statistical description of the random fields $u_1, \ldots, u_7$ we have to obtain all the finite-dimensional probability distributions for the variables $u_{j_1}(M_1), \ldots, u_{j_n}(M_n)$ at all possible finite sets of space-time points $M_1 = (\mathbf{x}_1, t_1), \ldots, M_n = (\mathbf{x}_n, t_n)$ (for details see §3 in Monin and Yaglom, 1965). The probability densities of these finite-dimensional distributions will be designated as $p_{M_1 \ldots M_n}(u_{j_1}, \ldots, u_{j_n})$. Instead of the probability densities it is sometimes convenient to use their Fourier transforms with respect to $u_{j_1}, \ldots, u_{j_n}$, which are the *characteristic functions*

$$\varphi_{M_1 \ldots M_n}(\theta_{j_1}, \ldots, \theta_{j_n}) = \overline{\exp\left[i \sum_{m=1}^{n} \theta_{j_m} u_{j_m}(M_m)\right]} = \int \exp\left[i \sum_{m=1}^{n} \theta_{j_m} u_{j_m}\right] p_{M_1 \ldots M_n}(u_{j_1}, \ldots, u_{j_n})\, du_{j_1} \ldots du_{j_n}. \tag{2.6}$$

As stated before, a bar over symbols denotes statistical averaging, i.e., mathematical expectation calculations. In the statistical description of the turbulent flows the probability distribution functions (2.6) for the hydrodynamic fields at finite sets of space–time points are studied, rather than particular cases of those fields.

The hydrodynamic equations (2.1)–(2.5) yield evolution equations for the finite-dimensional probability distributions (2.6) (see Monin, 1967a, 1967b, and §19.7 in Monin and Yaglom, 1975). Consider, for instance, the characteristic function $\varphi_{M_1 \ldots M_n}(\theta_1, \ldots, \theta_n)$ for the velocity vectors $\mathbf{u}_1 = \mathbf{u}(M_1), \ldots, \mathbf{u}_n = \mathbf{u}(M_n)$ at the points $M_1, \ldots, M_n$ with coordinates $\mathbf{x}_1, \ldots, \mathbf{x}_n$ and times $t_1, \ldots, t_n$. In this case,

$$\frac{\partial}{\partial t_k} \varphi_{M_1 \ldots M_n}(\theta_1, \ldots, \theta_n) = \overline{i\theta_{k\alpha}[\partial u_\alpha(M_k)/\partial t_k] \exp\left[i \sum_{m=1}^{n} \theta_m \cdot \mathbf{u}(M_m)\right]} \tag{2.7}$$

Substitution of $\partial u_\alpha(M_k)/\partial t_k$, determined from (2.1), into the right-hand side of (2.7) makes it possible to represent this side by finite-dimensional characteristic functions. In this way we obtain

$$\frac{\partial}{\partial t_k}\varphi_{M_1\ldots M_n}(\theta_1,\ldots,\theta_n)=i\frac{\partial^2}{\partial x_{k\alpha}\partial\theta_{k\alpha}}\varphi_{M_1\ldots M_n}(\theta_1,\ldots,\theta_n)-$$

$$-\theta_{k\alpha}E(\mathbf{x}_k,\mathbf{x})\frac{1}{\rho_0}\frac{\partial}{\partial x_\alpha}\left[\frac{\partial}{\partial\theta_4}\varphi_{M_1\ldots M_nM}(\theta_1,\ldots,\theta_n,\theta_4)\right]_{\theta_4=0}+$$

$$+\nu\theta_{k\alpha}E(\mathbf{x}_k,\mathbf{x})\frac{\partial^2}{\partial x_\beta\partial x_\beta}\left[\frac{\partial}{\partial\theta_\alpha}\varphi_{M_1\ldots M_nM}(\theta_1,\ldots,\theta_n,\theta)\right]_{\theta=0}+$$

$$+g\theta_{k3}E(\mathbf{x}_k,\mathbf{x})\frac{1}{\rho_0}\left[\frac{\partial}{\partial\theta_5}\varphi_{M_1\ldots M_nM}(\theta_1,\ldots,\theta_n,\theta_5)\right]_{\theta_5=0}+$$

$$+2e_{\alpha\beta\gamma}\theta_{k\alpha}\omega_{k\gamma}\frac{\partial}{\partial\theta_{k\beta}}\varphi_{M_1\ldots M_n}(\theta_1,\ldots,\theta_n). \tag{2.8}$$

Here, the arguments of the characteristic functions θ, θ_4 and θ_5 correspond to the random values $\mathbf{u}=\mathbf{u}(M)$, $u_4=p(M)$ and $u_5=\rho(M)$. Also, $M=(\mathbf{x},t_k)$ and $E(\mathbf{x}_k,\mathbf{x})$ is the operator that substitutes $\mathbf{x}_k$ for $\mathbf{x}$. The Fourier transform of (2.8) with respect to θ_1, $\ldots$, θ_n yields an equation for the n-point probability density $f_n=p_{M_1,\ldots,M_n}(\mathbf{u}_1,\ldots,\mathbf{u}_n)$, which can be reduced to the elegant form

$$\partial f_n/\partial t_k=-(\partial u_{k\alpha}f_n/\partial x_{k\alpha}+\partial b_{k\alpha}f_n/\partial u_{k\alpha});$$

$$b_{k\alpha}=E(\mathbf{x}_k,\mathbf{x})\left[-\frac{1}{\rho_0}\frac{\partial\langle p\rangle}{\partial x_\alpha}+\nu\Delta\langle u_\alpha\rangle+g\frac{\langle\rho\rangle}{\rho_0}\delta_{\alpha3}+2e_{\alpha\beta\gamma}\omega_\gamma u_\beta\right]. \tag{2.9}$$

Here, the square brackets denote the *conditional* mathematical expectation of the random values of the hydrodynamic fields variables at the point $(\mathbf{x},t_k)$, provided that $\mathbf{u}_1=\mathbf{u}(M_1),\ldots,\mathbf{u}_n=\mathbf{u}(M_n)$ are fixed.

Sometimes the instantaneous probability distributions are of interest, although they afford a less adequate description of turbulent flows. The evolution equations for those probabilities can be obtained from (2.8) and (2.9) by summation over all k; thereafter, one should set $t_1=\cdots=t_n=t$ and take into account that $\Sigma_k\,\partial/\partial t_k=\partial/\partial t$.

All equations of the type (2.8)–(2.9) for finite-dimensional probability distributions derived from the hydrodynamic equations are linear. However, they are not closed, since the time derivatives of the n-point distribution functions depend on the values of the $(n+1)$-point functions. Therefore, these equations form an infinite linked chain as in the case of equations for many-particle distribution functions in the kinetic theory of gases. Typical of turbulence theory would be an attempt to obtain a system of equations for the distribution functions of lower orders, by presenting the distribution functions of subsequent orders in terms of those of lower order using some closure hypothesis. Nevertheless, the 'closure problem' of turbulence theory in terms of distribution functions has not yet been thoroughly investigated (see §19.7 in Monin and Yaglom, 1975).

The description of turbulence can be significantly simplified if we abandon attempts at a complete description of finite-dimensional probability distributions and confine ourselves to the determination of their simplest and, in practice, most important characterististics: the low-order moments. Here we think primarily of the first and the second moments. The first moments $\overline{u_{jk}}$ are the average hydrodynamic variables in a turbulent flow; the fluctuations $u'_{jk} = u_{jk} - \overline{u_{jk}}$ describe the turbulence proper. The quadratic mean values of $\sigma_{jk} = [\overline{u'^2_{jk}}]^{1/2}$ present the intensity of the fluctuations u_{jk}. The one-point moments of second order $\tau_{jk} = -\rho_0 \overline{u'_j u'_k}$, $m_k = \overline{\rho' u'_k}$, $q_k = c_p \rho_0 \overline{T' u'_k}$ and $I_k = \rho_0 \overline{S' u'_k}$ are the components of the turbulent fluxes of momentum, mass, heat, and salt. The two-point moments of second order (*correlation functions*) $\overline{u'_{j_1}(M_1) u'_{j_2}(M_2)}$ describe statistical relations between the fluctuations u'_{j_1} and u'_{j_2} at different points in space and time. When divided by $\sigma_{j_1}(M_1)\sigma_{j_2}(M_2)$, these moments are correlation coefficients. They characterize the statistical structure of the hydrodynamic variables in turbulent flow.

A general equation for the moments of the velocity field can be obtained if (2.9) is multiplied by $F(\mathbf{u}_1, \ldots, \mathbf{u}_n)$ and subsequently integrated over all $\mathbf{u}_1, \ldots, \mathbf{u}_n$:

$$\frac{\partial \overline{F}}{\partial t_k} = -\frac{\partial \overline{u_{k\alpha} F}}{\partial x_{k\alpha}} + \overline{b_{k\alpha} \frac{\partial F}{\partial u_{k\alpha}}}. \tag{2.10}$$

These equations for the moments of the velocity field are called Friedman–Keller equations. Similar equations can be readily derived for moments that include fluctuations of velocity and of other hydrodynamic variables. All these equations are linear but not closed, since the time derivatives of the n-th order moments depend on the moments of the $(n+1)$th order. A closed system of equations for the moments of the first n orders can be obtained only in an approximate way, if the $(n+1)$th order moments in these equations are expressed in terms of those of lower order with some 'closure hypothesis' (see §§5, 6 and §§16–19 in Monin and Yaglom, 1965, 1967).

The equations for the first moments are derived by direct averaging of the hydrodynamic equations. This yields

$$\begin{aligned}
\partial \rho_0 \bar{u}_i / \partial t &= -\partial/\partial x_\alpha (\rho_0 \bar{u}_i \bar{u}_\alpha + p\delta_{i\alpha} - \tau_{i\alpha} - \bar{\sigma}_{i\alpha}) + g\bar{\rho}\delta_{i3} + 2\rho_0 e_{i\alpha\beta}\omega_\beta \bar{u}_\alpha;\\
\partial \bar{u}_\alpha / \partial x_\alpha &= 0; \quad \bar{\rho} - \rho_0/\rho_0 \simeq -\alpha(\overline{T} - T_0) + \beta(\overline{S} - S_0);\\
\frac{\partial (c_p \rho_0 \overline{T})}{\partial t} &= -\frac{\partial}{\partial x_\alpha}\left(c_p \rho_0 \overline{T} \bar{u}_\alpha + q_\alpha - c_p \rho_0 \chi \frac{\partial \overline{T}}{\partial x_\alpha} \right) + c_p \rho_0 (g\rho_0 \Gamma \overline{w});\\
\frac{\partial (\rho_0 \overline{S})}{\partial t} &= -\frac{\partial}{\partial x_\alpha}\left(\rho_0 \overline{S} \bar{u}_\alpha + I_\alpha - \rho_0 D \frac{\partial \overline{S}}{\partial x_\alpha} \right).
\end{aligned} \tag{2.11}$$

These equations are usually referred to as the Reynolds equations. They differ from the non-averaged hydrodynamic equations (2.1)–(2.5) by additional unknowns. Because of the unknown Reynolds stresses τ_{ji} and turbulent fluxes of heat q_k and salt I_k, the Reynolds equations are not closed. The closure of this Friedman–Keller system of equations on the first-moment level requires that these new unknowns (the one-point moments of second order) be expressed as functions of the mean hydrodynamic variables.

In the so-called *semi-empirical* turbulence theory (which will be discussed in §15) these functions are assumed to be linear.

Multiplying the first equation in (2.11) by $\bar{u}_i$ and summing over i, we get the following equation for the kinetic energy $E_s = \frac{1}{2}|\bar{\mathbf{u}}|^2$ of the mean motion per unit mass:

$$\partial \rho_0 E_s/\partial t = -\partial/\partial x_\alpha(\rho_0 E_s \bar{u}_\alpha + \bar{p}\bar{u}_\alpha - \tau_{\alpha\beta}\bar{u}_\beta - \bar{\sigma}_{\alpha\beta}\bar{u}_\beta) + \\ + g\bar{\rho}\bar{w} - \tau_{\alpha\beta}(\partial \bar{u}_\beta/\partial x_\alpha) - \rho_0 \epsilon_s; \qquad (2.12)$$

where $\epsilon_s = \nu/2(\partial \bar{u}_\alpha/\partial x_\beta + \partial \bar{u}_\beta/\partial x_\alpha)^2$ is the viscous dissipation rate of the kinetic energy of the mean motion per unit mass.

In the following, we will frequently use a statistically stationary and horizontally homogeneous model (SSHH model) of turbulence, in which all finite-dimensional probability distributions (2.6) are invariant with respect to all shifts of the time origin and to horizontal shifts of the system of space coordinates. In that case all one-point statistical characteristics, the averaged pressure gradient included, depend solely on the depth z of the point of observation, and the average flow is horizontal: $\bar{u}_1 = \bar{u}(z)$, $\bar{u}_2 = \bar{v}(z)$, $\bar{u}_3 = 0$. For the SSHH model the Reynolds equations can be written in the form:

$$\frac{1}{\rho_0}\frac{\partial}{\partial z}\left(\tau_{xz} + \rho_0 \nu \frac{\partial \bar{u}}{\partial z}\right) + f\left(\bar{v} - \frac{1}{f\rho_0}\frac{\partial \bar{p}}{\partial x}\right) = 0;$$

$$\frac{1}{\rho_0}\frac{\partial}{\partial z}\left(\tau_{yz} + \rho_0 \nu \frac{\partial \bar{v}}{\partial z}\right) - f\left(\bar{u} + \frac{1}{f\rho_0}\frac{\partial \bar{p}}{\partial y}\right) = 0;$$

$$\frac{\partial}{\partial z}(\rho_0 \sigma_w^2 + \bar{p}) = g\bar{\rho} + 2\rho_0 \omega_y \bar{u};$$

$$q_z - c_p \rho_0 \chi \frac{\partial \bar{T}}{\partial z} = \text{const}; \qquad I_z - \rho_0 D \frac{\partial \bar{S}}{\partial z} = \text{const}. \qquad (2.13)$$

Here, the first two equations describe the so-called Ekman boundary layer, while the third one shows that the average SSHH flow is, strictly speaking, not hydrostatic. The last two equations indicate that the total vertical heat and salt fluxes in SSHH flow are independent of depth. The total heat flux remains constant if the radiant heat flux is also taken into account.

Subtracting the first of the Reynolds equations (2.11) from the non-averaged Navier–Stokes equations (2.1), we obtain an equation for the fluctuations u_i' of the velocity field. We neglect fluctuations of the Coriolis force, because their effect on small-scale turbulence is negligible. The equation for u_i' readily yields the following equation for the Reynolds-stress tensor:

$$-\frac{\partial \tau_{ij}}{\partial t} = -\frac{\partial}{\partial x_\alpha}\left[-\tau_{ij}\bar{u}_\alpha + \rho_0 \overline{u_i' u_j' u_\alpha'} + (\overline{p' u_i'}\delta_{j\alpha} + \overline{p' u_j'}\delta_{i\alpha}) - \right. \\ \left. - (\overline{u_i'\sigma_{j\alpha}'} + \overline{u_j'\sigma_{i\alpha}'})\right] + \overline{p'\left(\frac{\partial u_i'}{\partial x_j} + \frac{\partial u_j'}{\partial x_i}\right)} + \left(\tau_{i\alpha}\frac{\partial \bar{u}_j}{\partial x_\alpha} + \tau_{j\alpha}\frac{\partial \bar{u}_i}{\partial x_\alpha}\right) - \\ - \left(\overline{\sigma_{i\alpha}'\frac{\partial u_j'}{\partial x_\alpha}} + \overline{\sigma_{j\alpha}'\frac{\partial u_i'}{\partial x_\alpha}}\right) + g(m_i\delta_{j3} + m_j\delta_{i3}), \qquad (2.14)$$

where, according to (2.3), $m_k = -(\alpha/c_p)g_k + \beta I_k$. Setting $i = j$ and summing over this index, we obtain the following equation for the kinetic energy of the turbulence per unit mass, $b^2 = \frac{1}{2}\overline{|\mathbf{u}'|^2}$:

$$\frac{\partial \rho_0 b^2}{\partial t} = -\frac{\partial}{\partial x_\alpha}(\rho_0 b^2 \bar{u}_\alpha + \overline{p'u'_\alpha} + \tfrac{1}{2}\rho_0 \overline{u'_\beta u'_\beta u'_\alpha} - \overline{u'_\beta \sigma'_{\alpha\beta}}) + \\ + gm_3 + \tau_{\alpha\beta}\frac{\partial \bar{u}_\beta}{\partial x_\alpha} - \rho_0 \epsilon, \tag{2.15}$$

where

$$\epsilon = \frac{\nu}{2}\overline{\left(\frac{\partial u'_\alpha}{\partial x_\beta} + \frac{\partial u'_\beta}{\partial x_\alpha}\right)^2}$$

is the dissipation rate of turbulent kinetic energy per unit mass. Note that the term $\tau_{\alpha\beta}\partial\bar{u}_\beta/\partial x_\alpha$ has different signs in the right-hand sides of (2.12) and (2.15). Hence, it describes direct kinetic energy exchange between the mean flow and the turbulence. In the case of small-scale three-dimensional turbulence, this term is almost always positive, i.e., it describes the generation of turbulence from the kinetic energy of the mean motion.

In the SSHH model, the diagonal elements of the Reynolds stress tensor are governed by the following simplified version of (2.14):

$$\begin{aligned} \frac{1}{2}\frac{\partial \rho_0 \sigma_u^2}{\partial t} &= -\frac{1}{2}\frac{\partial}{\partial z}\left(\rho_0 \overline{u'^2 w'} - \rho_0 \nu \frac{\partial \sigma_u^2}{\partial z}\right) + \tau_{xz}\frac{\partial \bar{u}}{\partial z} - \\ &\quad - \overline{p'\frac{\partial v'}{\partial y}} - \overline{p'\frac{\partial w'}{\partial z}} - \rho_0 \epsilon_u = 0; \\ \frac{1}{2}\frac{\partial \rho_0 \sigma_v^2}{\partial t} &= -\frac{1}{2}\frac{\partial}{\partial z}\left(\rho_0 \overline{v'^2 w'} - \rho_0 \nu \frac{\partial \sigma_v^2}{\partial z}\right) + \tau_{yz}\frac{\partial \bar{v}}{\partial z} + \\ &\quad + \overline{p'\frac{\partial v'}{\partial y}} - \rho_0 \epsilon_v = 0; \\ \frac{1}{2}\frac{\partial \rho_0 \sigma_w^2}{\partial t} &= -\frac{1}{2}\frac{\partial}{\partial z}\left(\rho_0 \overline{w'^3} + 2\overline{p'w'} - \rho_0 \nu \frac{\partial \sigma_w^2}{\partial z}\right) + \\ &\quad + \overline{p'\frac{\partial w'}{\partial z}} + gm_z - \rho_0 \epsilon_w = 0, \end{aligned} \tag{2.16}$$

where $\epsilon_i = \nu\overline{|\nabla u'_i|^2}$ is the dissipation rate of the component energy $\frac{1}{2}\overline{(u_i'^2)}$. These equations indicate that only horizontal velocity fluctuations can be directly generated by the mean flow, while vertical fluctuations can be induced only by the potential energy of the stratification, provided it is unstable, i.e., $m_z > 0$. On the other hand, in stable stratification, when $m_z < 0$, vertical fluctuations lose their energy to counteract the buoyancy forces. The terms containing the pressure fluctuations, $\overline{p'\partial v'/\partial y}$ and $\overline{p'\partial w'/\partial z}$, are of opposite sign in the equations for σ_u^2, σ_v^2 and σ_w^2. Hence, they describe the energy exchange between longitudinal and transverse fluctuations. To clarify their role we consider the particular case of a plane-parallel mean flow, aligned with the X-axis and with neutral stratification ($\bar{v} = m_z = 0$). Integrating (2.16) over a volume that has zero turbulent kinetic energy fluxes at its boundaries, we obtain

$$\int \overline{p' \frac{\partial v'}{\partial y}} \, dV = \int \rho_0 \epsilon_v \, dV > 0, \qquad \int \overline{p' \frac{\partial w'}{\partial z}} \, dV = \int \rho_0 \epsilon_w \, dV > 0;$$

$$\int \tau_{xz} \frac{\partial \bar{u}}{\partial z} \, dV = \int \overline{p' \frac{\partial v'}{\partial y}} \, dV + \int \overline{p' \frac{\partial w'}{\partial z}} \, dV + \tag{2.17}$$

$$+ \int \rho_0 \epsilon_u \, dV = \int \rho_0 \epsilon \, dV > 0.$$

The energy of the mean motion is seen to be transferred in a direct way to the longitudinal fluctuations u' only, while the transverse fluctuations v' and w' receive their energy from the longitudinal ones by means of pressure fluctuations. This implies that the pressure fluctuations contribute toward the partition of energy among the fluctuations in various directions, thus creating the possibility that the turbulence becomes isotropic.

The turbulent energy equation for the SSHH model can be derived by summing the three component equations in (2.16):

$$-\frac{\partial}{\partial z}\left(\rho_0 \overline{\frac{u'^2 + v'^2 + w'^2}{2} w'} + \overline{p'w'} - \rho_0 \nu \frac{\partial b^2}{\partial z}\right) +$$

$$+ \left(\tau_{xz} \frac{\partial \bar{u}}{\partial z} + \tau_{yz} \frac{\partial \bar{v}}{\partial z}\right)(1 - \mathrm{Rf}) - \rho_0 \epsilon = 0, \tag{2.18}$$

where

$$\mathrm{Rf} = -g m_z \left(\tau_{xz} \frac{\partial \bar{u}}{\partial z} + \tau_{yz} \frac{\partial \bar{v}}{\partial z}\right)^{-1}. \tag{2.19}$$

Here the ratio of the work done by the buoyancy forces (with a minus sign) to that performed by the Reynolds stresses, Rf, is called the flux Richardson number. This number is negative for unstable stratification and positive for stable stratification. In the latter case, the turbulence obtains energy from the averaged motion, i.e., $\tau_{xz}(\partial \bar{u}/\partial z) + \tau_{yz}(\partial \bar{v}/\partial z) > 0$, but loses it again by counteracting the buoyancy forces ($g m_z < 0$). The joint effect of these terms can compensate for the viscous energy dissipation only if $1 - \mathrm{Rf} > 0$. In other words, $\mathrm{Rf} < 1$ is a necessary condition for SSHH turbulence. This is called the Richardson criterion. It is rather approximate, a more realistic criterion being $\mathrm{Rf} < \mathrm{Rf}_{cr} < 1$. Note that the first term in (2.18) is a flux divergence of turbulent energy, which represents energy partition in space. At rather low values of Rf this term is sometimes set proportional to $g m_z$. With the proportionality coefficient designated as $\sigma - 1$ (>0), the Richardson criterion takes the form $1 - \sigma \mathrm{Rf} > 0$, i.e., $\mathrm{Rf}_{cr} = 1/\sigma$.

In the SSHH model, the equations for the non-diagonal elements of the Reynolds stress tensor assume the form, obtained from (2.14):

$$
\begin{aligned}
-\frac{\partial \tau_{xy}}{\partial t} &= -\frac{\partial}{\partial z}\left(\rho_0 \overline{u'v'w'} + \nu \frac{\partial \tau_{xy}}{\partial z}\right) + \overline{p'\left(\frac{\partial u'}{\partial y} + \frac{\partial v'}{\partial x}\right)} + \\
&\quad + \left(\tau_{xz}\frac{\partial \bar{v}}{\partial z} + \tau_{yz}\frac{\partial \bar{u}}{\partial z}\right) = 0; \\
-\frac{\partial \tau_{xz}}{\partial t} &= -\frac{\partial}{\partial z}\left(\rho_0 \overline{u'w'^2} + \overline{p'u'} + \nu \frac{\partial \tau_{xz}}{\partial z}\right) + \\
&\quad + \overline{p'\left(\frac{\partial u'}{\partial z} + \frac{\partial w'}{\partial x}\right)} - \rho_0 \sigma_w^2 \frac{\partial \bar{u}}{\partial z} + g m_x = 0; \\
-\frac{\partial \tau_{yz}}{\partial t} &= -\frac{\partial}{\partial z}\left(\rho_0 \overline{v'w'^2} + \overline{p'v'} + \nu \frac{\partial \tau_{yz}}{\partial z}\right) + \overline{p'\left(\frac{\partial v'}{\partial z} + \frac{\partial w'}{\partial y}\right)} - \\
&\quad - \rho_0 \sigma_w^2 \frac{\partial \bar{v}}{\partial z} + g m_y = 0.
\end{aligned}
\tag{2.20}
$$

Equations for temperature and salinity fluctuations are derived in a similar way. Within the SSHH model they are:

$$
\begin{aligned}
\frac{1}{2}\frac{\partial c_p \rho_0 \sigma_T^2}{\partial t} &= -\frac{1}{2}\frac{\partial}{\partial z}\left(c_p \rho_0 \overline{T'^2 w'} - c_p \rho_0 \chi \frac{\partial \sigma_T^2}{\partial z}\right) - \\
&\quad - q_z\left(\frac{\partial \bar{T}}{\partial z} - g\rho_0 \Gamma\right) - c_p \rho_0 \epsilon_T = 0; \\
\frac{1}{2}\frac{\partial \rho_0 \sigma_S^2}{\partial t} &= -\frac{1}{2}\frac{\partial}{\partial z}\left(\rho_0 \overline{S'^2 w'} - \rho_0 D \frac{\partial \sigma_S^2}{\partial z}\right) - \\
&\quad - I_z \frac{\partial \bar{S}}{\partial z} - \rho_0 \epsilon_S = 0,
\end{aligned}
\tag{2.21}
$$

where $\epsilon_T = \chi\overline{|\nabla T'|^2}$ and $\epsilon_S = D\overline{|\nabla S'|^2}$ are the rates at which temperature and salinity inhomogeneities are dissipated by molecular diffusion. The second terms on the right-hand side of (2.21) describe the generation of temperature and salinity fluctuations. To make these pulsations stationary, these terms must compensate for the dissipation; hence, they must be positive: $q_z(\partial\bar{T}/\partial z - g\rho_0\Gamma) < 0$ and $I_z(\partial\bar{S}/\partial z) < 0$, which means that the heat and salt fluxes must be directed against the gradients of the corresponding mean fields.

In the framework of the SSHH model, let us find expressions for the other one-point moments of second order, that is for the turbulent heat and salt fluxes q_k and I_k, and for the correlation $\overline{T'S'}$. For instance, the equations for q_k become

$$\frac{\partial q_x}{\partial t} = -\frac{\partial}{\partial z}\left(c_p\rho_0\overline{T'u'w'} - c_p\rho_0\nu\overline{T'\frac{\partial u'}{\partial z}} - c_p\rho_0\chi\overline{u'\frac{\partial T'}{\partial z}}\right) -$$

$$-q_z\frac{\partial\bar{u}}{\partial z} + c_p\tau_{xz}\left(\frac{\partial\bar{T}}{\partial z} - g\rho_0\Gamma\right) + c_p\overline{p'\frac{\partial T'}{\partial x}} = 0;$$

$$\frac{\partial q_y}{\partial t} = -\frac{\partial}{\partial z}\left(c_p\rho_0\overline{T'v'w'} - c_p\rho_0\nu\overline{T'\frac{\partial v'}{\partial z}} - c_p\rho_0\chi\overline{v'\frac{\partial T'}{\partial z}}\right) -$$

$$-q_z\frac{\partial\bar{v}}{\partial z} + c_p\tau_{yz}\left(\frac{\partial\bar{T}}{\partial z} - g\rho_0\Gamma\right) + c_p\overline{p'\frac{\partial T'}{\partial y}} = 0; \tag{2.22}$$

$$\frac{\partial q_z}{\partial t} = -\frac{\partial}{\partial z}\left(c_p\rho_0\overline{T'w'^2} + c_p\overline{T'p'} - c_p\rho_0\nu\overline{T'\frac{\partial w'}{\partial z}} -\right.$$

$$\left.-c_p\rho_0\chi\overline{w'\frac{\partial T'}{\partial z}}\right) - c_p\rho_0\sigma_w^2\left(\frac{\partial\bar{T}}{\partial z} - g\rho_0\Gamma\right) + c_p\overline{p'\frac{\partial T'}{\partial z}} +$$

$$+ c_p g\rho_0(-\alpha\sigma_T^2 + \beta\overline{T'S'}) - c_p\rho_0(\nu+\chi)\overline{\nabla T'\cdot\nabla w'} = 0.$$

The equations for I_k become

$$\frac{\partial I_x}{\partial t} = -\frac{\partial}{\partial z}\left(\rho_0\overline{S'u'w'} - \rho_0\nu\overline{S'\frac{\partial u'}{\partial z}} - \rho_0 D\overline{u'\frac{\partial S'}{\partial z}}\right) -$$

$$-I_z\frac{\partial\bar{u}}{\partial z} + \tau_{xz}\frac{\partial\bar{S}}{\partial z} + \overline{p'\frac{\partial S'}{\partial x}} = 0;$$

$$\frac{\partial I_y}{\partial t} = -\frac{\partial}{\partial z}\left(\rho_0\overline{S'v'w'} - \rho_0\nu\overline{S'\frac{\partial v'}{\partial z}} - \rho_0 D\overline{v'\frac{\partial S'}{\partial z}}\right) -$$

$$-I_z\frac{\partial\bar{v}}{\partial z} + \tau_{yz}\frac{\partial\bar{S}}{\partial z} + \overline{p'\frac{\partial S'}{\partial z}} = 0; \tag{2.23}$$

$$\frac{\partial I_z}{\partial t} = -\frac{\partial}{\partial z}\left(\rho_0\overline{S'w'^2} + \overline{S'p'} - \rho_0\nu\overline{S'\frac{\partial w'}{\partial z}} - \rho_0 D\overline{w'\frac{\partial S'}{\partial z}}\right) -$$

$$-\rho_0\sigma_w^2\frac{\partial\bar{S}}{\partial z} + \overline{p'\frac{\partial S'}{\partial z}} + g\rho_0(-\alpha\overline{T'S'} + \beta\sigma_S^2) -$$

$$-\rho_0(\nu+D)\overline{\nabla S'\cdot\nabla w'} = 0.$$

Finally, the correlation $\overline{T'S'}$ is governed by

$$\frac{\partial \overline{T'S'}}{\partial t} = -\frac{\partial}{\partial z}\left(\overline{T'S'w'} - \chi \overline{S'\frac{\partial T'}{\partial z}} - D\overline{T'\frac{\partial S'}{\partial z}}\right) -$$

$$-\frac{1}{\rho_0} I_z \left(\frac{\partial \overline{T}}{\partial z} - g\rho_0 \Gamma\right) - \frac{1}{c_p \rho_0} q_z \frac{\partial \overline{S}}{\partial z} - (\chi + D)\overline{\nabla T' \cdot \nabla S'} \qquad (2.24)$$

Equations for one-point moments of second order similar to (2.16), (2.20)–(2.22) were first considered in the context of the atmospheric boundary layer (Monin, 1965a, 1965b, 1965c).

Note that, in the SSHH model, turbulence can be considered locally homogeneous and locally symmetric in the vertical direction (Monin, 1965a). The probability distributions of the *differences* in the hydrodynamic variables at sufficiently close-spaced points then depend *explicitly* on the *differences* between the coordinates of the observation points, rather than on the coordinates themselves. The probability distributions can implicitly depend on the depths of the observation points through the parameters ϵ, ϵ_T and ϵ_S. They must be invariant under rotations of the coordinate system around the vertical axis and also under mirror reflections in any vertical plane. If the spatial *correlation functions* (the two-point instantaneous moments) of the fluctuations can be characterized by the general expression

$$\overline{u'_{j_1}(\mathbf{x}_1, t) u'_{j_2}(\mathbf{x}_2, t)} = B_{j_1}^{j_2}(\mathbf{r}_h, z_1, z_2), \qquad (2.25)$$

where $\mathbf{r}_h = \mathbf{x}_{2h} - \mathbf{x}_{1h}$ is the horizontal distance between the observation points, then the spatial structure functions (the two-point moments of spatial fluctuation differences) can be represented by a more specific expression, derived on the basis of local turbulence properties:

$$\overline{[u'_{j_1}(\mathbf{x}_2, t) - u'_{j_1}(\mathbf{x}_1, t)]\,[u'_{j_2}(\mathbf{x}_2, t) - u'_{j_2}(\mathbf{x}_1, t)]} = D_{j_1 j_2}(\mathbf{r}), \qquad (2.26)$$

where $\mathbf{r} = \mathbf{x}_2 - \mathbf{x}_1$. The scalar structure functions then depend on the arguments r_h and r_z, while the structure functions with tensor indices are linear combinations of constant tensors and those composed of the vectors r_i and λ_i (λ_i is a vertically directed unit vector). Since the one-point moments $\overline{\nabla u'_{j_1} \cdot \nabla u'_{j_2}}$, used in (2.16), (2.20)–(2.23), are expressed through the derivatives of the structure functions at zero, the formulated properties of the structure functions lead to the following equations:

$$\begin{aligned} \overline{\nabla u'_i \cdot \nabla u'_j} &= \overline{|\nabla u'|^2}\,\delta_{ij} + (\overline{|\nabla w'|^2} - \overline{|\nabla u'|^2})\lambda_i \lambda_j; \\ \overline{\nabla T' \cdot \nabla u'_i} &= \overline{\nabla T' \cdot \nabla w'}\,\lambda_i. \end{aligned} \qquad (2.27)$$

Hence, $\overline{|\nabla v'|^2} = \overline{|\nabla u'|^2}$, $\overline{\nabla u'_i \cdot \nabla u'_j} = 0$ for $i \neq j$ and $\overline{\nabla T' \cdot \nabla u'_j} = 0$ for $j \neq 3$ (here S' can be substituted for T'). The terms of this type that tend towards zero have already been omitted from (2.20), (2.22)–(2.23).

3. MECHANISMS OF TURBULENCE GENERATION IN THE OCEAN

Possible mechanisms of turbulence generation can be found in Monin (1970a). We consider them in turn.

3.1. Instability of Vertical Velocity Gradients in Drifting Flow

These are induced by the direct effect of the wind on the ocean surface. This effect is felt throughout the upper mixed layer (UML), which is quasi-homogeneous, i.e., neutrally stratified because of the mixing. The instability of drift currents and of neutrally stratified shear flows is determined by the Reynolds criterion $\mathrm{Re} > \mathrm{Re}_{cr}$. In the Reynolds number (1.2) the length scale L should be replaced by the vertical integral scale (external scale) of the turbulence, expressed here by the UML depth δ, while the velocity scale U should be replaced by the velocity difference across this layer, which is the drift velocity U at the surface. Taking into account the exact (though not total) analogy between the UML and the boundary layer on a flat plate, Re_{cr} can be set equal to $\mathrm{Re}_{\delta\, cr} \simeq 3000$ for the boundary layer, so that turbulence is generated by drifting flows if

$$\mathrm{Re} = \frac{U\delta}{\nu} > \mathrm{Re}_{cr} \simeq 3000. \tag{3.1}$$

The depth δ can be easily estimated for a stationary UML under summer conditions, which compensate for mechanical mixing effects. In this case (Kitaigorodsky, 1960)

$$\delta = 2\,\frac{\rho_0 u_*^3}{g|m_z|}, \tag{3.2}$$

where u_* and m_z are the friction velocity and the vertical mass flow at the surface, respectively. Giving u_* the typical value 1 cm s^{-1} and m_z the value 10 g cm^{-2} year^{-1}, by Monin's estimate (Monin, 1970b) we obtain $\delta \simeq 60$ m. A map of average annual m_z values (Agafonova and Monin, 1972) is shown in Figure 3.1. With u_* and δ known, the velocity U can be estimated by an empirical formula from boundary layer theory

$$U/u_* = 8.74(u_*\delta/\nu)^{1/7}. \tag{3.3}$$

Hence, at $u_* = 1$ cm s^{-1}, $\delta = 60$ m and $\nu = 10^{-2}$ cm^2 s^{-1}, we have $U \simeq 30$ cm s^{-1}. The Reynolds number (3.1) is then approximately 1.8×10^7, so that (3.1) is satisfied with a wide margin.

3.2. Overturning of Surface Waves

This, and (to a lesser extent) hydrodynamic instability of UML wave motions induced by surface waves, is the most common and, apparently, the most powerful mechanism of UML turbulence. It has not yet been quantitatively estimated. Note, however, data by Longuet-Higgins (1969), which show that turbulence consumes about 10^{-4} of the wave energy per cycle. The total energy of surface gravity waves per unit area is $\frac{1}{2}(g\rho_0 a^2)$, where a is a typical wave amplitude. Therefore, the turbulence energy generation rate per unit water mass is

$$\epsilon = ga^2/2\tau\delta \times 10^{-4}, \tag{3.4}$$

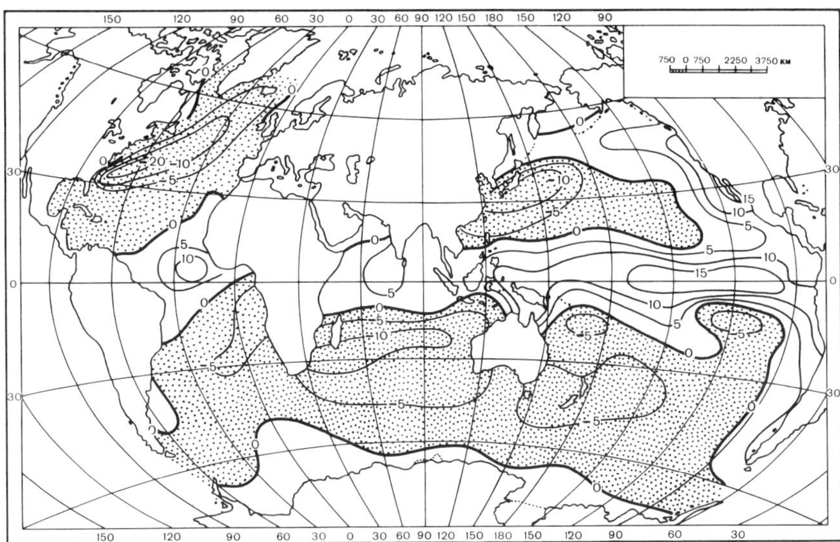

Fig. 3.1. Distribution of the mass flux M in the World Ocean (g cm^2 °C units) (Agafonova and Monin, 1972).

where τ is a typical wave period. For a = 3 m, τ = 15 s and δ = 60 m, we have $\epsilon \simeq 5 \times 10^{-2}$ cm^2 s^{-3}. The dependence of the wave-generated turbulence energy on the wave characteristics and on the depth was analyzed by Benilov (1973) within the framework of similarity theory.

3.3. Instability of Vertical Velocity Gradients in Stratified Large-Scale Oceanic Flows

To derive the criterion of hydrodynamic instability for such flows, let us consider the linearized equations for adiabatic (i.e., isentropic and isohaline (Monin, 1973b) disturbances in SSHH flows. These have the form

$$\frac{\bar{d}\mathbf{u}'}{dt} + w'\frac{\partial \bar{\mathbf{u}}}{\partial z} = -\frac{\nabla \pi'}{\rho_*} - N^2 \zeta' \lambda; \qquad \frac{\bar{d}\zeta'}{dt} = w'; \qquad \operatorname{div} \mathbf{u}' = 0, \tag{3.5}$$

where $\bar{d}/dt = \partial/\partial t + \bar{\mathbf{u}} \cdot \nabla$. Also, $\pi' = \rho_* p'/\rho_0$ is the standardized pressure disturbance; $\zeta' = g(p' - c^2 \rho')/\rho_0 c^2 N^2$ is the vertical shift of liquid particles; and N is the so-called Brunt–Väisälä frequency, determined by

$$N^2 = \frac{g}{\rho_*}\frac{\partial \rho_*}{\partial z} = \frac{g}{\rho_0}\left(\frac{\partial \rho_0}{\partial z} - \frac{g\rho_0}{c^2}\right). \tag{3.6}$$

Here, ρ_* is the so-called *potential density*, i.e, the density adiabatically reduced to a standard pressure. For $\partial\rho_*/\partial z > 0$ or $N^2 > 0$ the buoyancy force counteracts the particle shifts in the vertical direction so that the stratification becomes stable; while for $\partial\rho_*/\partial z < 0$ or $N^2 < 0$ the stratification is unstable. Expressing the disturbances as elementary plane waves propagating, say, along the X-axis (so that they depend on x and t by the law $\exp[ik(x - ct)]$ and do not depend on y, due to the continuity equation $\partial u'/\partial x$ +

$\partial w'/\partial z = 0$) one can derive the stream function ψ by setting $u' = -\partial\psi/\partial z$ and $w' = \partial\psi/\partial x$. In this case, its complex amplitude $\psi(z)$ can be derived from (3.5). This yields

$$\frac{\partial}{\partial z}\left(\rho_* U \frac{\partial \Psi}{\partial z}\right) - \left\{\frac{1}{2}\frac{\partial}{\partial z}\left(\rho_* \frac{\partial \bar{u}}{\partial z}\right) + k^2 \rho_* U + \right.$$
$$\left. + \frac{\rho_*}{U}\left[\frac{1}{4}\left(\frac{\partial \bar{u}}{\partial z}\right)^2 - N^2\right]\right\} \Psi = 0, \tag{3.6'}$$

where $U = \bar{u} - c$ and $\Psi = \psi U^{-1/2}$. Assuming the phase velocity of the disturbance waves, c, to have a non-zero imaginary part, we now multiply (3.6) by the complex conjugate Ψ^* and integrate it over the ocean depth (with the boundary conditions $w' = 0$ and hence $\Psi = 0$ at $z = 0, H$). The imaginary part of the relation obtained in this way is

$$(\mathrm{Im}\, c)\int_0^H \left\{\left|\frac{\partial \Psi}{\partial t}\right|^2 + k^2|\Psi|^2 - \left[\frac{1}{4}\left(\frac{\partial \bar{u}}{\partial z}\right)^2 - N^2\right]\left|\frac{\Psi}{U}\right|^2\right\} \rho_*\, \mathrm{d}z = 0. \tag{3.7}$$

If the expression in square brackets is not positive throughout, i.e., if

$$\mathrm{Ri} = N^2\left(\frac{\partial \bar{u}}{\partial z}\right)^{-2} \geqslant \frac{1}{4}, \tag{3.8}$$

then (3.7) holds only at $\mathrm{Im}\, c = 0$, i.e., under the condition that all elementary wave disturbances are neutral. Thus, the inequality (3.8) proves to be a *sufficient* condition for hydrodynamic *stability* of stratified SSHH flows. Consequently, the fulfilment of the condition $\mathrm{Ri} < \frac{1}{4}$ in some oceanic layers is a *necessary* prerequisite of hydrodynamic *instability* in stratified flows. This criterion was first obtained by Miles (1961) and Howard (1961). The quantity Ri in (3.8) is called the *Richardson number*. Sometimes it is also called the *gradient* or *local* Richardson number as opposed to the flux Richardson number (2.19).

In order to establish the relationship between the gradient and flux Richardson numbers, let us consider (2.20) for τ_{xz}, (2.22) for q_x and (2.23) for I_x, neglecting the terms with vertical flux divergences of these second moments, or assuming the divergences to be proportional to the terms

$$\overline{p'\left(\frac{\partial u'}{\partial z} + \frac{\partial w'}{\partial x}\right)}, \qquad \overline{c_p p' \frac{\partial T'}{\partial x}} \quad \text{and} \quad \overline{p' \frac{\partial S'}{\partial x}},$$

respectively, and employing the semi-empirical equations

$$\begin{aligned}
\overline{p'\left(\frac{\partial u_i'}{\partial x_j} + \frac{\partial u_j'}{\partial x_i}\right)} &= c_1 \frac{b}{l}(\tau_{ij} + \tfrac{2}{3}\rho_0 b^2 \delta_{ij}) - \\
&\quad - \tilde{c}_1 \rho_0 \frac{b^3}{l}(\lambda_i \lambda_j - \tfrac{1}{3}\delta_{ij}); \\
\overline{p' \frac{\partial T'}{\partial x_k}} &= -c_2 \frac{b}{l}\frac{q_k}{c_p} - \tilde{c}_2 \frac{\rho_0 b^2 \sigma_T}{l}\lambda_k; \\
\overline{p' \frac{\partial S'}{\partial x_k}} &= -c_2 \frac{b}{l} I_k - \tilde{c}_2 \frac{\rho_0 b^2 \sigma_S}{l}\lambda_k.
\end{aligned} \tag{3.9}$$

Here, l is the turbulence scale, while c_1, c_2, $\tilde{c}_1$ and $\tilde{c}_2$ are numerical constants (semi-empirical formulas of this type are discussed in Monin, 1965a, b, c). Equations (3.9) then take the form

$$c_1 \frac{b}{l} \tau_{xz} + g m_x = \rho_0 \sigma_w^2 \frac{\partial \bar{u}}{\partial z};$$

$$\tau_{xz} \left(\frac{\partial \bar{T}}{\partial z} - g \rho_0 \Gamma \right) - c_2 \frac{b}{l} \frac{q_x}{c_p} = \frac{q_z}{c_p} \frac{\partial \bar{u}}{\partial z}; \tag{3.10}$$

$$\tau_{xz} \frac{\partial \bar{S}}{\partial z} - c_2 \frac{b}{l} I_x = I_z \frac{\partial \bar{u}}{\partial z}.$$

Similar equations are also valid if τ_{yz}, m_y, q_y, I_y, $\partial \bar{v}/\partial z$ are substituted for τ_{xz}, m_x, q_x, I_x, $\partial \bar{u}/\partial z$. Taking into account that

$$N^2 = g \left[-\alpha \left(\frac{\partial \bar{T}}{\partial z} - g \rho_0 \Gamma \right) + \beta \frac{\partial \bar{S}}{\partial z} \right] \quad \text{and} \quad m_k = -\alpha \frac{q_k}{c_p} + \beta I_k,$$

we derive the following expressions for τ_{xz} and τ_{yz}:

$$\tau_{xz} = \rho_0 K \frac{\partial \bar{u}}{\partial z}; \qquad \tau_{yz} = \rho_0 K \frac{\partial \bar{v}}{\partial z};$$

$$K = \left(c_2 \frac{b}{l} \sigma_w^2 + g \frac{m_z}{\rho_0} \right) \left(c_1 c_2 \frac{b^2}{l^2} + N^2 \right)^{-1}. \tag{3.11}$$

Here, K is the kinematic eddy viscosity. Note that the quantities m_x, q_x and I_x are also proportional to $\partial \bar{u}/\partial z$, while m_y, q_y and I_y are proportional to $\partial \bar{v}/\partial z$ and τ_{xy} can be proved to be proportional to the product $(\partial \bar{u}/\partial z) \cdot (\partial \bar{v}/\partial z)$. Let us also introduce a similar eddy diffusivity for mass, K_ρ, by

$$g m_z = -\rho_0 K_\rho N^2. \tag{3.12}$$

Substituting τ_{xz}, τ_{yz} and $g m_z$ from (3.11), (3.12) into the definition (2.19) of the flux Richardson number, we obtain

$$\text{Rf} = \alpha_\rho \text{Ri}; \quad \alpha_\rho = \frac{K_\rho}{K}; \quad \text{Ri} = N^2 \left[\left(\frac{\partial \bar{u}}{\partial z} \right)^2 + \left(\frac{\partial \bar{v}}{\partial z} \right)^2 \right]^{-1}. \tag{3.13}$$

This definition of Ri is a straightforward generalization of (3.8).

Large-scale oceanic flows are *quasi-hydrostatic*. Except in a narrow equatorial zone they are also *quasi-geostrophic* everywhere, i.e.,

$$\bar{u} \simeq -\frac{1}{f\bar{\rho}} \frac{\partial \bar{p}}{\partial y}; \qquad \bar{v} \simeq \frac{1}{f\bar{\rho}} \frac{\partial \bar{p}}{\partial x}. \tag{3.14}$$

Hence

$$\frac{\partial \bar{u}}{\partial z} \simeq -\frac{g}{f\bar{\rho}} \frac{\partial \bar{\rho}}{\partial y} - \frac{\bar{u}}{\bar{\rho}} \frac{\partial \bar{\rho}}{\partial z} \simeq -\frac{g}{f\bar{\rho}} \frac{\partial \bar{\rho}}{\partial y};$$

$$\frac{\partial \bar{v}}{\partial z} \simeq \frac{g}{f\bar{\rho}} \frac{\partial \bar{\rho}}{\partial x} - \frac{\bar{v}}{\bar{\rho}} \frac{\partial \bar{\rho}}{\partial z} \simeq \frac{g}{f\bar{\rho}} \frac{\partial \bar{\rho}}{\partial x}. \tag{3.15}$$

The corresponding equations in atmospheric physics are called the thermal wind equations. Hence

$$\left|\frac{\partial \bar{\mathbf{u}}}{\partial z}\right| \simeq \frac{g}{f\bar{\rho}} |\nabla_h \bar{\rho}| \simeq \frac{\eta N^2}{f},$$

where $\eta \simeq |\nabla_h \bar{\rho}| \cdot (\partial\bar{\rho}/\partial z)^{-1}$ is the slope of the isopycnic surface with respect to the horizontal plane. Substituting this value of $|\partial\bar{\mathbf{u}}/\partial z|$ into (3.13) we obtain the Richardson criterion for the instability of large-scale flow:

$$\mathrm{Ri} = \left(\frac{f}{N\eta}\right)^2 < \mathrm{Ri}_{\mathrm{cr}}. \tag{3.16}$$

The rate of turbulent kinetic energy generation per unit mass by large-scale flows (scales of the order of several thousand kilometers) is in all probability $\epsilon \simeq 10^{-5}$ cm^2 s^{-3}.

3.4. Hydrodynamic Instability of Quasi-Horizontal Meso-Scale Non-Stationary Flows

These are induced, for example, by tidal and inertial oscillations at scales of tens of kilometers. In all likelihood, here $\epsilon \simeq 10^{-3}$ cm^2 s^{-3}. Empirical evidence for the generation of small-scale turbulence of this kind can be found, e.g., in Lemmin *et al.* (1975).

3.5. Instability of Local Velocity Gradients in Internal Waves

This, as well as the overturning of these waves, is a universal, and apparently basic, mechanism of small-scale turbulence generation throughout the depth of the ocean (below the upper mixed layer, and especially in the pycnocline where the greatest energy density of internal waves is concentrated).

A large fraction of internal wave energy is usually observed in the lowest mode. To obtain the criterion for its hydrodynamic instability, let us calculate the Richardson number (3.8) for the case of a wave which has a large wavelength compared with the depth h of the pycnocline (density jump layer) but still has a frequency ω_0 substantially exceeding the inertial frequency f. In such a wave the pycnocline oscillates practically as a whole, so that the derivative $\partial\zeta/\partial z$ is small, and the internal wave equation (1.12) assumes the form

$$\frac{\partial^2 \zeta}{\partial z^2} + k^2 \frac{N^2 - \omega_0^2}{\omega_0^2} \zeta = 0.$$

Differentiating (1.14) for u with respect to z, we obtain the horizontal velocity shear in the form

$$\frac{\partial u}{\partial z} \simeq \frac{k_1 \omega_0}{k^2} \frac{\partial^2 \zeta}{\partial z^2} \simeq k_1 \frac{\omega_0^2 - N^2}{\omega_0} \zeta.$$

Hence, the Richardson number $\mathrm{Ri} = N^2 (\partial u/\partial z)^{-2}$ can be written as

$$\mathrm{Ri} \simeq \left(\frac{\omega_0}{N} - \frac{N}{\omega_0}\right)^{-2} (k_1 \zeta)^{-2},$$

and the criterion Ri $< \frac{1}{4}$ becomes (Phillips, 1967)

$$k_1 \zeta > 2 \left(\frac{N}{\omega_0} - \frac{\omega_0}{N} \right)^{-1}. \tag{3.17}$$

If this requirement is satisfied, the vertical velocity gradients in internal waves lose their stability first in the regions of the wave crests and hollows, thereby giving rise to turbulence. Ivanov *et al.* (1974), Morozov (1974), and Sanford (1975) reported that the Richardson criterion was satisfied in their experiments. Direct investigations of internal wave instability and the resulting small-scale turbulence were performed by Beliayev *et al* (1975). The hydrodynamic stability limit of internal waves (Ri = $\frac{1}{4}$ or, according to (3.17), $\zeta \simeq 2\omega_0/(Nk)$) will determine the limiting two-dimensional wave spectrum $E(k) \simeq \zeta^2/k^2 \simeq 4\omega_0^2/(N^2 k^4)$. The frequency ω_0 of the lowest mode of internal waves is determined by the relation

$$\omega_0^2 = \frac{\Delta\rho}{\rho_0} gk(1 + \coth k\delta)^{-1},$$

where δ, as above, is the UML depth, $\Delta\rho$ is the vertical density difference in the pycnocline, and $h \simeq (g/N^2) \cdot (\Delta\rho/\rho_0)$ is its thickness. Hence, for the limiting two-dimensional spectrum we have

$$E(k) \simeq h(1 + \coth k\delta)^{-1} k^{-3}. \tag{3.18}$$

This spectrum is proportional to k^{-3} for comparatively short waves ($k\delta \gg 1$, $\omega_0^2 \simeq k$) and to k^{-2} for long waves ($k\delta \ll 1$, $\omega_0^2 \simeq k^2$). The one-dimensional spectrum is obtained from the two-dimensional one by multiplying the latter by k. The frequency spectrum $E(\omega) \simeq \zeta^2/\omega$ is proportional to ω^{-3} for short waves and to ω^{-1} for long waves. Internal wave spectra of this kind are often observed in the ocean.

Internal waves can turn over when their local acceleration $\omega^2\zeta$ is comparable with g. This is, however, improbable because of their low frequencies. However, Orlanski and Bryan (1969) detected that internal waves propagating in a shear flow can be overturned by convection, which pulls down their crests. The criterion for this process is the inequality $|u| \geqslant \omega/k_h$, where u is the orbital particle velocity in a wave and ω/k_h is the phase wave velocity. With the use of

$$\left|\frac{\partial u}{\partial z}\right| = k_z|u|, \qquad \omega^2 = \frac{k_h^2 N^2}{k_h^2 + k_z^2}, \qquad \text{Ri} = N^2 k_z^{-2} |u|^{-2},$$

this criterion can be written in terms of the Richardson number:

$$\text{Ri} \leqslant 1 + k_h^2/k_z^2. \tag{3.19}$$

This criterion is much less strict than the condition $\omega^2\zeta \simeq g$, especially for waves propagating downwards, e.g., those generated by colliding surface waves (Brekhovskih *et al.*, 1972). Overturning of internal waves characterized by wavelengths of several meters, periods of a few minutes and phase velocities of several centimeters per second, was observed by Woods (1968a, b, c), who took underwater pictures of coloured microstructure layers (see Figure 3.2).

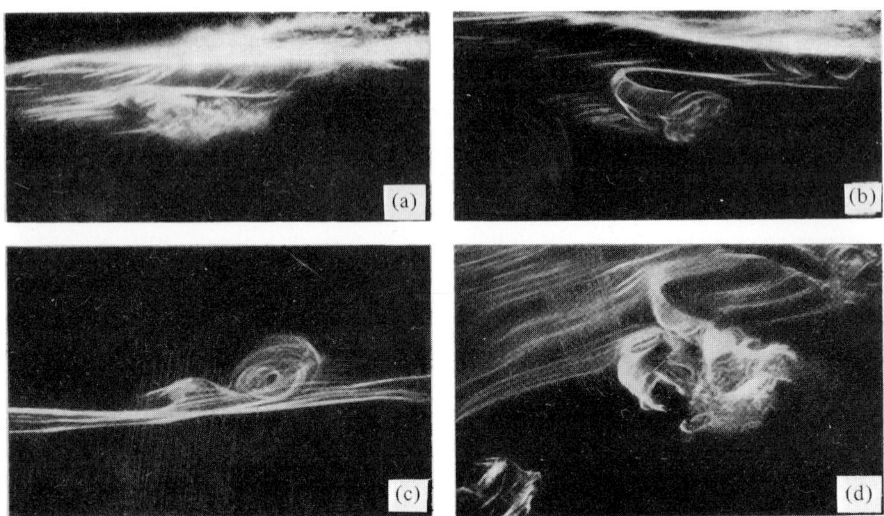

Fig. 3.2. Development of one of four breakers in the bulk ocean (Turner, 1965). (a) Initial steep wave, (b) developed breaker, (c) final stage of a rotating breaker, while the secondary breaker is seen, (d) neighboring breakers seen from above.

Garrett and Munk (1972) estimated the efficiency of turbulence generation by internal waves by employing the climatological energy spectrum of internal waves:

$$E(\mathbf{k}, \omega) = \frac{E}{2j\pi^4} \rho_{00} M^{-4} N_0^3 f \omega^{-1} (\omega^2 - f^2)^{-1} k^{-1};$$
$$f \leqslant \omega \leqslant N_0, \quad 0 \leqslant k \leqslant j\pi M N_0^{-1} (\omega^2 - f^2)^{1/2}. \tag{3.20}$$

Here, $E = 2\pi \times 10^{-5}$, $M = 1.22 \times 10^{-6}$ cm^{-1}, $N_0 = 0.83 \times 10^{-3}$ s^{-1}, and $j = 20$ is the equivalent number of modes. Calculated from this spectrum, the root-mean-square isopycnal slope is $\eta \simeq \frac{1}{38}$, the vertical velocity gradient is $c = |\partial \mathbf{u}/\partial z| = 1.84 \times 10^{-3} (N/2\pi N_0)^{3/2}$ s^{-1} and the Richardson number $\mathrm{Ri} = N^2 c^{-2} = 8.1 (N/2\pi N_0)^{-1}$. Thus, the mean climatological isopycnal slope proves to be one thirty-eighth of the critical slope $\eta = 1$ at which the waves turn over. The mean climatological velocity gradient, however, appears to be only $5.7 (N/2\pi N_0)^{-1/2}$ times less than the value corresponding to the critical Richardson number $\mathrm{Ri} = \frac{1}{4}$. Therefore, at not very low values of N, hydrodynamic instability of internal waves is much more probable than overturning. This conclusion becomes more obvious in the presence of vertical microstructures. As follows from (1.14), the quantities $c = |\partial \mathbf{u}/\partial z|$ and $N^2 \simeq \partial \rho/\partial z$ are approximately proportional. Therefore, in the microstructure the ratio c'/N^2 equals the climatological c/N^2. Hence, Ri' is N'^2/N^2 times lower, i.e., closer to the critical value, than the climatological value of Ri ($N'^2/N^2 \simeq 20$ according to Gregg and Cox, 1972).

Garrett and Munk tried to calculate the rate of turbulent energy generation by internal waves from the expression

$$\rho_0 \epsilon = \int (\tfrac{1}{12} \rho_0 N^2 \Delta^3) Q(\Delta)\, \mathrm{d}\Delta = \rho_0 K_\rho N^2. \tag{3.21}$$

Here Δ is the thickness of turbulent spots, which is determined by the levels at which Ri is equal to the 'Thorpe number', Th, below which decaying turbulence vanishes (Th = 0.27 ± 0.01 according to laboratory measurements by Thorpe, 1973). Also, $Q(\Delta)\,\mathrm{d}\Delta$ is the number of spots with a thickness falling within $\Delta \pm \frac{1}{2}\,\mathrm{d}\Delta$ per unit area in the plane (z, t), $\frac{1}{12}(\rho_0 N^2 \Delta^3)$ is the potential energy change per unit area, and K_ρ is the turbulent mass exchange coefficient given by (3.12). Here, K_ρ is equal to $\frac{1}{12}(\Delta_0^3 n)$, where Δ_0 is obtained from the condition

$$c_{\mathrm{Th}} = \Delta_0^{-1} \int_{-\Delta_0/2}^{+\Delta_0/2} c\,\mathrm{d}z$$

and n is the number of supercritical peaks c per unit area in (z, t).

Setting $c = c_{\mathrm{cr}}(1 - (\pi^2 r^2/2L_0^2))$, we obtain

$$\Delta_0 = \frac{2\sqrt{6}}{\pi} L_0 \left[1 - \left(\frac{\mathrm{Ri}_{\mathrm{cr}}}{\mathrm{Th}}\right)^{1/2}\right]^{1/2} \simeq 0.3 L_0 .$$

Let us take $n = 1/L_0\tau_0$, where τ_0 is the interval between peaks that have a vertical separation of L_0. In this case, $K_\rho = 0.0023 L_0^2 \tau_0^{-1}$. For L_0 Garrett and Munk chose a certain quantity inversely proportional to the root-mean-square vertical wavenumber (based on the velocity shear spectrum). This quantity appeared to be $27(N/2\pi N_0)^{-1}$ m. In this case, to achieve the standard value of $K_\rho = 1\ \mathrm{cm}^2\ \mathrm{s}^{-3}$, there must appear one peak every $0.2(N/2\pi N_0)^{-2}$ days. Independent calculations of τ_0 require data on the probability distribution of c. Garrett and Munk showed by rough estimates that K_ρ rapidly (and exponentially) decreases with N (i.e., with depth). The instability of internal waves in the microstructure increases the turbulence by a very small amount, which only results in a slight dissipation.

3.6. Convection in Layers with Unstable Density Stratification

This results mainly from cooling of the ocean surface during cold seasons (Bulgakov, 1975). Sometimes, convection in these layers can also be induced by salt accumulation in surface waters during periods of intensive evaporation, by bottom heating in locations characterized by a considerable heat flux, or, finally, by lateral intrusions of low-density water, as is often observed in microstructure layers (see Chapter 4 in Fedorov, 1976). Experimental data on the statistical characteristics of turbulent convection are still unavailable. In steady-state conditions, the energy generation rate in turbulent convection induced by buoyancy forces can be calculated by

$$\epsilon = \frac{g|m_z|}{\rho_0}. \tag{3.22}$$

This equals $3 \times 10^{-4}\ \mathrm{cm}^2\ \mathrm{s}^{-3}$ for $|m_z| = 10\ \mathrm{g\ cm}^{-2}\ \mathrm{y}^{-1}$. Convection in microstructure layers will be discussed in the next section.

3.7. Instability of Vertical Velocity Gradients in a Bottom Boundary Layer (BBL)

In many respects, this layer is similar to the atmospheric boundary layer (ABL) over the mainland. In particular, the thickness of the BBL, h, can be estimated, as in the case of ABL, by

$$h = au_*/f. \tag{3.23}$$

Here, u_* is the friction velocity at the surface and the numerical coefficient a depends on the BBL stratification (it decreases with growing stability and increases with growing instability). Due to the continuity of the vertical momentum flux at the ocean surface, we have $\rho_0 u_*^2 = \rho_a u_{*a}^2$, where the index a refers to the values in the atmosphere. Hence, the value of u_* in the UML is $(\rho_0/\rho_a)^{1/2} \simeq 28$ times smaller than that of u_{*a}. If the value of u_* in the BBL is the same, the BBL thickness is 28 times smaller than the ABL thickness for the same stratification. Indeed, according to measurements, $h \simeq 10$ m (see, e.g., Nihoul, 1977) and Re $= 4.5 \times 10^6$ in the BBL, with the Reynolds number based on (3.3). The turbulent energy dissipation (and generation) rate ϵ is proportional to u_*^3. Thus, if $\epsilon \simeq 5$ cm² s⁻³ in the ABL, then $\epsilon \simeq 2 \times 10^{-4}$ cm² s⁻³ in the BBL. In fact, it is somewhat higher since internal waves near the bottom also contribute to turbulence generation in the BBL. Note also that the BBL can be substantially non-stationary in tidal flows.

From the viewpoint of turbulence generation, the ocean can be divided into three layers: (1) the UML, with turbulence maintained largely by overturning surface waves and $\epsilon \simeq 10^{-1} - 10^{-2}$ cm² s⁻³; (2) the bulk ocean, with turbulence generated primarily by internal wave instabilities and $\epsilon \simeq 10^{-5}$ cm² s⁻³; and (3) the BBL, with turbulence created chiefly by Ekman layer instability and $\epsilon \simeq 10^{-4}$ cm² s⁻³.

The following statistical information is available on these layers. The depth of the UML, in which the temperature varies negligibly with depth, is about 100 m in the tropics, 10–20 m at high latitudes in summer and several hundred meters (sometimes reaching the bottom) in winter. The UML has a distinct lower boundary, with an irregular shape which is determined by larger-scale turbulent vortices of the order of the UML thickness and by internal waves. The layer of discontinuity (seasonal thermocline), several meters thick, lies below the UML; its temperature decreases by several degrees with depth. Further down is the main thermocline, whose temperature gradually decreases and whose lower boundary is found at about 1500 m depth (with temperatures of about 10°C at 300 m, 4.5°C at 1000 m and 2.7–3.2°C at 1500 m). With further increasing depth the temperature changes negligibly, reaching 1–1.5°C at the bottom (in the Atlantic Ocean from 2.5° in the North to –0.5°C in the South). The only exception is in polar waters, where a very thin upper layer, heated up in summer, covers the coldest subsurface water. This is followed at 1–2 km by a warmer layer with gradually decreasing temperature and, finally, by the isothermal zone.

The vertical salinity profile between 100 and 200 m depth consists of a subsurface layer with a high salinity (the greatest value along the entire vertical profile), followed at 600–1200 m by an intermediate layer of low salinity (the lowest value along the entire vertical profile), with a deep-water layer of approximately constant salinity located still deeper. However, various ocean areas are characterized by vertical profiles different from those typical of equatorial-tropical waters. In temperate tropical waters, the surface salinity minimum vanishes. In subpolar waters, the salinity increases monotonically with depth and in polar waters the increase is especially rapid in the uppermost layer. Of regional importance are the North-Atlantic type, characterized by a monotonic salinity decrease with depth; the Mediterranean type, with salinity maxima at the surface and at a depth of 500–1000 m; and the Indo–Malay type, with a single salinity maximum at 500 m. This variety of salinity and temperature profiles gives rise to quite different shapes of the so-called T–S-curves, which are plotted on the coordinates S and T, with depths indicated by dots (see Figure 3.3).

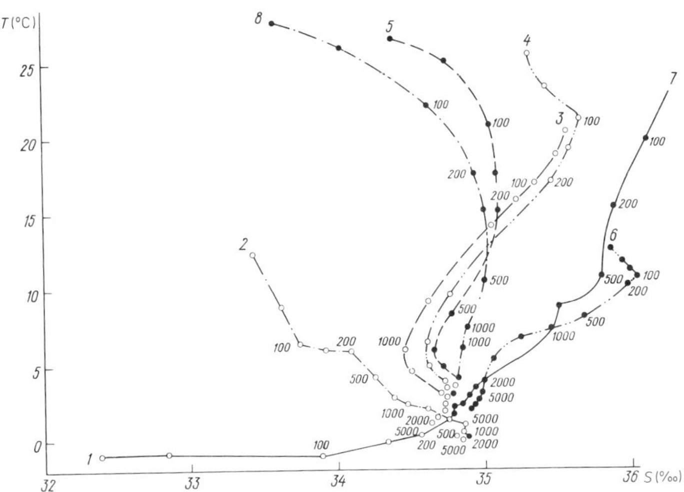

Fig. 3.3. Typical T–S-curves for ocean water (Stepanov, 1965). (1) polar, (2) subpolar, (3) moderate latitudes, (4) tropics, (5) equatorial, (6) North Atlantic Ocean, (7) Mediterranean, (8) Indo–Malay waters. Figures on the T–S-curves show the depths in meters.

Sea water densities reduced to atmospheric pressure at constant T and S can be readily measured in $\sigma_t = 1000(\rho - 1)$ units. The density at pressure p will be $(1 + 10^{-3}\sigma_t)(1 - \mu p)^{-1}$, where $\mu \simeq (4\text{–}5) \times 10^{-5}$ atm^{-1} is the water compressibility coefficient, which decreases slightly with pressure. Both the total density and σ_t increase with depth almost everywhere. Therefore, σ_t is smallest at the ocean surface and throughout the UML. The most rapid density growth, which can be $\delta\sigma_t \simeq 1$ per 10 m of depth, is observed in the layer of discontinuity. The increase in σ_t slows down at still greater depths. The value of σ_t at the bottom is 27.80–27.82 in the northern part of the Pacific Ocean, 27.87 in the Antarctic Ocean and 28.10 in the Arctic Ocean. Instead of $\partial\rho/\partial z$ it is convenient to consider the Brunt–Väisälä frequency N, defined by (3.6), which usually increases with depth from the ocean surface up to the layer of discontinuity, where the period $2\pi/N$ is some 10 min. From the layer of discontinuity down to the bottom this frequency decreases by dozens of units. The smooth curve is superposed by a sequence of pronounced maxima in the microstructure sublayer.

In the UML the turbulent velocity fluctuations are, as a rule, of the order 1 cm s^{-1} and decrease rapidly with depth. At the ocean surface ϵ is about $10\text{–}10^{-1}$ cm^2 s^{-3}, while in the discontinuity layer it reduces, on average, to $10^{-3}\text{–}10^{-4}$. Turbulent temperature fluctuations first decrease rapidly with depth, but they become greatest (of the order 10^{-1} °C) in the seasonal thermocline, where the vertical temperature gradient is very high. It is likely that the fluctuations of salinity, electric conductivity (some

10^{-4}–10^{-6} Ω^{-1} cm^{-1}) and the speed of sound behave in the same manner. In the UML ϵ_T varies, in all likelihood, within 10^{-3}–10^{-8} (°C)^2 s^{-1}, and $\epsilon_S \simeq 10^{-7}$–$10^{-8}$ (‰)^2 s^{-1}.

Throughout the bulk of the ocean, turbulence is distributed not uniformly but in patches that arise from internal wave instability and in microstructure layers that result from the spreading of these patches. This turbulence distribution can be characterized by the intermittency coefficient $p(z)$, i.e., the average area fraction occupied by turbulence at depth z. The coefficient $p(z)$ was measured by Grant *et al.* (1963) from a submarine, and was found to be unity in a 50 m thick UML. It decreased to 0.05 at 100 m depth, with subsequent negligible changes down to 300 m depth. Most likely, in the bulk of the ocean $p(z)$ is about 10^{-2}. To characterize the intermittency of turbulence in more detail, Kolmogorov demonstrated the variability of structure functions obtained by the moving-average method. As an example, Figure 3.4 depicts the time evolution of the structure function $D_{\sigma\sigma}(r)$ of electrical conductivity at horizontal shifts r of 1 and 30 cm. Examples of this type demonstrate that the small-scale structure of turbulence varies with periods characteristic of internal waves (periods of several minutes).

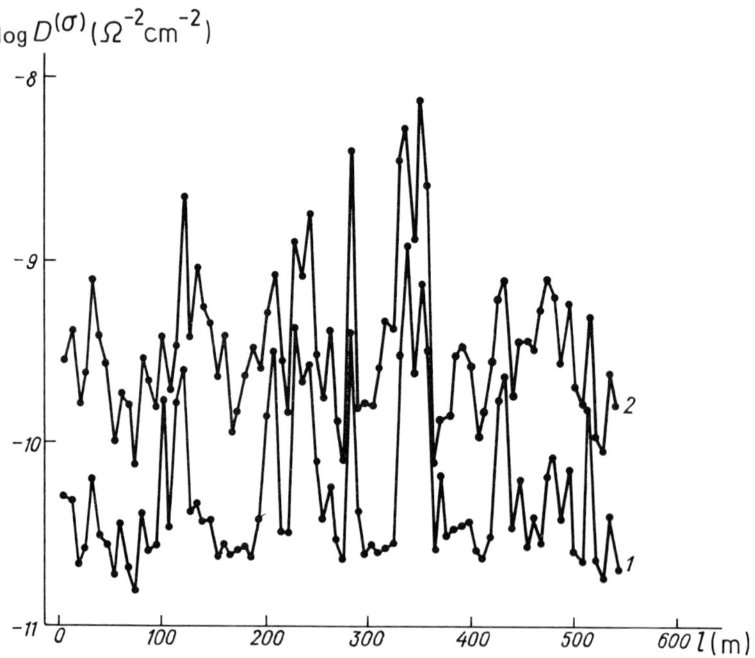

Fig. 3.4. Varying structure function of electric conductivity fluctuations, plotted from measurements taken using a towed device for (1) a 1 cm interval, and (2) a 30 cm interval.

Monin *et al.* (1970) have determined that, in the bulk ocean, the Brunt–Väisälä frequency N decreases with depth, as $N = w_*/z$ where w_* is a constant approximately equal to 2.2 m s^{-1} which varies negligibly in different ocean regions (Figure 3.5). On the other hand, they also found that next to the 'law of depth', $Nz = w_*$, there is, in the bulk ocean, also a 'law of distance from the bottom', $N = (\Gamma/L)(H - z)$, where H

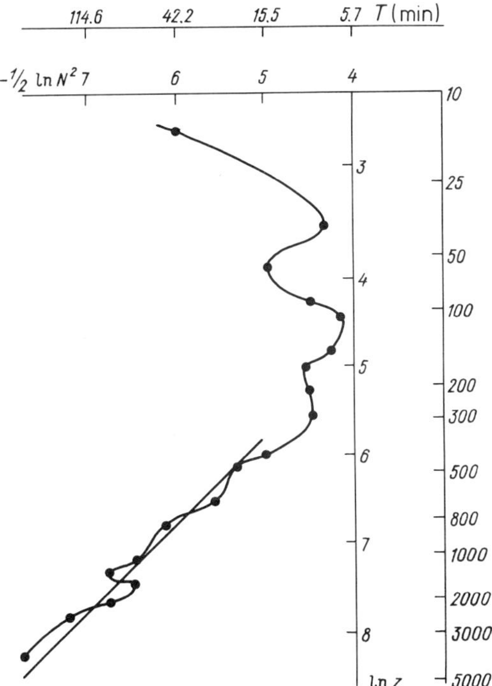

Fig. 3.5. Vertical profile of the Brunt–Väisälä frequency N (T is the Väisälä period) taken at the station 'Vityaz', No. 4311 ($\varphi = 20°03'$ N, $\lambda = 151°49'$ W, 5293 m depth).

is the ocean depth, $\Gamma = \partial\bar{u}/\partial z$, and L is the integral scale of the turbulence. Here, Γ/L varyes in the range $(1–9) \times 10^{-7}$ m^{-1} s^{-1} (see Figure 3.6). Following Long (1969), we can interpret this law as the condition of resonance between the frequency $(H - z)(\partial\bar{u}/\partial z)L^{-1}$ of the L-scale inhomogeneity transfer with the velocity $(H - z)(\partial\bar{u}/\partial z)$, and the frequency N, this condition ensuring internal wave propagation throughout the bulk ocean. This law can also be treated as an asymptotic expression for the turbulent boundary layer with stable stratification. Here, at increasing $H - z$ the velocity shear $\partial\bar{u}/\partial z$ tends to a constant, while the gradients of temperature and salinity, and hence N, are proportional to $(H - z)^2$, with $\alpha_\rho = K_\rho/K \simeq (H - z)^{-2}$. This approach yields

$$\frac{\Gamma}{L} \simeq \frac{1}{\mathrm{Ri}_m}\left(\frac{gm_z}{\rho}\right)^2 u_*^{-5}, \tag{3.24}$$

where $\mathrm{Ri}_m \simeq 0.1$ is the maximum Richardson number in stable stratification. Equating the expression for N in the 'law of depth' to that in the 'law of distance from the bottom' in the middle of the ocean (say, at $z = H/2$), one can estimate that $m_z \simeq \rho u_*^3/(gL) \simeq 1$ g cm^{-2} y^{-1} (Monin, 1970b). In this case, $\rho'/\rho \simeq 10^{-6}$ and $T' \simeq \rho'/(\alpha\rho) \simeq 10^{-2}$ °C, so that the density and temperature fluctuations prove to be comparable to, or even stronger than, those in the atmosphere under conditions of stable stratification.

Deepwater turbulence measurements in the BBL are still scarce. It is necessary

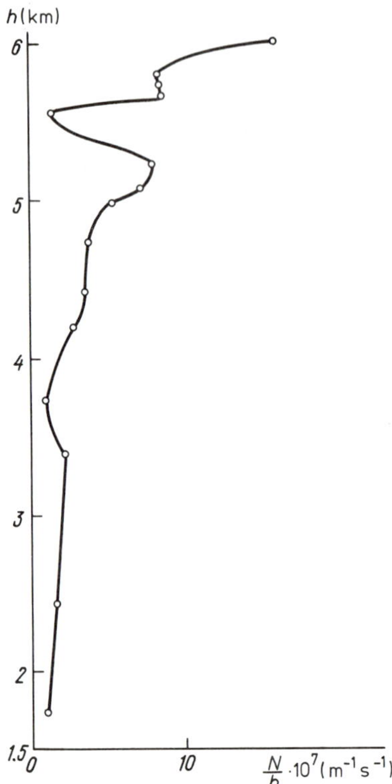

Fig. 3.6. Vertical profile of the reduced Brunt–Väisälä frequency taken at the station 'Vityaz', No. 4371 (φ = 27°06′9 N, λ = 153°45′ E, 6020 m depth).

to mention the measurements of convective fluctuations at frequencies of 10^{-1}–10^2 cycles h^{-1}, made by Munk and Wimbush (1971) who successfully interpreted the results within the framework of the similarity theory of turbulent convection. This theory is the subject of the next section.

4. STRATIFICATION EFFECTS

Turbulence in stratified statistically stationary and horizontally homogeneous (SSHH) flows (except for its small-scale components, which are affected strongly by molecular viscosity, thermal conductivity and diffusivity) can be described by the similarity theory first developed for the atmospheric surface layer by Monin and Obukhov (1953, 1954) (see also Chapter IV in Monin and Yaglom, 1965) and for the atmospheric boundary layer by Monin (1950); see also Kazansky and Monin, (1960, 1961).

The basic formulation of the similarity theory is that turbulence in a layer with SSHH flow is fully characterized by *five* external parameters: the layer depth h, the roughness height z_0 of the surface, the buoyancy parameter g/ρ_0, and the differences U of velocity and potential density $\delta_{\rho *}$ across the layer. If it is also necessary to determine the characteristics of the temperature and salinity fluctuations, we have to use the differences in temperature and salinity across the layer, δT and δS, instead of $\delta \rho_*$ (note that $\delta \rho_* = -\rho_0 \alpha \delta T + \rho_0 \beta \delta S$). In the same way, we have to use the parameters αg and βg, where α and β are the compressibility coefficients determined by (2.3), instead of g/ρ_0.

These external parameters determine, in particular, such important internal characteristics as the friction velocity u_* at the rough surface and the vertical turbulent fluxes of mass (m_z), heat (q_z) and salt (I_z) at the surface ($m_z = -(\alpha/c_p)q_z + \beta I_z$). Hence, they determine the depth L of the constant-flux layer and the density R_*, temperature T_* and salinity S_* scales, defined by

$$L = -\frac{u_*^3}{\kappa (g/\rho_0) m_z}; \qquad R_* = -\frac{m_z}{\kappa u_*};$$
$$T_* = -\frac{q_z}{\kappa c_p \rho_0 u_*}; \qquad S_* = -\frac{I_z}{\kappa \rho_0 u_*}. \tag{4.1}$$

In this case, $R_* = -\rho_0 \alpha T_* + \rho_0 \beta S_*$, where $\kappa \simeq 0.4$ is the Kármán constant, used to simplify some expressions. In stable density stratification, $m_z < 0$, $L > 0$, and $R_* > 0$. If the stratification is unstable, $m_z > 0$, $L < 0$, and $R_* < 0$. The same external parameters are also employed to determine the turning angle γ of the velocity across the layer, the friction coefficient $\eta = u_*/U$ and the internal stratification parameter $\mu = h/L$ (positive for stable, and negative for unstable stratification).

The similarity theory shows in particular that the one-point probability densities of $\mathbf{u}$, ρ, T, and S must have the form

$$p(\mathbf{u}, \rho, T, S) = u_*^{-3} |R_* T_* S_*|^{-1} F \times$$
$$\times \left(\frac{\mathbf{u}}{u_*}, \frac{\rho}{R_*}, \frac{T}{T_*}, \frac{S}{S_*}; \frac{z}{h}, \mu, \frac{z_0}{h} \right), \tag{4.2}$$

where F is a universal function of its arguments, and z is measured from the surface. Our first application is the UML, in which the depth z is measured from the equilibrium level of the ocean surface. The theory applies even better to the BBL, where the height from the bottom, $z_1 = H - z$, is more convenient to use than z. It is also worth trying to apply the similarity theory to the whole depth of the ocean or to particular layers in its microstructure.

Note that in the equations below only the additive constants can depend on z_0:

$$\bar{u}(z) = \bar{u}(z_0) + \frac{u_*}{\kappa}\left[f_u\left(\frac{z}{h}, \mu\right) - f_u\left(\frac{z_0}{h}, \mu\right) \right];$$
$$\bar{v}(z) = \bar{v}(z_0) + \frac{u_*}{\kappa}\left[f_v\left(\frac{z}{h}, \mu\right) - f_v\left(\frac{z_0}{h}, \mu\right) \right];$$
$$\bar{\rho}_*(z) = \bar{\rho}_*(z_0) + R_*\left[f_\rho\left(\frac{z}{h}, \mu\right) - f_\rho\left(\frac{z_0}{h}, \mu\right) \right]; \tag{4.3}$$
$$\bar{T}(z) = \bar{T}(z_0) + T_*\left[f_T\left(\frac{z}{h}, \mu\right) - f_T\left(\frac{z_0}{h}, \mu\right) \right];$$
$$\bar{S}(z) = \bar{S}(z_0) + S_*\left[f_S\left(\frac{z}{h}, \mu\right) - f_S\left(\frac{z_0}{h}, \mu\right) \right].$$

while the vertical gradients of the average fields and the probability distributions of the fluctuations $\mathbf{u}', \rho', T', S'$ are z_0-independent.

For $z \ll h$ the explicit dependence on h in (4.2) and (4.3) must vanish so that the universal functions then depend not on the three arguments taken separately but on the two products $\xi = \mu z/h = z/L$ and $\xi_0 = \mu z_0/h = z_0/L$. In particular, since $z_0 \ll h$, we obtain $f_u(z_0/h, \mu) \simeq f_u(z_0/L)$. This also holds for $f_v, \ldots, f_s$ in (4.3).

Consider now the condition $z \ll |L|$. This can be satisfied by decreasing z or by increasing $|L|$. The length $|L|$ is decreased by reducing $|m_z|$, i.e., by approaching neutral stratification. A sufficiently low z thus corresponds to a sublayer with neutral stratification. Here, the influence of the parameter μ must vanish, so that instead of the three arguments z/h, μ, and z_0/h or the two products z/L and z_0/L, only the z/z_0-dependence can be of importance. In this sublayer, the functions $f_u, \ldots, f_s$ in (4.3) must assume the logarithmic form

$$\bar{u}(z) \simeq U \cos\gamma + \frac{u_*}{\kappa} \ln\frac{z}{z_0}; \qquad \bar{v}(z) \cong U \sin\gamma;$$

$$\bar{\rho}_*(z) \simeq \bar{\rho}_*(z_0) + \frac{R_*}{\alpha_\rho} \ln\frac{z}{z_0}; \qquad \bar{T}(z) \simeq \bar{T}(z_0) + \frac{T_*}{\alpha_T} \ln\frac{z}{z_0}; \tag{4.4}$$

$$\bar{S}(z) \simeq \bar{S}(z_0) + \frac{S_*}{\alpha_S} \ln\frac{z}{z_0}.$$

The first two of these expressions are written for a coordinate system in which the x-axis is directed along the shear stress vector at $z = z_0$. The velocity amplitude at this level is designated as U. The logarithmic laws (4.4) can also be derived by integrating the asymptotic expressions for the mean gradients concerned. For example,

$$\frac{\partial \bar{u}}{\partial z} = \frac{u_*}{\kappa h} f_u'\left(\frac{z}{h}, \mu\right) \simeq \frac{u_*}{\kappa z} \varphi_u\left(\frac{z}{L}\right) \simeq \frac{u_*}{\kappa z}, \tag{4.5}$$

provided $\varphi_u(0) = 1$, which is equivalent to the definition of the Kármán constant κ.

Along with the surface (or boundary)-layer laws (4.4) for the mean variables, it is possible to formulate the so-called defect laws. For instance, at the lower UML boundary (or at the upper BBL boundary, or the axis of a microstructure layer) we have

$$\bar{u}(z) = \frac{u_*}{\kappa} \psi_u\left(\frac{z}{h}, \mu\right); \qquad \bar{v}(z) = \frac{u_*}{\kappa} \psi_v\left(\frac{z}{h}, \mu\right);$$

$$\bar{\rho}(z) - \bar{\rho}(h) = R_* \psi_\rho\left(\frac{z}{h}, \mu\right); \qquad \bar{T}(z) - \bar{T}(h) = T_* \psi_T\left(\frac{z}{h}, \mu\right); \tag{4.6}$$

$$\bar{S}(z) - \bar{S}(h) = S_* \psi_S\left(\frac{z}{h}, \mu\right).$$

At low values of z these defect laws must hold simultaneously with the logarithmic surface-layer laws. Having equated them, let us designate the following z-independent dimensionless quantities:

$$\frac{\kappa U \cos\gamma}{u_*} - \ln\frac{z_0}{h} = \psi_u\left(\frac{z}{h},\mu\right) - \ln\frac{z}{h} \xrightarrow[z/h\to 0]{} A(\mu);$$

$$\frac{\kappa U \sin\gamma}{u_*} = \psi_v\left(\frac{z}{h},\mu\right) \xrightarrow[z/h\to 0]{} B(\mu);$$

$$\alpha_\rho\frac{\delta\rho_*}{R_*} + \ln\frac{z_0}{h} = -\alpha_\rho\psi_\rho\left(\frac{z}{h},\mu\right) + \ln\frac{z}{h} \xrightarrow[z/h\to 0]{} C(\mu); \tag{4.7}$$

$$\alpha_T\frac{\delta T}{T_*} + \ln\frac{z_0}{h} = -\alpha_T\psi_T\left(\frac{z}{h},\mu\right) + \ln\frac{z}{h} \xrightarrow[z/h\to 0]{} D(\mu);$$

$$\alpha_S\frac{\delta S}{S_*} + \ln\frac{z_0}{h} = -\alpha_S\psi_S\left(\frac{z}{h},\mu\right) + \ln\frac{z}{h} \xrightarrow[z/h\to 0]{} E(\mu).$$

We now obtain the following expressions for the momentum, mass, heat, and salt exchange. These determine the drag coefficient u_*/U, the turning angle γ and the coefficients of mass exchange $R_*/(\delta\rho_*)$, heat transfer $T_*/(\delta T)$ and salt exchange $S_*/(\delta S)$ in terms of the external paramters of stratified SSHH flow. The expressions are:

$$U/u_* = \frac{1}{\kappa}\left\{B^2(\mu) + \left[A(\mu) - \ln\frac{h}{z_0}\right]^2\right\}^{1/2};$$

$$\sin\gamma = \frac{1}{\kappa}\frac{u_*}{U}B(\mu); \qquad \frac{\delta\rho_*}{R_*} = \frac{1}{\alpha_\rho}\left[\ln\frac{h}{z_0} + C(\mu)\right]; \tag{4.8}$$

$$\frac{\delta T}{T_*} = \frac{1}{\alpha_T}\left[\ln\frac{h}{z_0} + D(\mu)\right]; \qquad \frac{\delta S}{S_*} = \frac{1}{\alpha_S}\left[\ln\frac{h}{z_0} + E(\mu)\right].$$

Finally, the internal stratification parameter can be expressed in terms of the external Richardson number Ri and other external parameters:

$$\mathrm{Ri} = \frac{gh\delta\rho_*}{\rho_0 U^2} = \frac{\mu}{\alpha_\rho}\left[\ln\frac{h}{z_0} + C(\mu)\right]\left\{B^2(\mu) + \left[A(\mu) - \ln\frac{h}{z_0}\right]^2\right\}^{-1}. \tag{4.9}$$

The set (4.8)–(4.9) determines the turbulence characteristics of a stratified SSHH flow from external parameters.

The depth of the Ekman boundary layer (EBL) is determined by $h = h_0\zeta(\mu_0)$, where $h_0 = \kappa u_*/f$ is the neutrally stratified EBL depth, f is the Coriolis parameter, $\mu_0 = h_0/L$, and $\zeta(\mu_0)$ is a universal function. Equations (4.2)–(4.3) and (4.6)–(4.9) are valid here, with h_0 and μ_0 substituted for h and μ, while the similarity theory for the EBL can be obtained by the substitution of h by h_0, i.e., by using the Coriolis parameter f. In this case it is convenient to set $h_0/z_0 = \kappa(u_*/U)\,\mathrm{Ro}$ in (4.8). Here, $\mathrm{Ro} = U/(fz_0)$ is the so-called Rossby number, which is solely determined by the external parameters. This was derived by Rossby as long ago as 1932.

The EBL depth, determined from the vertical extent of appreciable momentum flux, can be obtained from $h \simeq (K/f)^{1/2}$, where K is the effective eddy viscosity (3.11). In neutral stratification, $K \leqslant \kappa u_* h$ and hence $h \simeq h_0 = \kappa u_*/f$. If the stratification is highly unstable, the EBL characteristics lose their dependence on u_* asymptotically (see below). In this case, $K \lessapprox (gm_z/\rho_0)^{1/3} h^{4/3}$, and thus $h \simeq (gm_z/\rho_0)^{1/2} f^{-3/2} \simeq h_0 |\mu|^{1/2}$.

In highly stable stratification the flux Richardson number

$$\mathrm{Rf} \simeq \frac{g|m_z|}{\rho_0}\left(\frac{\tau}{\rho_0} \cdot \frac{\partial \bar{u}}{\partial z}\right)^{-1} \simeq \frac{g|m_z|}{\rho_0} \cdot \frac{K}{u_*^4}$$

should not exceed its critical value $R \simeq 10^{-1}$. Therefore, $K \lessapprox R u_*^4 (g|m_z|/\rho_0)^{-1}$, whence $h \simeq h_0 \mu_0^{-1/2}$. Since $|L|$ decreases and $|\mu_0|$ increases with both stability and instability, the expressions obtained here demonstrate that the depth of appreciable momentum loss (i.e., the depth of the layer in which the momentum flux approaches zero) increases with decreasing stability.

However, under conditions of severe instability (either geothermal convection in the BBL or winter convection in the UML) the greatest change in velocity with depth occurs near the interface from which the convection develops. The further z_1 is from the interface, the smoother is the flow profile. In free convection this gives a velocity defect equal to $C_u \kappa^{-4/3} u_*^2 (gm_z z_1/\rho_0)^{-1/3}$, where C_u is a numerical constant of order unity. As a result, the depth where the flow deviates less from geostrophic than the small quantity au_* is given by

$$z_1 \simeq \left(\frac{C_u}{\kappa a}\right)^3 \frac{h_0}{|\mu_0|}.$$

This quantity reduces with increasing instability, although the vertical momentum flux induced by convection penetrates to still greater depths, which are proportional to $|\mu_0|^{1/2}$.

In a similar way one can find the intermediate asymptotics for the universal functions under highly unstable or highly stable stratification at $z \ll h$. Here the explicit dependence on h in (4.2)–(4.3) vanishes. These asymptotics demonstrate the effects of stratification on turbulence most vividly. Highly unstable stratification corresponds to large negative values of the dimensionless vertical coordinate $\xi = z/L$. This can be also obtained with fixed z and $m_z > 0$ for $u_* \to 0$. Here the asymptotic state is free convection. The turbulence then gains its energy not from the mean motion but from the potential energy available in unstable stratification and appears as water threads that start from various points at the level where the convection originates; these threads mix very little. In this case, the parameter u_* in (4.2)–(4.3) can be ignored. Since the remaining parameters, g/ρ_0 and m_z, do not form a fixed length scale, the condition of free convection appears to be self-similar. The universal functions of $\xi = z/L$ then become power laws and (4.3) becomes

$$\begin{aligned}
\bar{u}(z_2) - \bar{u}(z_1) &= -\frac{Cu_*}{\kappa}(|\xi_2|^{-1/3} - |\xi_1|^{-1/3});\\
\bar{\rho}_*(z_2) - \bar{\rho}_*(z_1) &= \frac{C|R_*|}{\alpha_\rho(-\infty)}(|\xi_2|^{-1/3} - |\xi_1|^{-1/3});\\
\bar{T}(z_2) - \bar{T}(z_1) &= \frac{C|T_*|}{\alpha_T(-\infty)}(|\xi_2|^{-1/3} - |\xi_1|^{-1/3});\\
\bar{S}(z_2) - \bar{S}(z_1) &= \frac{C|S_*|}{\alpha_S(-\infty)}(|\xi_2|^{-1/3} - |\xi_1|^{-1/3}).
\end{aligned} \tag{4.10}$$

Here, C is a numerical constant, $\xi_{1,2} = z_{1,2}/L$, and $\alpha_\rho(-\infty)$, $\alpha_T(-\infty)$ and $\alpha_S(-\infty)$ are the ratios K_ρ/K, K_T/K and K_S/K in free convection. The probability density for the fluctuations $\mathbf{u}'$, ρ', T', S' in free convection becomes

$$p(\mathbf{u}', \rho', T', S') \simeq u_*^{-3} |R_* T_* S_*|^{-1} F_1 \times$$
$$\times \quad \frac{\mathbf{u}'}{u_* |\xi|^{1/3}} ; \frac{\rho'}{R_* |\xi|^{-1/3}} ; \frac{T'}{T_* |\xi|^{-1/3}} ; \frac{S'}{S_* |\xi|^{-1/3}} , \tag{4.11}$$

where no scale is seen to contain u_*. Hence, the following expressions for the z-dependence of the root-mean-square fluctuations can be derived:

$$\sigma_{u_j} \simeq u_* |\xi|^{1/3} = \left[\kappa m_z \frac{gz}{\rho_0}\right]^{1/3} ;$$
$$\sigma_\rho \simeq |R_*| \cdot |\xi|^{-1/3} = \left[\kappa^4 m_z^{-2} \frac{gz}{\rho_0}\right]^{-1/3} ;$$
$$\sigma_T \simeq |T_*| \cdot |\xi|^{-1/3} = \left[\kappa^4 m_z \left(\frac{|q_z|}{c_p \rho_0}\right)^{-3} \frac{gz}{\rho_0}\right]^{-1/3} ; \tag{4.12}$$
$$\sigma_S \simeq |S_*| \cdot |\xi|^{-1/3} = \left[\kappa^4 m_z \left(\frac{|I_z|}{\rho_0}\right)^{-3} \frac{gz}{\rho_0}\right]^{-1/3} ,$$

so that the velocity fluctuations increase while σ_ρ, σ_T and σ_z decrease with increasing z.

We now consider unstable stratification in a layer that is characterized by neither extremely small nor extremely large $|\xi|$ (say $-1 < \xi < -0.1$), and assume that there is a mean flow. The longitudinal velocity fluctuations $\mathbf{u}'$ then gain their energy mainly from the interaction between the mean flow and the Reynolds stresses, while the vertical fluctuations ω' obtain their energy chiefly from the potential energy of unstable stratification, released by the buoyancy forces. The energy exchange between longitudinal and vertical pulsations is then negligible. One may expect that in this 'convective layer with velocity shear' the hydrodynamic equations are invariant with respect to affine transformations of the coordinates with different horizontal and vertical stretching factors, so that it is expedient here to assign different length scales, L_h and L_v, to the horizontal and vertical dimensions. For example, $u_*^2 = -\overline{u'w'}$ will have the dimension $L_h L_v t^{-2}$, where t is time. It can be proved easily that the scales for measuring the fluctuations w', p', T' and S' will be the quantities (4.12), while those for measuring u' and v' will be u_*^2/σ_w. Therefore,

$$p(\mathbf{u}', \rho', T', S') \simeq u_*^{-4} \sigma_w^2 \sigma_\rho^{-1} \sigma_T^{-1} \sigma_S^{-1} F_2 \times$$
$$\times \left(\frac{\sigma_w \mathbf{u}_h'}{u_*^2}, \frac{w'}{\sigma_w}, \frac{\rho'}{\sigma_\rho}, \frac{T'}{\sigma_T}, \frac{S'}{\sigma_S}\right), \tag{4.13}$$

where σ_w, σ_ρ, σ_T, and σ_S are given by (4.12). Thus, for instance, in the convective layer with velocity shear σ_u decreases proportionally to $u_*^2/\sigma_w \simeq z^{-1/3}$ with increasing z, while a further increase of z in free convection results in an increase of σ_u proportional to $\sigma_w \simeq z^{1/3}$.

Under highly stable stratification the flux Richardson number

$$\mathrm{Rf} = g|m_z| \left(\rho u_*^2 \frac{\partial \bar{u}}{\partial z}\right)^{-1}$$

approaches its limiting value R so that the average velocity profile $\bar{u}(z)$ appears to be asymptotically linear, with a gradient

$$\frac{\partial \bar{u}}{\partial z} \simeq \frac{g|m_z|}{\rho_0 u_*^2 R} = \frac{u_*}{\kappa RL} \tag{4.14}$$

In this case, the gradients of density, temperature and salinity are given by the asymptotic expressions

$$\frac{\partial \bar{\rho}_*}{\partial z} \simeq \frac{R_*}{\alpha_\rho(\xi)RL}\,; \quad \frac{\partial \bar{T}}{\partial z} \simeq \frac{T_*}{\alpha_T(\xi)RL}\,; \quad \frac{\partial \bar{S}}{\partial z} \simeq \frac{S_*}{\alpha_S(\xi)RL}\,. \tag{4.15}$$

Here, $\alpha_\rho \simeq \xi^{-2}$. The same might be also expected for the functions α_T and α_S (Monin *et al.*, 1970). Finally, since highly stable stratification makes the existence of large-scale turbulent eddies impossible, the turbulent exchange between different fluid layers is very much reduced. The turbulence this acquires a local character, i.e., its characteristics cease to depend on z. In this case,

$$p(\mathbf{u}', \rho', T', S') \simeq u_*^{-3} |R_* T_* S_*|^{-1} F_3\left(\frac{\mathbf{u}'}{u_*}, \frac{\rho'}{R_*}, \frac{T'}{T_*}, \frac{S'}{S_*}\right). \tag{4.16}$$

A qualitative resumé of this quantitative information on the effects of stratification turbulence can be presented as follows. The velocity fluctuation intensity, i.e., the turbulence proper, increases with increasing instability. The only exception occurs in convection with velocity shear when the horizontal velocity fluctuations decrease, while the vertical ones still increase. The density fluctuations depend on stratification in a more complex way. They are negligible in highly stable stratification because the turbulence is extremely weak. The turbulence gains strength with decreasing stability but the mean density gradient decreases and the velocity fluctuations become larger when reaching a certain moderate stability. Thereafter, they decrease to a minimum at neutral stratification. Subsequently, both the mean density gradient and the turbulence increase, which results in a rapid increase of the density fluctuations. However, under conditions of high instability, due to strong mixing, the mean potential density becomes uniform and the increase in the density fluctuations slows down. Probably it stops or even reverses in direction. The behavior of temperature and salinity fluctuations can be even more complex because they depend on the signs of the contributions of the temperature and salinity gradients to the density gradients.

In line with the foregoing, when describing the effects of stratification on turbulence, we can take the scale L from (4.1) as a length scale, but similarity makes it possible to employ other length scales composed of external and internal parameters of turbulence. For instance, Panchev (1975) made use of the five local parameters αg, ϵ, ϵ_T, $\partial\bar{u}/\partial z$, and $\partial\bar{T}/\partial z$ (the effects of salinity on density were neglected) and arranged them into the most general expression with the dimension of length:

$$L_* = \epsilon^{1/2}\left(\alpha g\frac{\partial\bar{T}}{\partial z}\right)^{-3/4}\varphi\left[\frac{\alpha g\epsilon_T}{\epsilon}\left(\frac{\partial\bar{T}}{\partial z}\right)^{-1}, \left(\alpha g\frac{\partial\bar{T}}{\partial z}\right)^{-1/2}\frac{\partial\bar{u}}{\partial z}\right]. \tag{4.17}$$

Choosing a power law for φ and omitting some of the five determining parameters in succession, we can express L_* in seven different ways:

$$\begin{aligned} &L_* = (\alpha g)^{-3/2}\epsilon^{5/4}\epsilon_T^{-3/4}; \qquad L_* = (\alpha g)^{-3/4}\epsilon^{1/2}\left(\frac{\partial\bar{T}}{\partial z}\right)^{-3/4};\\ &L_* = (\alpha g)^{-1/4}\epsilon_T^{1/2}\left(\frac{\partial\bar{T}}{\partial z}\right)^{-5/4}; \qquad L_* = \alpha g\epsilon_T^{1/2}\left(\frac{\partial\bar{u}}{\partial z}\right)^{-5/2};\\ &L_* = \epsilon^{-1/4}\epsilon_T^{3/4}\left(\frac{\partial\bar{T}}{\partial z}\right)^{-3/2}; \qquad L_* = \epsilon^{1/3}\left(\frac{\partial\bar{u}}{\partial z}\right)^{-3/2};\\ &L_* = \epsilon_T^{1/2}\left(\frac{\partial\bar{T}}{\partial z}\right)^{-1}\left(\frac{\partial\bar{u}}{\partial z}\right)^{-1/2}. \end{aligned} \tag{4.18}$$

The first of these was introduced by Obukhov (1959) in order to describe buoyancy effects on the inertial subrange of a turbulence spectrum (see §5 below). The second was introduced by Ozmidov (1965). For the ocean, the more general form $L_* = \epsilon^{1/2}N^{-3/2}$ applies, where N is the Brunt-Väisälä frequency. It is expedient to employ the smallest of the buoyancy scales (4.18).

Special attention should be paid to turbulence in stable stratification, when turbulent mixing is hampered by energy losses caused by counteracting the buoyancy forces. Under natural conditions, it then does not work throughout the bulk of the water for a sufficiently long time period, but is concentrated in individual turbulent layers. These layers are vertically homogeneous because of mixing and they are separated by extremely thin interlayers or ('sheets'), which are characterized by microjumps of temperature, electrical conductivity, speed of sound, salinity, density, diffraction coefficient, and other thermodynamic parameters – and sometimes by microjumps of flow velocity. This thin-layer vertical structure is called the *microstructure* or the *vertical fine structure*. It manifests itself as inhomogeneities ('steps') in the vertical profiles of temperature and other thermodynamic variables or, more markedly, as numerous peaks in the vertical profiles of their gradients. Repeated measurements using the method of continuous probing carried out during a number of cruises by the Institute of Oceanology of the U.S.S.R. Academy of Sciences, and also by some foreign research groups, have shown that the microstructure occurs throughout the oceans of the world, except for regions of microconvection (which are quite rare, at least in low and temperate latitudes). Figure 4.1 exemplifies the reproducibility of the microstructure 'steps' by repeated probing.

After smoothing, at least visually, the microstructure 'steps' in the profile of a thermodynamic variable, one obtains a smooth curve that characterizes the large-scale ocean stratification (bulk stratification). Note that, from the viewpoint of the Richardson criterion (3.8), the bulk stratification is nearly always stable, i.e., the Richardson number $\mathrm{Ri}(z)$ calculated by this parameter substantially exceeds its critical value of $\frac{1}{4}$. In what

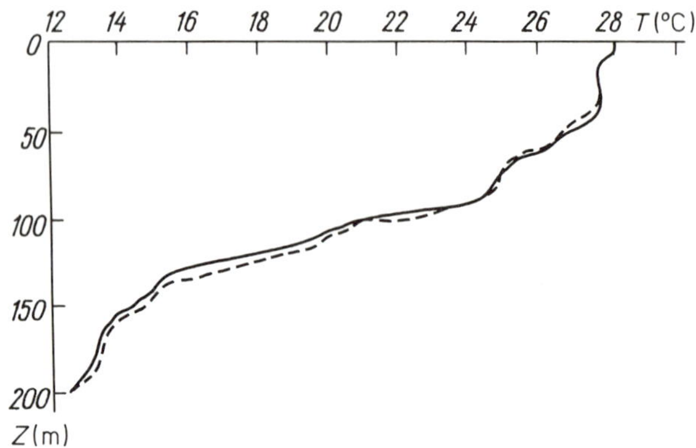

Fig. 4.1. Specimen of two temperature probings with a low-inertia device in the Indian Ocean (φ = 0°02′ S, λ = 75°44′ E, 13 February 1972).

way, then, does turbulence arise? The function Ri(z), plotted such that the microstructure 'steps' are taken into account (see Figure 4.2), shows that Ri $< \frac{1}{4}$ in some microstructure layers. Here, in all probability, large-scale turbulence was generated at the moment of probing, while in all the other layers (with Ri(z) $> \frac{1}{4}$) the turbulence decays with time. Internal waves can create conditions for the local generation of turbulence in stable bulk stratification (see point (5) and (3.17) in the preceding section). In the vicinity of the crests and troughs of these local waves the local Richardson number can reduce to its critical value of $\frac{1}{4}$, which results in spots of turbulence.

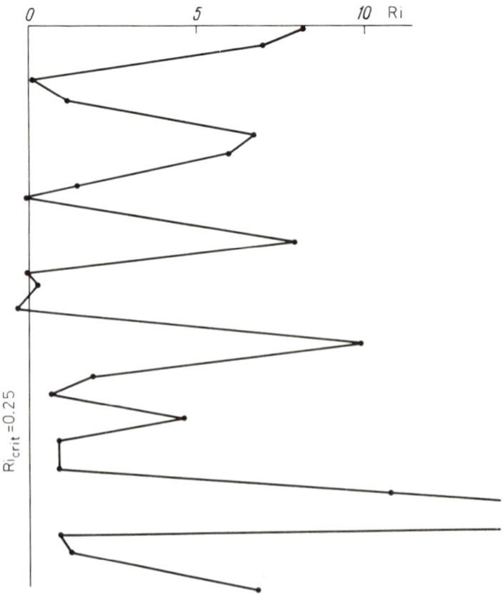

Fig. 4.2. Vertical distribution of the Reynolds number calculated from data obtained by synchronous probing of flow velocity and thermochalinic structure in the Indian Ocean (φ = 0°02′ S, λ = 75°44′ E, 13 February 1972).

The evolution of a newly-generated turbulence spot follows a distinct pattern. The turbulence-induced mixing makes the spot vertically quasi-homogeneous, so that the water density within its limits becomes constant. In conditions of stable stratification, when the density increases with depth, the density in the upper half of the spot then exceeds that at the same level in its environment, while in the lower half of the spot the density is less than that of the environment. The buoyancy forces then push the upper half of the spot down, whilst the lower half floats up. The spot thus becomes flattened: it 'collapses'. It spreads sideways because of mass conservation and turns into a thin 'pancake'. Its intrusion into the surrounding stratified fluid gives rise to a new microstructure layer.

If the initial internal wave has both a long period and a long wavelength (for instance, in the case of internal waves with tidal periods induced by the tide-generating forces and by the tides themselves), then the turbulence spots that arise are large and the microstructure layers resulting from them are rather thick. Such layers can give rise to internal waves with smaller periods and lengths, forming smaller turbulence spots and thinner microstructure layers, and so on, up to internal waves with the smallest periods and lengths, the smallest turbulence spots, and the thinnest microstructure layers. Thus, the solution of the question of the order of evolution can be found in the cascade process: internal waves → turbulence spots → microstructure layers → internal waves, and so on. This cascade process can result in the formation of a quasi-stationary spectrum of internal waves, intermittent turbulence, and microstructure layers. In Nature, however, other processes that affect real spectra can also occur, including storms and quasi-stationary horizontal inhomogeneities of geographic and dynamic origin.

In the initial stages of evolution (as studied experimentally by Wu (1969) and theoretically by Kao (1976)), a turbulence spot expands by way of intrusion, at first readily counteracting the resistance of the water and thus giving rise to internal waves, and then reaching steady-state conditions when the driving force of the intrusion is balanced by the sum of shape and wave resistances. These stages develop rapidly, within tens of N^{-1}, where N is the Brunt-Väisälä frequency. Then comes the long-term final stage (investigated by Barenblatt in 1978), when the driving force of the intrusion is balanced by viscous resistance. In the first stage of the evolution, the relative rate $S^{-1}\ \mathrm{d}S/\mathrm{d}t$ at which the horizontal area S increases is proportional to the inflow towards the central level of the spot, $N^2 t$, since the free-fall acceleration is proportional to N^2. *The horizontal* diameter L then increases as $(L - L_0)/L_0 \simeq N^2 t^2$ (Wu, 1969). In the second stage, $\mathrm{d}L/\mathrm{d}t \simeq Nh$, where h is the thickness of the intrusion 'tongue'; in the case of a round spot $h \simeq V/L^2$, where V is the spot volume which is independent of time, so that $L \sim [N_0 V(t - t_0)]^{1/3}$. Finally, in the viscous intrusion stage the mass conservation equation can be written as

$$\frac{\partial h}{\partial t} + \operatorname{div} h\mathbf{v} = 0. \tag{4.19}$$

The propagation velocity of the intrusion is obtained by equating the driving force, $(\delta S)\,\Delta(ph)$, and the resistance force $(\delta S)\,C\mu\mathbf{v}/h$:

$$\mathbf{v} = -\frac{h}{C\mu}\,\Delta(ph). \tag{4.20}$$

Here, μ is the viscosity, C is a numerical constant ($C = 12$ for the case of a viscous flow between plane walls), and p is the excess pressure in the mixed fluid. In the intrusion 'tongue' the pressure is distributed vertically by the hydrostatic law $p = p_1 - g\rho_1(z - z_1) + \frac{1}{8}\rho_1 N^2 h^2$, and in the environment by $p = p_1 - g\rho_1(z - z_1) + \frac{1}{2}\rho_1 N^2 (z - z_1)^2$, so that the difference in mean pressures throughout the cross-section of the intrusion 'tongue' results in an excess pressure of $\frac{1}{12}(\rho_1 N^2 h^2)$. Substituting this into (4.20) we reduce (4.19) to the form

$$\frac{\partial h}{\partial t} - \kappa \Delta h^5 = 0, \qquad \kappa = \frac{\rho_1 N^2}{20 C \mu}. \tag{4.21}$$

For the axially symmetric case this yields the Barenblatt expressions

$$h = \left[\frac{V}{2\pi\kappa(t - t_1)}\right]^{1/5} f(\zeta); \qquad \zeta = r\left[\frac{\kappa V^4}{16\pi^4}(t - t_1)\right]^{-1/10};$$
$$f(\zeta) = \left(\frac{10^{1/5}}{6}\right)^{1/4}\left(1 - \frac{\zeta^2}{\zeta_0^2}\right)^{1/4}; \qquad 0 \leqslant \zeta \leqslant \zeta_0 = \frac{10^{3/5}}{2}, \tag{4.22}$$

where $f(\zeta) = 0$ at $\zeta \geqslant \zeta_0$. The thickness of the corresponding 'pancake' varies very little almost up to the edge, where it sharply dwindles to nothing. In this case, the edge of the intrusion, $\zeta = \zeta_0$, propagates according to

$$r = \zeta_0\left[\frac{\kappa V^4}{16\pi^4}(t - t_1)\right]^{1/10}. \tag{4.23}$$

This is extremely slow. Figure 4.3 demonstrates an empirical confirmation of (4.23) in laboratory experiments (Zatsepin *et al.*, 1978) Note that turbulence spots formed by plane internal waves are not axially symmetric (round), but are cylindrical with their horizontal axes directed along the y-axis. In this case, instead of (4.22), Barenblatt obtained

$$h = \left[\frac{V^2}{4\kappa H^2(t - t_1)}\right]^{1/6} f_1(\zeta_1); \qquad \zeta_1 = x\left[\frac{\kappa V^4}{16 H^4}(t - t_1)\right]^{-1/6};$$
$$f_1(\zeta_1) = (\zeta_0^2/15)^{1/4}(1 - \zeta^2/\zeta_0^2)^{1/4}; \qquad 0 \leqslant \zeta \leqslant \zeta_0 = (15)^{1/6}\,\frac{[2\Gamma(5/4)\Gamma(1/2)]^{2/3}}{\Gamma(7/4)} \tag{4.24}$$

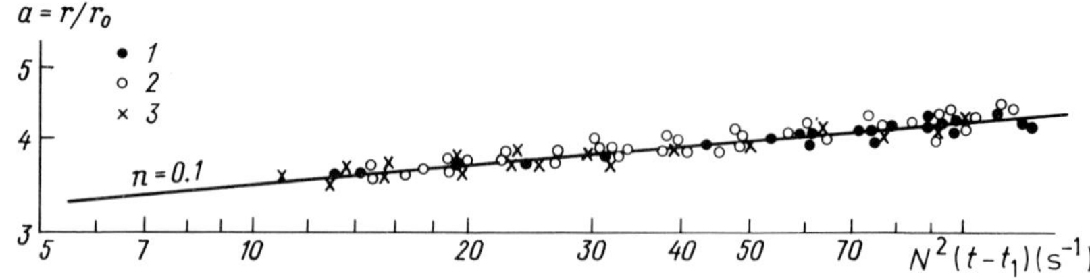

Fig. 4.3. Increase in the relative size of a turbulence spot in the final stage of its development (straight line – theory, dots – experiment). N is the Brunt–Väisälä frequency (Zatsepin *et al.*, 1978). (1) $N = 1.00\ s^{-1}$, (2) $N = 0.63\ s^{-1}$, (3) $N = 0.58\ s^{-1}$.

Instead of (4.23) the formula of propagation of the intrusion front edge now follows the equation

$$x = \zeta_0 \left[\frac{\kappa V^4}{16H^4} (t - t_1) \right]^{1/6}. \tag{4.25}$$

Here, the maximum intrusion thickness decreases with time according to $h \simeq (t - t_1)^{-1/6}$, where H is the length of the cylindrical spot along the y-axis.

The idea that the horizontal expansion of turbulence spots generated by internal waves results in microstructure formation was not accepted immediately. The first hypothesis on the existence of microstructure in the oceans was put forward after an investigation (Stommel *et al.*, 1956) into the possibility of the evolution of microconvection in a layer of salt water with stable density stratification but with temperature and salinity gradients of opposite signs. This is due to 'double diffusion', i.e., due to the difference in the diffusivities of heat and salt in water. It should be remembered that the thermal diffusivity in sea water is 100 times higher than the salt diffusivity. Note, however, that the hypothesis that turbulence in conditions of highly stable stratification evolves only in individual thin layers, i.e., in turbulent 'pancakes', was put forward by Kolmogorov as far back as 1948–49 with no association with the processes of double diffusion whatsoever. This hypothesis was qualitatively confirmed by the very first turbulence measurements in the free atmosphere. These were carried out by the Institute of Geophysics of the U.S.S.R. Academy of Sciences in the beginning of the nineteen-fifties using free-floating balloons. This work lead to the discovery of 'turbulence clouds'. Later, the microstructure in the free atmosphere was studied using radar observations, see, e.g., Phillips (1967) and Ludlam (1967).

The idea of the double diffusion of heat and salt was later developed by Stern (1960) and Stommel (1962), and this gave rise to a series of theoretical and laboratory investigations of thermohaline microconvection. Laboratory experiments by Turner and Stommel (1964) and Turner (1965) showed that the occurrence of cold, and relatively fresh, water over denser, warm salt water results in a succession of convective layers in which convection is induced by fast upward heat diffusion. The penetrative convection through the upper boundary of each layer is delayed by a density microjump created at this boundary by mixing in a stable gradient. Together with the convective vertical heat flux q_z, this type of stratification also causes a vertical salt flux I_z. When divided by $c_p \rho_0 \chi \delta T/h$ and $\rho_0 D \delta S/h$, respectively, where δT and δS are the vertical differences in temperature and salinity and h is the layer width, and thus made dimensionless, these fluxes become proportional to $\mathrm{Ra}^{1/3}$, where $\mathrm{Ra} = \alpha g h^3 \delta T/\nu\chi$ is the Rayleigh number in (1.6) with proportionality coefficients that depend on the ratio $\beta\delta S/\alpha\delta T$ of the contributions from salinity and temperature to the vertical density difference. Experiments show that the ratio $\beta I_z/\alpha q_z$ of the potential energy changes induced by salt and heat transfer first rapidly falls with increasing $\beta\delta S/\alpha\delta T$ and then, when $\beta\delta S/\alpha\delta T > 2$, becomes constant and approximately equal to 0.15. Thus, when the salinity makes a substantial contribution to the density gradient, 15% of the potential energy released by thermal convection is used to elevate salt.

Turner (1967) and Stern and Turner (1969) found that when warm salt water overlies cold, denser, but less salty water, laminar convection arises in a number of layers in the form of long, narrow vertical cells called *salt fingers*. This is because, while the salinity anomalies are preserved because salt diffusion is slow, the relatively fast horizontal smoothing of the temperature anomalies results in density anomalies which induce convection. Experiments by Turner demonstrated that, in this case, the ratio $\alpha q_z/\beta I_z$ is almost independent of the parameter $\alpha\delta T/\beta\delta S$, and approximately equals 0.56. In other words, more than half of the potential energy released by salinity convection is spent on heat transfer. Salt fingers prove, therefore, to be an effective mechanism for the vertical transfer not only of salt, but also of heat.

Measurements under advantageous stratification conditions in the sea have revealed, in a number of cases, vertical microstructure created by double diffusion. For instance, the stratification that arises when cold, fresher water overlies warm, saltier water was observed in Lake Wanda in the Antarctic, above pits filled with hot salt water at the bottom of the Red Sea, over the warm, salty Red Sea waters in the Gulf of Aden, over the Mediterranean waters in the east Atlantic, along the coast of Somalia, and under the ice of the Arctic. See Figure 4.4, which is based on measurements by Neal, Neshyba

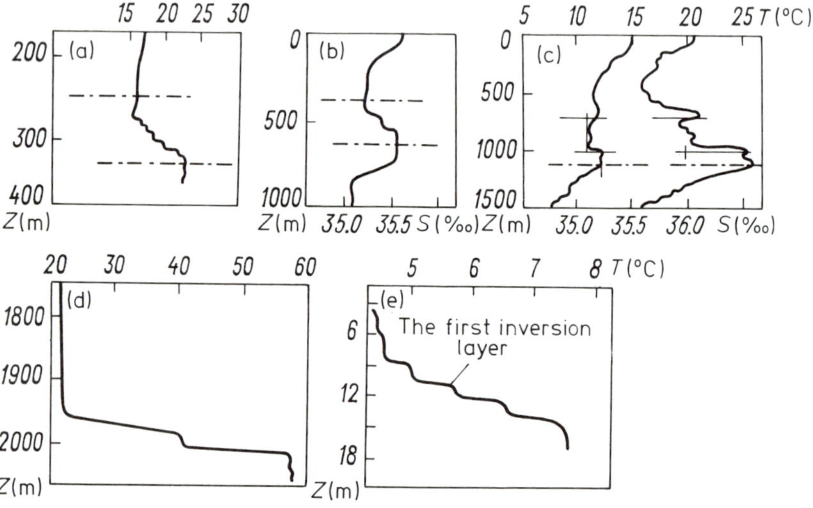

Fig. 4.4. Stepwise thermochaline structure of temperature inversions: (a) at the 'Meteor' station, No. 49, in the Gulf of Aden, (b) at the 'Meteor' station, No. 130, offshore Somalia, (c) at the 'Meteor' station, No. 52, in the East Atlantic, (d) at the 'Meteor' station, No. 384, in the Red Sea, (e) in Lake Wanda in the Antarctic.

and Denner (1969). The layers with salt fingers were observed where warm salt water overlies cold fresher water in the main thermocline of the Sargasso Sea in the vicinity of the Bermuda Islands, in the North trade wind flow in the Atlantic, and under the Mediterranean water in the east Atlantic. See Figure 4.5, based on measurements by Tait and Howe, 1971. Williams (1974) managed to observe and photograph the cell layer structure with salt fingers in the Atlantic using the Schlieren technique.

Nevertheless, the type of stratification in which the temperature and salinity gradients make opposing contributions to the density gradient, and which is characterized by double diffusion of heat and salt, is not typical of the ocean. Generally, the temperature

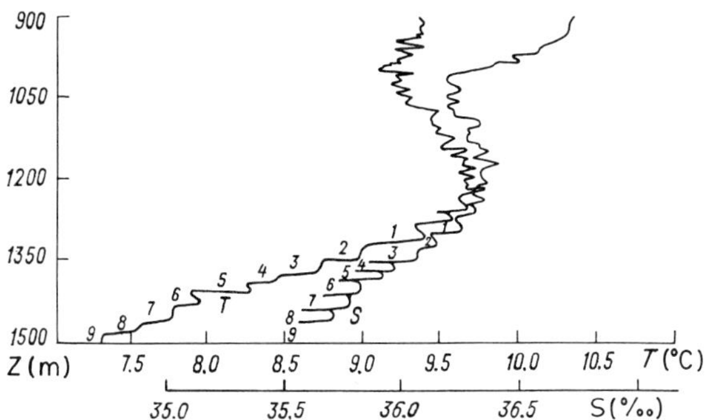

Fig. 4.5. Stepwise microstructure of temperature and salinity beneath the Mediterranean water in the Atlantic at 'Discovery' station, No. 15.

decreases and the salinity increases with depth. The stratification then proves to be stable with respect to both temperature and salinity, making the formation of microconvection layers by double diffusion impossible. At the same time, repeated measurements have shown that in this typical stratification there usually exists a microstructure. Moreover, Simpson and Woods (1970) detected a stepwise microstructure in the temperature profiles of the fresh water in Loch Ness, Scotland (Figure 4.6), where the mechanism of double diffusion is not in evidence because of the absence of salt.

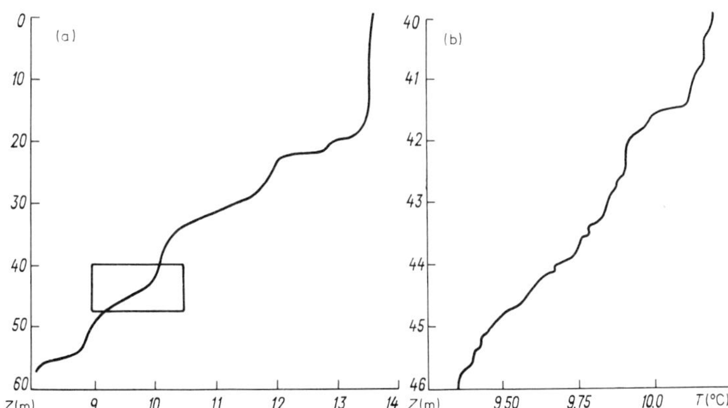

Fig. 4.6. Temperature structure of Loch Ness. The vertical step size was 6.5 cm, with a resolution of 6×10^{-3} °C Figure (a). Figure (b) is the detailed structure of the part of the temperature profile in (a) that is marked with a rectangle.

However, there still exists the possibility of the double diffusion of heat and momentum. Remember that in water the momentum diffusivity, i.e., the kinematic viscosity, is 7–13 times larger than the thermal diffusivity, i.e., the kinematic heat conductivity. Indeed, McIntyre (1969) developed a theory of the double diffusion of heat and momentum (or the moment of momentum, as he termed it). Baller (1972) attempted to apply this theory to the generation of microstructure in the ocean. In this context, it is useful to remember that thin-layer microstructure can be traced not only in the thermodynamic variables but also in the velocity. Various water layers thus have, generally speaking, different vector velocities. This was confirmed by direct measurements carried out by Monin *et al.* (1973), who employed continuous velocity probing with a 'cross-beam' device. This technique is based on the detection of the Doppler shear of the sound waves which are diffracted by natural inhomogeneities within a restricted volume limited by acoustic beams (see Figure 4.7). The same data were obtained by Sanford (1975) with

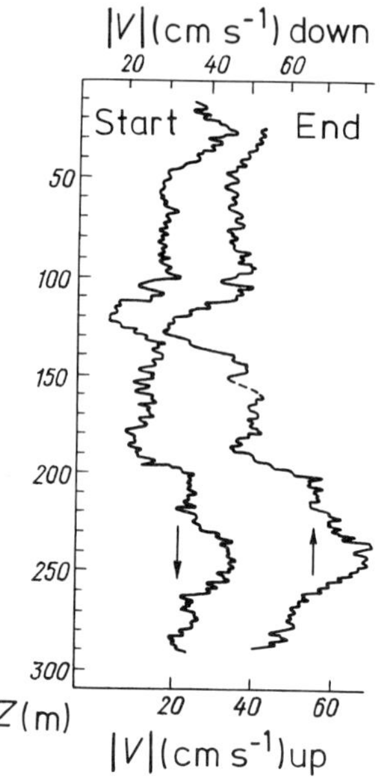

Fig. 4.7. Specimen of the root-mean-square velocity measured by lowering and lifting a crossbeam probe during the seventh expedition of the 'Dmitriy Mendeleyev', station 495, 16 March 1972. The probing was started at 20.15 and finished at 21.15.

a small electromagnetic flow meter. Note that, in some cases, Sanford observed that the flow direction reversed within half an inertial period (see Figure 4.8). This demonstrates that the velocity microstructure cannot always be attributed to double diffusion of heat and momentum but, for instance, can also be associated with inertial oscillations or other internal waves.

Assuming that double diffusion conditions occur very seldom, we now take the isentropic and isohaline layer-by-layer inflow of water from the side, i.e., *lateral convection*, to be the typical mechanism of microstructure generation. For instance, if there are two adjacent columns of differently colored water, then the layer-by-layer convection (which is much faster than molecular diffusion) mixes them into a single column with alternating layers of the two colors. In the ocean, this mutual penetration of fluids of different origin can be traced over substantial distances. For example, Stommel and Fedorov (1967) observed a saline layer, generated by intensive evaporation over the North Australian Shelf, protruding hundreds of kilometers into the Timor Sea. The Mediterranean

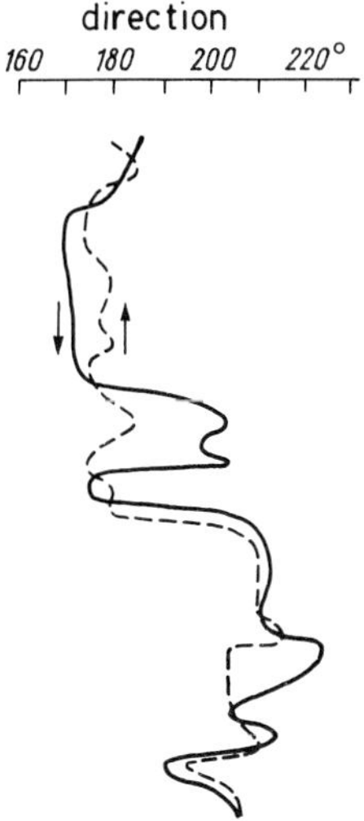

Fig. 4.8. Specimen of flow direction measurements taken by lowering (solid line) and lifting (dashed line) a crossbeam probe during the seventh expedition of the 'Dmitriy Mendeleyev', station 595, 16 March 1972. The probing was started at 20.15 and finished at 21.15.

water in the Atlantic Ocean, and the Red Sea water in the Indian Ocean, serve as further examples of this phenomenon. Microstructure layers can thus arise from lateral convection in horizontally inhomogeneous conditions caused by the geographical environment and, at the scale of internal waves, from internal wave instabilities that are non-uniformly distributed in space and lead to the generation of turbulence spots.

5. THEORY OF TURBULENCE SPECTRA

Since the turbulence in oceanic SSHH flows can be assumed to be locally homogeneous (or even homogeneous in horizontal planes), the spatial structure functions $D_{j_1j_2}(\mathbf{r})$ – see (2.26) – of the hydrodynamic variables do not explicitly depend on the coordinates of the observation points $\mathbf{x}_1$ and $\mathbf{x}_2$, but only on the difference $r = \mathbf{x}_2 - \mathbf{x}_1$. They can be written as integrals of the type

$$D_{j_1j_2}(\mathbf{r}) = 2\int(1 - \cos\mathbf{k}\cdot\mathbf{r})F_{j_1j_2}(\mathbf{k})\,\mathrm{d}\mathbf{k}. \tag{5.1}$$

Here, the integral refers to the entire space of wave-vectors $\mathbf{k}$, except for the point $\mathbf{k} = 0$; the *three-dimensional spectral density* matrix $\|F_{j_1j_2}(\mathbf{k})\|$ is Hermitian, i.e., $F_{j_1j_2}(\mathbf{k}) = F^*_{j_1j_2}(-\mathbf{k}) = F^*_{j_2j_1}(\mathbf{k})$ and non-negative for all $\mathbf{k} \neq 0$, while for any j and $k_0 > 0$ its diagonal elements obey the integration requirements $\int_{k<k_0} k^2 F_{jj}(\mathbf{k})\,\mathrm{d}\mathbf{k} < \infty$ and $\int_{k>k_0} F_{jj}(\mathbf{k})\,\mathrm{d}\mathbf{k} < \infty$. These spectral densities will be the subject of this section.

Since buoyancy forces act on the density fluctuations of fluid particles of all spatial scales, including the smallest ones, SSHH turbulence is not locally isotropic but is locally symmetric relative to the vertical direction, as noted in §2. Therefore, the scalar spectral functions, exemplified by the spectral density of the kinetic energy per unit mass, $F(\mathbf{k}) =$

$\frac{1}{2}F_{\alpha\alpha}(k)$, depend not only on the modulus of the wave vector (the wavenumber $k = |\mathbf{k}|$) but also on the horizontal and vertical wavenumbers $k_h = (k_1^2 + k_1^2)^{1/2}$ and k_3 separately. Integrating the three-dimensional spectrum over all wave-vector directions – over a sphere of radius k in wave space – we obtain the wavenumber spectrum

$$E(k) = \int F(\sqrt{k_1^2 + k_2^2}, k_3)\, dk_1\, dk_2\, dk_3, \quad \sqrt{k_1^2 + k_2^2 + k_3^2} = k. \tag{5.2}$$

Let us also determine the one-dimensional spectrum along a horizontal straight line L that is parallel, say, to the x_1-axis:

$$F_1(k_1) = \int F(\sqrt{k_1^2 + k_2^2}, k_3)\, dk_2\, dk_3. \tag{5.3}$$

For large wavenumbers k_1 (i.e., for small scales of inhomogeneities along the straight line L), the one-dimensional spectrum (5.3) can be derived from the readily-measurable frequency spectrum $E_1(\omega)$ of the fluctuations at a fixed point on the straight line L in terms of G. I. Taylor's 'frozen turbulence' hypothesis:

$$F_1(k_1) = \bar{u}E_1(\bar{u}k_1). \tag{5.4}$$

Here, $\bar{u}$ is the mean velocity component along the straight line L. In the case of locally isotropic turbulence, the functions (5.2) and (5.3) can be readily expressed in terms of each other since they are related by $E(k) = -2k(dF_1(k)/dk)$. Hence, the frequency spectrum $E_1(\omega)$ can be used to construct the wavenumber spectrum $E(k)$. As a result, we shall confine ourselves to a discussion of the wavenumber spectra.

According to Kolmogorov's theory (1941, see also §21.1–21.3 in Monin and Yaglom, 1967) the small-scale spectral components of fully developed turbulence at extremely high Reynolds numbers arise because all large turbulent vortices are hydrodynamically unstable. They disintegrate into smaller vortices, supplying these with kinetic energy. In stationary conditions this process is characterized by a constant energy flux per unit mass, ϵ, throughout the spectrum. Only the smallest vortices are stable; they dissipate their kinetic energy directly (at the same rate ϵ), since they have to overcome the friction caused by the kinematic viscosity ν. For locally isotropic turbulence, we can now begin our discussion of the shape of the spectra of all variables involved by formulating Kolmogorov's first similarity hypothesis: *the statistical characteristics of the small-scale components of fully developed turbulence are determined by the two dimensional parameters ϵ and ν.* These parameters define the following *internal scales* (Kolmogorov microscales) of length η, time τ_η, and velocity $v_\eta = \eta/\tau_\eta$:

$$\eta = \left(\frac{\nu^3}{\epsilon}\right)^{1/4}; \quad \tau_\eta = \left(\frac{\nu}{\epsilon}\right)^{1/2}; \quad v_\eta = (\nu\epsilon)^{1/4}. \tag{5.5}$$

These can be interpreted as typical dimensions, lifetimes, and internal velocities for the smallest turbulent vortices. The statistical characteristics of fully developed, locally isotropic turbulence measured within these scales will be universal, i.e., the same for all turbulent flows. According to Kolmogorov's first similarity hypothesis, the wavenumber spectrum of turbulent kinetic energy must have the form

$$E(k) = (\epsilon\nu^5)^{1/4}\varphi(\eta k), \tag{5.6}$$

where $\varphi(\eta k)$ is some universal function. This function can be predicted with certainty in part of the spectrum, since the direct effects of viscosity are essential only for the smallest turbulent vortices. Kolmogorov's second similarity hypothesis reads: *the statistical characteristics of the small-scale components of fully developed turbulence with length and time scales that greatly exceed η and τ_η are fully determined by the single dimensional parameter, ϵ.* The range of scales that are much larger than η and τ_η, but much smaller than the external (integral) scales L and τ, i.e., the scales of the flow as a whole, is referred to as the *inertial range*. In this range, viscous forces are negligible. The statistical conditions are thus determined exclusively by the inertial forces, which results in energy transfer from larger to smaller scales. According to the second similarity hypothesis, turbulence is self-similar in the inertial range. In the inertial range of wavenumbers, $1/L \ll k \ll 1/\eta$, the universal function $\varphi(\eta k)$ in (5.6) therefore must be chosen such that ν does not occur in the kinetic energy spectrum, i.e., $\varphi(\eta k) \simeq (\eta k)^{-5/3}$. Hence, the kinetic energy spectrum in the inertial range must have the form

$$F(k) = C_1 \epsilon^{2/3} k^{-5/3}. \tag{5.7}$$

This is the so-called *5/3 power law* of Kolmogorov and Obukhov. For the structure functions of velocity, this corresponds to the 2/3 power law

$$D_{je}(\mathbf{r}) = \frac{27}{55} \Gamma\left(\frac{1}{3}\right)\left(-\frac{1}{3}\frac{r_j r_e}{r^2} + \frac{4}{3}\delta_{je}\right) C_1 \epsilon^{2/3} \mathbf{r}^{2/3}. \tag{5.8}$$

The expressions (5.6)–(5.8) are supported by numerous measurements at sea and in the laboratory. For instance, Figure 9.5 (p. 113) depicts data concerning ocean turbulence spectra in tidal flows at high Reynolds number, $\mathrm{Re} = 3 \times 10^8$ (Grant and Moilliet, 1962; Grant *et al.*, 1962; Stewart and Grant, 1962). The universal function $\varphi(\eta k)$ is shown in Figure 9.5, with $k\eta$ plotted along the x-axis on a logarithmic scale. The logarithmic scale used for $\varphi(k\eta)$ can disguise the spread of empirical points. This spread is negligible, as seen in Figure 9.15 (p. 123), which presents (on a linear scale) the same function $\varphi(k\eta)$, multiplied by $(k\eta)^2$, which yields the energy dissipation spectrum $k^2 E(k)$ in dimensionless form. The maximum energy dissipation occurs at the wavenumber $k \simeq \frac{1}{8}\eta$. The most reliable value of the coefficient C_1 in (5.7), estimated from these and some other data, is $C_1 \simeq 1.4$. The inertial range of wavenumbers, in which the 5/3-power law (5.7) is valid, increases with Reynolds number. This can be seen from Figure 5.1, which shows curves obtained in laboratory measurements. At low Reynolds numbers, of the order of 3000 or less, the spectrum has no inertial range.

If we want to derive the function $\varphi(k\eta)$ theoretically, we have to use some semi-empirical turbulence theory that eliminates the problems associated with the unclosed Friedman–Keller equations for the moments of the velocity field. We now consider the approach first developed by Batchelor (1959) and applied to passive scalar contaminants at high Prandtl numbers ν/χ. This theory was used by Novikov (1961) to study vorticity fields. Let us consider turbulent inhomogeneities of the vorticity field $\omega_j(\mathbf{x}, t)$ with spatial scales smaller than η. In an area this small the velocity field can be taken to depend linearly on the relative coordinates, i.e., $u_j(\mathbf{x}_0 + \mathbf{r}, t) \simeq u_j(\mathbf{x}_0, t) + a_{j\alpha} r_\alpha$. Here $\mathbf{x}_0$ is a fixed point in the interior of the particle under discussion, and $a_{j\alpha} = \partial u_j/\partial x_\alpha$ can be assumed to be constant over short space and time intervals. The vorticity equation for

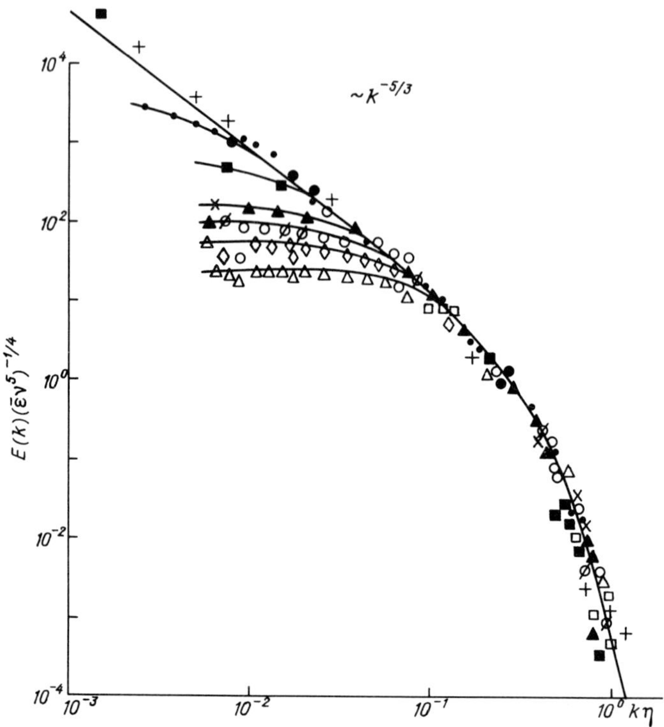

Fig. 5.1. Normalized longitudinal velocity spectrum according to laboratory measurements carried out by different investigators at various Reynolds numbers.

incompressible fluids, $\partial\omega_j/\partial t + u_\beta\partial\omega_j/\partial x_\beta - \omega_\beta\partial u_j/\partial x_\beta = \nu\Delta\omega_j$ can then be linearized and written in a system of coordinates that moves and rotates together with the fluid particle and is connected to its principal axes of deformation:

$$\frac{\partial\omega_j}{\partial t} + a_\beta x_\beta \frac{\partial\omega_j}{\partial x_\beta} - a_j\omega_j = \nu\Delta\omega_j. \tag{5.9}$$

Here, a_β is the principal deformation rate, $0 < a_1 \geqslant a_2 \geqslant a_3 < 0$, and $a_1 + a_2 + a_3 = 0$ due to incompressibility. The solution of this equation, with the initial condition $\omega_j(\mathbf{x}, 0) = iA_j(0)\exp[i\mathbf{k}(0)\cdot\mathbf{x}]$, is assumed to be of the form $\omega_j(\mathbf{x}, t) = iA_j(t)\exp[i\mathbf{k}(t)\cdot\mathbf{x}]$, where A_j and $\mathbf{k}$ are real. Substituting this function into (5.9), differentiating it, cancelling the term $\exp[i\mathbf{k}\cdot\mathbf{x}]$, and separating the real and imaginary parts of the resulting equation, we obtain $\partial A_j/\partial t - a_jA_j = -\nu k^2A_j$, and $\partial k_j/\partial t + a_jk_j = 0$. Hence $k_j(t) = k_j(0)\exp(-a_jt)$, so that at large t the vector $\mathbf{k}(t)$ becomes asymptotically parallel to the axis of maximum compression, x_3. Its length becomes $\mathbf{k}(t) \simeq k_3(t) = k_3(0)\exp(|a_3|t)$. Also $A_j(t) = A_j(0)\exp(a_jt - \nu\int_0^t k^2\,\mathrm{d}t)$, so that the vector $\boldsymbol{\omega}(\mathbf{x}, t)$ becomes asymptotically parallel to the axis of maximum expansion, x_1. Its length becomes $\boldsymbol{\omega}(\mathbf{x}, t) \simeq |\omega_1(\mathbf{x}, t)| \simeq |A_1(0)|\exp(a_1t - \nu\int_0^t k^2\,\mathrm{d}t)$.

Now, let $\delta\omega = \delta\omega_1(t)$ be the spectral vorticity component that exists at time t in the interval δk around the point k on the k_3 axis. In the time interval $\mathrm{d}t$ the wavenumber and vorticity both increase as $\mathrm{d}(\delta k) = |a_3|\delta k\,\mathrm{d}t$ and $\mathrm{d}(\delta\omega) = (a_1 - \nu k^2)\delta\omega\,\mathrm{d}t$. The energy spectrum, given by

$$E(k) = \frac{\delta\omega\delta\omega^*}{2k^2\delta k},$$

then increases as

$$\begin{aligned}\mathrm{d}E &= \frac{\delta\omega\delta\omega^*}{2k^2\delta k}\left[\frac{\mathrm{d}(\delta\omega)}{\delta\omega} + \frac{\mathrm{d}(\delta\omega^*)}{\delta\omega^*} - \frac{2\mathrm{d}k}{k} - \frac{\mathrm{d}(\delta k)}{\delta k}\right] \\ &\simeq E[2(a_1 - \nu k^2) - 3|a_3|]\,\mathrm{d}t \\ &= E(2a_1|a_3|^{-1} - 3 - 2|a_3|^{-1}\tau_\eta^{-1}\eta^2k^2)\,\frac{\mathrm{d}k}{k}.\end{aligned}$$

Hence,

$$\begin{aligned}&E(k) = C_2\epsilon^{2/3}k^{-5/3}(\eta k)^{2\sigma - 4/3}\exp[-(\gamma\eta k)^2], \\ &\text{with } \sigma = \overline{a_1|a_3|^{-1}}, \qquad \gamma^2 = \tau_\eta^{-1}\overline{|a_3|^{-1}}.\end{aligned} \tag{5.10}$$

Novikov estimated that $1/2 < \sigma < 1$ and $\gamma^2 \geqslant 3$. For $\sigma = 2/3$ and $C_2 = \gamma^{1/3}/\Gamma(2/3)$, (5.10) obeys the normalization requirement $2\nu\int_0^\infty k^2E\,\mathrm{d}k = \epsilon$ and agrees with the 5/3-power law (5.7).

For passive scalar contaminants, which include the temperature T if the medium is neutrally stratified, one can introduce (Obukhov, 1949) the inhomogeneity measure $\frac{1}{2}\int(T - \overline{T})^2\rho\,\mathrm{d}V$ and the inhomogeneity dissipation rate per unit mass $\epsilon_T = \chi\overline{(\Delta T')^2}$. For passive scalars, the first similarity hypothesis can now be formulated as follows: *for the fully developed turbulence at extremely large Reynolds* ($\mathrm{Re} = L\delta U/\nu$) *and Peclet* ($\mathrm{Pe} = L_T\delta_T U/\chi$) *numbers, the statistical characteristics of the small-scale components of T are fully determined by the four parameters ϵ, ν, ϵ_T, and χ*. Consistent with this similarity hypothesis, the wavenumber spectrum of the field T must have the form

$$E_T(k) = (\epsilon_T/\epsilon)(\epsilon\chi^5)^{1/4}\varphi_T(\eta k, \mathrm{Pr}), \tag{5.11}$$

where $\mathrm{Pr} = \nu/\chi$ is the Prandtl number. We confine ourselves below to situations involving large Prandtl numbers, which correspond to the diffusion of heat and salt in sea water. In this case, the internal scale of the contaminant field, $\eta_T = (\chi^3/\epsilon)^{1/4}$, is smaller than η and the second similarity hypothesis can be formulated as follows: *in fully developed turbulence, the statistical characteristics of the small-scale components of T are fully determined by the two dimensional parameters ϵ and ϵ_T for length and time scales that substantially exceed η and τ_η*. In the *inertial-convective* range of wavenumbers, $1/L \ll k \ll 1/\eta$, the function $\varphi_T(\eta k, \mathrm{Pr})$ in (5.11) therefore must have a form which ensures that the parameters ν and χ do not occur. In order to obtain this form, the function φ_T must be independent of the Prandtl number in this range and must be made proportional to $(\eta k)^{-5/3}$. Hence, in the inertial-convective range, the spectrum of T must take the form

$$E_T(k) = B_1\epsilon_T\epsilon^{-1/3}k^{-5/3}. \tag{5.12}$$

This is the so-called 5/3-power law of Obukhov–Corrsin. Here, B_1 is a numerical constant, shown by experiment to be close to a value of 1.1.

The form of the function $\varphi_T(\eta k, \mathrm{Pr})$ at a high Pr can be specified by Batchelor's hypothesis (1959), which assumes that the spectrum $E_T(k)$, at very high wavenumbers, $k \gg 1/\eta$, cannot depend on ϵ *directly* (since at high Pr most of the energy dissipation occurs at smaller wavenumbers), but only through the typical deformation rates $\tau_\eta^{-1} = (\epsilon/\nu)^{1/2}$. Deformation produces convective mixing by rotating and bringing together the isothermal (or isohaline) surfaces in the interior of fluid particles. In accordance with Batchelor's hypothesis, the function $\varphi_T(\eta k, \mathrm{Pr})$ in (5.11) must involve the parameter ϵ only in the combination ϵ/ν. Therefore, we must have

$$\varphi_T(\eta k, \mathrm{Pr}) = (\mathrm{Pr})^{3/4}\,\Phi[\eta k(\mathrm{Pr})^{-1/2}]. \tag{5.13}$$

As long as the argument of the function Φ is small, i.e., in the *viscous–convective* range to the spectrum, $1 \ll k \ll (\mathrm{Pr})^{1/2}$, then the diffusivity χ should have a negligible effect on the shape of the spectrum $E_T(k)$. This requirement is satisfied if $\Phi(x) \simeq x^{-1}$. Thus, the inertial-convective range has $E_T(k) \simeq k^{-5/3}$, the viscous–convective range exhibits $E_T(k) \simeq k^{-1}$, and, finally, the *viscous-diffusive* range $k\eta \gg (\mathrm{Pr})^{1/2}$ has a spectrum $E_T(k)$ that sharply decreases due to molecular diffusion. These assumptions were confirmed by the temperature fluctuation spectra recorded in the ocean by Grant *et al.* (1968). One of these spectra is presented in Figure 10.11.

At $\eta k \gg 1$, the function $\varphi_T(\eta k, \mathrm{Pr})$ can be constructed theoretically in terms of the approach used above to construct the kinetic energy spectrum (5.10). Instead of the vorticity equation (5.9) in a system of coordinates that moves and rotates with the fluid and lines up with the principal axes of deformation, we shall employ a similar equation for the temperature

$$\frac{\partial T}{\partial t} + a_\beta x_\beta \frac{\partial T}{\partial x_\beta} = \chi \Delta T,$$

which has a solution of the form

$$T(\mathbf{x}, t) = A(t)\exp[i\mathbf{k}(t)\cdot\mathbf{x}].$$

In this case,

$$k_j(t) = k_j(0)\exp(-a_j t); \qquad A(t) = A(0)\exp\left(-\chi\int_0^t k^2\,\mathrm{d}t\right),$$

so that at large values of t the vector $\mathbf{k}(t)$ will also be asymptotically parallel to the axis of maximum compression of the particle, x_3, with a length $k(t) \simeq k_3(0)\exp(|a_3|t)$. The isothermal surfaces T = const then become normal to the x_3-axis and draw closer together. In other words, the temperature gradients increase due to convection. The amplitude δT of the spectral component of the temperature field which is present at time t in an interval δk around k_3 then increases by $\mathrm{d}(\delta T) = -\chi k^2 \delta T\,\mathrm{d}t$ in the time step $\mathrm{d}t = \mathrm{d}k/(|a_3|k)$. The spectrum $E_T(k) = \overline{(\delta T)^2/(\delta k)}$ then increases by

$$\mathrm{d}E_T = \overline{\frac{(\delta T)^2}{\delta k}\left[2\frac{\mathrm{d}(\delta T)}{\delta T} - \frac{\alpha(\delta k)}{\delta k}\right]} \simeq E_T\overline{[-2\chi k^2 - |a_3|]}\,\mathrm{d}t$$

$$= E_T\left(-2\frac{\chi}{\nu}\eta^2 k^2 \tau_\eta^{-1}\,\overline{|a_3|^{-1}} - 1\right)\frac{\mathrm{d}k}{k}.$$

Hence,

$$E_T(k) = \frac{\gamma^2 \epsilon_T \tau_\eta}{k} \exp\left[-\frac{1}{\Pr}(\gamma\eta k)^2\right], \tag{5.14}$$

where γ is determined in the same way as the constant in (5.10), i.e., it is found from the normalization requirement: $2\chi \int_0^\infty k^2 E_T \, dk = \epsilon_T$. So far, (5.6)–(5.14) have ignored the effects that the buoyancy forces, and thus the stratification, have on the turbulence spectra. Let us first discuss the effects of buoyancy forces in the small-scale range of the spectrum, $\eta k \gg 1$, where the inertia forces are negligible and the buoyancy forces can be statistically balanced by the viscous forces. Following Nozdrin (1974), we introduce the buoyancy force $(\alpha T - \beta S)\mathbf{g}$ and consider the following vorticity equation

$$\frac{\partial \omega_j}{\partial t} + \alpha_\beta x_\beta \frac{\partial \omega_j}{\partial x_\beta} - a_j \omega_j = \nu \Delta \omega_j + (\mathbf{g} \times \nabla)_j (\alpha T - \beta S) \tag{5.15}$$

instead of (5.9).

This treatment must be supplemented with a diffusion equation for T and a similar equation for S. Employing the same reasoning as above, and taking into account that asymptotically $(\mathbf{g} \times \mathbf{k})_1 \simeq g_2 k_3$, $(\mathbf{g} \times \mathbf{k})_2 \simeq g_1 k_3$, $(\mathbf{g} \times \mathbf{k})_3 \simeq 0$, we obtain the following equations for the spectral components

$$\begin{aligned} &\mathrm{d}(\delta\omega) = [(a_1 - \nu k^2)\delta\omega + g_2 k(\alpha\delta T - \beta\delta S)] \, \mathrm{d}t; \\ &\mathrm{d}(\delta T) = -\chi k^2 \delta T \, \mathrm{d}t; \qquad \mathrm{d}(\delta S) = -D k^2 \delta S \, \mathrm{d}t; \\ &\mathrm{d}(\delta k) = |a_3| \delta k \, \mathrm{d}t. \end{aligned} \tag{5.16}$$

Hence, the spectra

$$E(k) = \frac{\overline{\delta\omega \cdot \delta\omega^*}}{2k^2 \delta k}; \qquad E_T(k) = \frac{\overline{(\delta T)^2}}{\delta k}; \qquad E_S(k) = \frac{\overline{(\delta S)^2}}{\delta k}$$

and the cross-spectra

$$E_{\omega^* T}(k) = \frac{\overline{\delta\omega^* \cdot \delta T}}{\delta k}; \qquad E_{\omega^* S}(k) = \frac{\overline{\delta\omega^* \cdot \delta S}}{\delta k}; \qquad E_{T^* S}(k) = \frac{\overline{\delta T^* \delta S}}{\delta k}$$

are governed by the following equations:

$$\begin{aligned} &k\frac{\mathrm{d}E_T}{\mathrm{d}k} = -E_T\left(2\frac{\chi}{\nu}\gamma^2\eta^2k^2 + 1\right); \qquad k\frac{\mathrm{d}E_S}{\mathrm{d}k} = -E_S\left(2\frac{D}{\nu}\gamma^2\eta^2k^2 + 1\right); \\ &k\frac{\mathrm{d}E_{T^*S}}{\mathrm{d}k} = -E_{T^*S}\left(\frac{\chi + D}{\nu}\gamma^2\eta^2k^2 + 1\right); \\ &k\frac{\mathrm{d}E_{\omega^*T}}{\mathrm{d}k} = E_{\omega^*T}\left(\sigma - 1 - \frac{\nu + \chi}{\nu}(\gamma\eta k)^2\right) + (\alpha E_T - \beta E_{TS^*})\frac{\mu\eta k}{\tau_\eta}; \\ &k\frac{\mathrm{d}E_{\omega^*S}}{\mathrm{d}k} = E_{\omega^*S}\left(\sigma - 1 - \frac{\nu + D}{\nu}(\gamma\eta k)^2\right) + (\alpha E_{T^*S} - \beta E_S)\frac{\mu\eta k}{\tau_\eta}; \\ &k\frac{\mathrm{d}E}{\mathrm{d}k} = E(2\sigma - 3 - 2(\gamma\eta k)^2) + \\ &\qquad + \left(\alpha\frac{E_{\omega^*T} + E_{\omega T^*}}{2} - \beta\frac{E_{\omega^*S} + E_{\omega S^*}}{2}\right)\frac{\mu v_\eta}{k}, \end{aligned} \tag{5.17}$$

where $\mu = \overline{v_\eta^{-1} g_2 |a_3|^{-1}}$. These equations are successively integrated by quadratures. The spectra $E_T(k)$, $E_S(k)$, and $E_{T*S}(k)$ appear to have the universal form (5.14). They differ from one another by the coefficients χ/ν, D/ν, and $\frac{1}{2}(\chi + D)/\nu$ in the exponents, as well as by the dimensional normalizing factors. The spectra $E_{\omega *T}(k)$, $E_{\omega *S}(k)$, and $E(k)$ contain terms of the type (5.10) and additional terms with the factor μ. The latter terms introduce corrections for the buoyancy forces that also affect the normalizing factors for these spectra. In general, these spectra fail to have a universal form but depend on the relations between their integral parameters ϵ, ϵ_T, ϵ_S, $\epsilon_{TS} = (\chi + D)\overline{\nabla T \cdot \nabla S}$, etc. Nozdrin (1974) analyzed these spectra in detail by integrating equations of the type (5.17) with generalized hypergeometric series. In the diffusion equation for temperature he took the term $\chi \Delta T$ into consideration along with the thermal diffusion term $\chi \Delta S$ derived from the describing the diffusive flux of contaminant in the entropy equation for the mixture (see Kamenkovich, 1973). The non-universal shape of certain spectra is illustrated by the spectrum of the electrical conductivity $C' \simeq -\alpha_1 T' + \beta_1 S'$. This is derived from (5.17):

$$E_C(k) = \frac{\gamma^2 \tau_\eta}{k} \left\{ \alpha_1^2 \epsilon_T \exp\left[-\frac{\chi}{\nu}(\gamma\eta k)^2\right] - \right.$$

$$\left. - 2\alpha_1 \beta_1 \epsilon_{TS} \exp\left[-\frac{1}{2}\frac{\chi + D}{\nu}(\gamma\eta k)^2\right] + \beta_1^2 \epsilon_S \exp\left[-\frac{D}{\nu}(\gamma\eta k)^2\right]\right\}. \tag{5.18}$$

Hence, the shape of the electrical conductivity spectrum will be different in various cases, depending on the relation between the parameters $\alpha_1^2 \epsilon_T$, $\alpha_1 \beta_1 \epsilon_{TS}$, and $\beta_1^2 \epsilon_S$, and also on the value of γ^2. This conclusion is confirmed by empirical data. It was Nasmyth (1972) who detected the effects of buoyancy forces on the small-scale end of kinetic energy spectra in ocean turbulence. He found that in the dissipation range these spectra, usually rise above the universal curve described by (5.6).

At the small-scale end of the inertial-convective range the buoyancy forces are negligible compared with the inertia forces, and the turbulence spectra are described by the 5/3-power laws (5.7) and (5.12). However, the buoyancy forces grow with scale, and at the large-scale end of the inertial-convective range they may become comparable with the inertia forces. The 'buoyancy scale' L_* at which their influence becomes essential can be composed of the buoyancy parameter αg and the parameters that determine the inertial-convective range, ϵ and ϵ_T. The salinity effect on the water density is neglected for the time being. As a result, we obtain $L_* = (\alpha g)^{-3/2} \epsilon^{5/4} \epsilon_T^{-3/4}$, which is the first equation given in the set (4.18). When this scale lies inside the inertial-convective range, according to the similarity hypothesis by Bolgiano (1959) and Obukhov (1959), then the *statistical characteristics of fully developed turbulence components in this range will be completely determined by the parameters* ϵ, ϵ_T, *and* αg. In particular, it follows that over the inertial-convective range the wavenumber spectra of kinetic energy and temperature obey (5.7) and (5.12). The factors C_1 and B_1 are then universal functions of the dimensionless wavenumber kL_* rather than constants.

The asymptotic form of these universal functions at the large-scale end of the inertial-convective range, i.e., at small kL_*, can be determined with Bolgiano's second hypothesis (Bolgiano, 1959, 1962). This states that in conditions of stable stratification the turbulence energy which is transferred through the spectrum from large to small scales mainly serves to counteract the buoyancy forces at large scales. Only a negligible portion of the energy reaches the dissipation range and is converted into heat. The viscous energy dissipation rate ϵ then proves to be extremely low and barely affects the steady-state condition at the large-scale end of the inertial-convective range. Hence, at this end of the range the parameter ϵ must disappear (5.7) and (5.12). These equations involve the coefficients C_1 and B_1, which are functions of kL_*. Therefore, these functions must have the form $C_1 \simeq (kL_*)^{-8/15}$ and $B_1 \simeq (kL_*)^{4/15}$. As a result, (5.7) and (5.12) reduce to

$$E(k) = C_2 (\alpha g)^{4/5} \epsilon_T^{2/5} k^{-11/5}; \qquad E_T(k) = B_2 (\alpha g)^{-2/5} \epsilon_T^{4/5} k^{-7/5}. \tag{5.19}$$

As the wavenumber increases the rate at which the kinetic energy and temperature spectra decrease are, respectively, greater, and less, than the 5/3-power law at the small-scale end of the inertial-convective range. Figure 5.2 furnishes an example of the spectra of longitudinal velocity u' and temperature T' plotted by Nozdrin (1975) from simultaneous measurements at a depth of 30 m in polygon no. 3 in the equatorial zone of the Pacific Ocean (164°30′ East) during the eleventh cruise of the 'Dmitriy Mendeleyev' in the winter of 1973–74. These spectra obey the buoyancy-range laws (5.19) at $k < 0.8$ cm^{-1} and the 5/3-power law at $k > 0.8$ cm^{-1}. The temperature spectrum follows Batchelor's minus-one power law.

As noted before, the buoyancy forces prevent the turbulence from being locally isotropic; it is only locally symmetric with respect to the vertical. As a result, the theory of locally isotropic random fields, which states that there is no correlation between scalar and solenoid vector fields, is not applicable here. Hence, the temperature field, generally speaking, correlates with the velocity field, i.e., the turbulent heat flux $\mathbf{q} = c_p \rho_0 \overline{T'\mathbf{u}'}$ is non-zero and any wave number k can make a non-zero contribution $c_p \rho_0 E_{Tuj}(k)$ to the

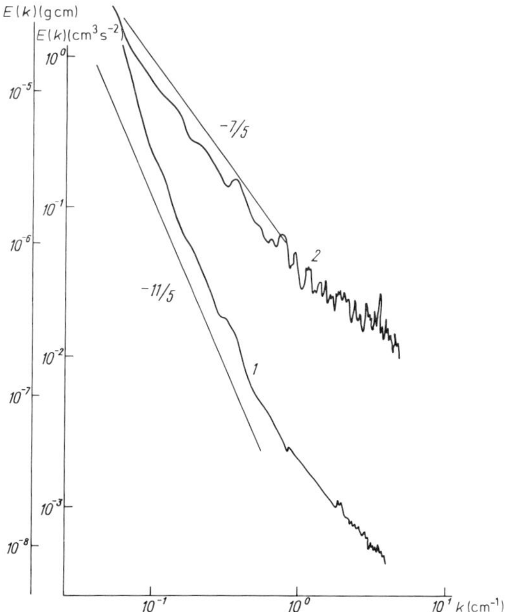

Fig. 5.2. Spectral densities of velocity (1) and temperature (2) fluctuations from simultaneous measurements taken during the eleventh cruise of the 'Dmitriy Mendeleyev', Polygon 3, 20 m depth (Nozdrin, 1975).

flux. The vertical turbulent heat flux q_z and its spectrum, which (apart from a constant factor) is given by the function $E_{Tw}(k)$, are of particular interest. In the buoyancy range this function is proportional to the geometric mean of the functions (5.19) for dimensional reasons:

$$E_{Tw}(k) \simeq (\alpha g)^{1/5} \epsilon_T^{3/5} k^{-9/5}. \tag{5.20}$$

At large values of k it obeys the 5/3-power law $E_{Tw} \simeq \epsilon_T^{1/2} \epsilon^{1/6} k^{-5/3}$ of the inertial-convective range.

Equations (5.19) and (5.20) are valid provided that the scale L_* is much smaller than the external scale (integral scale) L. At scales that exceed those in the inertial-convective range, the mean field gradients $\partial \bar{u}/\partial z$ and $\partial \overline{T}/\partial z$ will be of paramount importance if this is not the case. For instance, if αg and $\partial \overline{T}/\partial z$ are the determining parameters in this range, then the spectra can be expressed as follows

$$E(k) \simeq \alpha g \left|\frac{\partial \overline{T}}{\partial z}\right| k^{-3}; \qquad E_T(k) \simeq \left|\frac{\partial \overline{T}}{\partial z}\right|^2 k^{-3}. \tag{5.21}$$

Spectra of this type were discovered by turbulence measurements in the free atmosphere (Shur, 1962) and interpreted theoretically by Lumley (1964).

Equations (5.19) and (5.20) represent the asymptotic form of turbulence spectra in the buoyancy range, which is located at the large-scale end of the inertial-convective range. Interpolation formulas that describe the transition from the 5/3-power laws in the inertial-convective range to the laws (5.19) and (5.20) in the buoyancy range can be derived in terms of some semi-empirical theory. We will discuss an approach developed by Monin (1962) and employed by a number of investigators. Let us derive exact equations for turbulence spectra making use of the fact that, due to local inhomogeneities, the hydrodynamic fields that describe the turbulence can be written as Fourier–Stieltjes integrals of the form

$$u_j(\mathbf{x}, t) = u_j(0, t) + \int (e^{i\mathbf{k}\cdot\mathbf{x}} - 1) Z_j(\mathbf{dk}, z, t). \tag{5.22}$$

Here, Z_j is a random spectral measure, the integration is carried out over the whole wave-space except for its origin. Setting the coordinate origin $\mathbf{x} = 0$ on a solid surface, let us assume that $u_j(0, t) = 0$. Substituting spectral expansions of the type (5.22) into the hydrodynamic equations (2.1)–(2.5) and eliminating pressure by use of the continuity equation (2.2), one can readily derive the following equations for the random spectral measures of velocity, temperature, and salinity:

$$\begin{aligned}
\left(\frac{\partial}{\partial t} + \nu k^2\right) Z_j(\mathbf{dk}) &= -ik_\alpha \Delta_{j\beta}(\mathbf{k}) \int Z_\beta(\mathbf{dk}_1) Z_\alpha(\mathbf{dk} - \mathbf{dk}_1) + \\
&\quad + \Delta_{j3}(\mathbf{k})[-\alpha g Z_T(\mathbf{dk}) + \beta g Z_S(\mathbf{dk})]; \\
\left(\frac{\partial}{\partial t} + \chi k^2\right) Z_T(\mathbf{dk}) &= -ik_\alpha \int Z_T(\mathbf{dk}_1) Z_\alpha(\mathbf{dk} - \mathbf{dk}_1) + \\
&\quad + g\rho\Gamma Z_3(\mathbf{dk}); \\
\left(\frac{\partial}{\partial t} + Dk^2\right) Z_S(\mathbf{dk}) &= -ik_\alpha \int Z_S(\mathbf{dk}) Z_\alpha(\mathbf{dk} - \mathbf{dk}_1).
\end{aligned} \tag{5.23}$$

Here, $\Delta_{jl}(k) = \delta_{jl} - (k_j k_l/k^2)$ and, for the sake of compactness, the random spectral measures are represented only by their wave arguments. We now multiply the equation for $Z_j(\mathbf{dk})$ by $Z_j^*(\mathbf{dk})$ (assuming summation over the index j) and add the corresponding complex-conjugate equation. When we average the resulting equality, employ the statistical stability of turbulence, i.e., the time independence of the mean products of random spectral measures, and perform similar operations on the equations for $Z_T(\mathbf{dk})$ and $Z_S(\mathbf{dk})$, then we obtain

$$\nu k^2 \overline{Z_\gamma(\mathbf{dk})Z_\gamma^*(\mathbf{dk})} = k_\alpha \operatorname{Im}\left\{\overline{Z_\beta^*(\mathbf{dk}) \int Z_\beta(\mathbf{dk}_1)Z_\alpha(\mathbf{dk} - \mathbf{dk}_1)}\right\} +$$

$$+ \operatorname{Re}\{-\alpha g\overline{Z_3(\mathbf{dk})Z_T^*(\mathbf{dk})} + \beta g\overline{Z_3(\mathbf{dk})Z_S^*(\mathbf{dk})}\};$$

$$\chi k^2 \overline{Z_T(\mathbf{dk})Z_T^*(\mathbf{dk})} = k_\alpha \operatorname{Im}\left\{\overline{Z_T^*(\mathbf{dk}) \int Z_T(\mathbf{dk}_1)Z_\alpha(\mathbf{dk} - \mathbf{dk}_1)}\right\} + \qquad (5.24)$$

$$+ g\rho_0 \Gamma \operatorname{Re}\{\overline{Z_3(\mathbf{dk})Z_T^*(\mathbf{dk}_1)}\};$$

$$Dk^2 \overline{Z_S(\mathbf{dk})Z_S^*(\mathbf{dk})} = k_\alpha \operatorname{Im}\left\{\overline{Z_S^*(\mathbf{dk}) \int Z_S(\mathbf{dk}_1)Z_\alpha(\mathbf{dk} - \mathbf{dk}_1)}\right\}.$$

Having integrated these equations over the entire wave-vector space beyond a sphere of radius k, and employing conventional symbols for wavenumber spectra, we obtain:

$$E_3(k) - \alpha g \int_k^\infty E_{wT}(k)\, dk + \beta g \int_k^\infty E_{wS}(k)\, dk = 2\nu \int_k^\infty k^2 E(k)\, dk;$$

$$E_{uTT}(k) + g\rho_0\Gamma \int_k^\infty E_{wT}(k)\, dk = \chi \int_k^\infty k^2 E_T(k)\, dk; \qquad (5.25)$$

$$E_{uSS}(k) = D \int_k^\infty k^2 E_S(k)\, dk.$$

To determine the spectra $E(k)$, $E_T(k)$, and $E_S(k)$ from these equations, we have to express the functions $E_3(k)$, $E_{uTT}(k)$, $E_{uSS}(k)$, $E_{wT}(k)$, and $E_{wS}(k)$ in terms of some semi-empirical hypothesis. Following Monin (1962) let us take into account that the functions $E_3(k)$, $E_{uTT}(k)$ and $E_{uSS}(k)$ describe the transfer of kinetic energy and temperature and salinity variance from large-scale turbulence components (with wavenumbers below k) to small-scale components (with wavenumbers above k). In terms of Heisenberg's (1948) semi-empirical theory, these functions can be presented as products of the eddy viscosity (turbulent exchange coefficient) $K(k)$ that results from the small-scale turbulence components (with wavenumbers above k) and the squared mean gradients of velocity, temperature, and salinity in the large-scale flow. The latter consists of the mean flow and the turbulence components at wavenumbers below k, with a lower limit k_0 that corresponds to the turbulence scale $L_0 = 2\pi/k_0$, i.e., to the maximum scale allowed by the flow geometry. The resulting expressions for the spectral fluxes are:

$$E_3(k) = K(k)\left[\left(\frac{\partial \bar{u}}{\partial z}\right)^2 + \left(\frac{\partial \bar{v}}{\partial z}\right)^2 + \int_{k_0}^{k} 2k^2 E(k)\, dk\right];$$

$$E_{uTT}(k) = \alpha_T K(k)\left[\left(\frac{\partial \bar{T}}{\partial z}\right)^2 + \int_{k_0}^{k} k^2 E_T(k)\, dk\right]; \tag{5.26}$$

$$E_{uSS}(k) = \alpha_S K(k)\left[\left(\frac{\partial \bar{S}}{\partial z}\right)^2 + \int_{k_0}^{k} k^2 E_S(k)\, dk\right].$$

In a similar manner, the integrals $\int_k^\infty E_{wt}(k)\, dk$ and $\int_k^\infty E_{wS}(k)\, dk$, which are proportional to the vertical fluxes of heat and salinity carried by the small-scale components, can be represented as products of $K(k)$ and the squares of the mean gradients of temperature and salinity in the large-scale flow:

$$\int_k^\infty E_{wT}(k)\, dk = \pm \alpha_T K(k)\left[\left(\frac{\partial \bar{T}}{\partial z}\right)^2 + \int_{k_0}^{k} k^2 E_T(k)\, dk\right]^{1/2};$$

$$\int_k^\infty E_{wS}(k)\, dk = \pm \alpha_S K(k)\left[\left(\frac{\partial \bar{S}}{\partial z}\right)^2 + \int_{k_0}^{k} k^2 E_S(k)\, dk\right]^{1/2}. \tag{5.27}$$

Here and below, the top symbol in $\pm$ or $\mp$ denotes a positive and the bottom symbol a negative contribution from the vertical gradients of temperature and salinity to the stable density stratification, as defined by

$$\frac{\partial \rho}{\partial z} = -\rho\alpha \frac{\partial T}{\partial z} + \rho\beta \frac{\partial S}{\partial z}.$$

Finally, let us express $K(k)$ through the values of k_1 and $E(k_1)$ for all $k_1 > k$ in the form of an integral based on dimensional considerations. This can be realized in different ways. Let us use the expression proposed by Howells (1960):

$$K(k) = \gamma_0 \left[\int_k^\infty \frac{E(k_1)}{k_1^2}\, dk_1\right]^{1/2}, \tag{5.28}$$

where γ_0 is a numeric constant. Equations (5.26)–(5.28) and (5.25) form a closed system for the spectra $E(k)$, $E_T(k)$, and $E_S(k)$. This system of equations (with salinity neglected) was solved by Benilov and Lozovatsky (1974) by a method described by Monin (1962). A somewhat different, but related, system of equations (with salinity neglected) was constructed on the basis of a procedure suggested by Tchen (1953, 1954), who considered a shear flow without thermal stratification, and solved by Gisina (1966, 1969). Turbulence spectra in stratified flows were investigated by Tchen later (see Tchen, 1975).

Equations (5.25)–(5.28) can be used to determine turbulence spectra both in the buoyancy range and in the inertial-convective, viscous-convective and viscous-diffusive ranges. However, in the latter two ranges these equations do not seem to be sufficiently accurate. Just like in Heisenberg's (1948) theory, which was employed as the model to construct these equations, they yield power laws of the type $E(k) \simeq k^{-7}$ in the viscous-diffusive range instead of the exponential laws (5.14) that follow from the more accurate equations (5.17) for the small-scale spectral range. Therefore, we shall use (5.25)–(5.28) only to determine the turbulence spectra in the buoyancy and inertial-convective ranges. At values of k inside these ranges, the integrals from k to ∞ on the right-hand sides of (5.25) can be replaced by those from 0 to ∞ with fair accuracy, since it is the small-scale spectral range that makes the major contribution to these integrals. The right-hand sides can then be replaced by the quantities ϵ, ϵ_T, and ϵ_S. With (5.26)–(5.28), the equations then take the form:

$$\gamma_0 \left[\int_k^\infty \frac{E(k)}{k^2}\, dk \right]^{1/2} \left\{ \left(\frac{\partial \bar{u}}{\partial z} \right)^2 + \left(\frac{\partial \bar{v}}{\partial z} \right)^2 + \int_{k_0}^k 2k^2 E(k)\, dk \mp \right.$$

$$\mp \alpha_T \alpha g \left[\left(\frac{\partial \bar{T}}{\partial z} \right)^2 + \int_{k_0}^k k^2 E_T(k)\, dk \right]^{1/2} \mp$$

$$\left. \mp \alpha_S \beta g \left[\left(\frac{\partial \bar{S}}{\partial z} \right)^2 + \int_{k_0}^k k^2 E_S(k)\, dk \right]^{1/2} \right\} = \epsilon; \tag{5.29}$$

$$\alpha_T \gamma_0 \left[\int_k^\infty \frac{E(k)}{k^2}\, dk \right]^{1/2} \left[\left(\frac{\partial \bar{T}}{\partial z} \right)^2 + \int_{k_0}^k k^2 E_T(k)\, dk \right] \simeq \epsilon_T;$$

$$\alpha_S \gamma_0 \left[\int_k^\infty \frac{E(k)}{k^2}\, dk \right]^{1/2} \left[\left(\frac{\partial \bar{S}}{\partial z} \right)^2 + \int_{k_0}^k k^2 E_S(k)\, dk \right] = \epsilon_S.$$

The second of these equations ignores the small 'adiabatic correction', i.e., the term derived from the second term on the left-hand side of the second equation in the set (5.25). This correction is negligible, since the adiabatic temperature gradient $g\rho_0\Gamma$ is, as a rule, small compared with the real mean gradient. Notice that the preservation of this correction would not hamper the solution, but only make the subsequent expressions slightly more cumbersome. Setting $k = k_0$ and $K(k_0) = K$ in (5.29), one obtains macroscopic equations for the turbulent energy and the temperature and salinity dissipation rates:

$$K \left[\left(\frac{\partial \bar{u}}{\partial z} \right)^2 + \left(\frac{\partial \bar{v}}{\partial z} \right)^2 \right] \mp \alpha g \alpha_T K \left| \frac{\partial \bar{T}}{\partial z} \right| \mp \beta g \alpha_S K \left| \frac{\partial \bar{S}}{\partial z} \right| = \epsilon;$$

$$\alpha_T K \left(\frac{\partial \bar{T}}{\partial z} \right)^2 = \epsilon_T; \qquad \alpha_S K \left(\frac{\partial \bar{S}}{\partial z} \right)^2 = \epsilon_S. \tag{5.30}$$

These equations coincide with (2.18)–(2.19) and (2.21), apart from small terms. Together with the mean equations of motion and of heat and salt transfer they enable us, in principle to determine the mean velocity, temperature, and salinity, as well as the quantities ϵ, ϵ_T, ϵ_S, and K as functions of z. In calculating the spectra, it will be assumed that these parameters are known.

Let us now introduce the following normalized spectra:

$$
\begin{aligned}
E(k) &= (\epsilon/\gamma_0)^{2/3} L^{5/3} f(kL);\\
E_T(k) &= \left(\frac{\epsilon_T}{\alpha_T\gamma_0}\right)(\epsilon/\gamma_0)^{-1/3} L^{5/3} f_T(kL);\\
E_S(k) &= \frac{\epsilon_S}{\alpha_S\gamma_0}(\epsilon/\gamma_0)^{-1/3} L^{5/3} f_S(kL);\\
L &= \left| \pm \alpha g \left(\frac{\alpha_T\epsilon_T}{\gamma_0}\right)^{1/2} \pm \beta g \left(\frac{\alpha_S\epsilon_S}{\gamma_0}\right)^{1/2} \right|^{-3/2} (\epsilon/\gamma_0)^{5/4}.
\end{aligned}
\tag{5.31}
$$

In the absence of salinity, the scale L becomes equal to the scale L_* in the first equation of the set (4.18), apart from a numeric multiplier. Hence, L is a natural generalization of the buoyancy scale for turbulence in salt water. From (5.29) and (5.31) we now define the following dimensionless parameters:

$$
\begin{aligned}
\Gamma_u &= \frac{1}{2}\left(\frac{\epsilon}{\gamma_0}\right)^{-2/3} L^{4/3}\left[\left(\frac{\partial \bar{u}}{\partial z}\right)^2 + \left(\frac{\partial \bar{v}}{\partial z}\right)^2\right];\\
\Gamma_T &= \left(\frac{\epsilon_T}{\alpha_T\gamma_0}\right)^{-1}\left(\frac{\epsilon}{\gamma_0}\right)^{1/3} L^{4/3}\left(\frac{\partial \bar{T}}{\partial z}\right)^2;\\
\Gamma_S &= \left(\frac{\epsilon_S}{\alpha_S\gamma_0}\right)^{-1}\left(\frac{\epsilon}{\gamma_0}\right)^{1/3} L^{4/3}\left(\frac{\partial \bar{S}}{\partial z}\right)^2.
\end{aligned}
\tag{5.32}
$$

As a result, (5.29) becomes:

$$
\begin{aligned}
&2F^2\left[\Gamma_u + \int_{x_0}^{x} x^2 f(x)\,\mathrm{d}x\right] \mp F = 1;\\
&F^2\left[\Gamma_T + \int_{x_0}^{x} x^2 f_T(x)\,\mathrm{d}x\right] = 1;\\
&F^2\left[\Gamma_S + \int_{x_0}^{x} x^2 f_S(x)\,\mathrm{d}x\right] = 1,
\end{aligned}
\tag{5.33}
$$

where

$$
x = kL, \quad x_0 = k_0 L, \quad F = \left[\int_x^{\infty} \frac{f(x)\,\mathrm{d}x}{x^2}\right]^{1/4}
$$

and the functions $f_T(x)$ and $f_S(x)$ are eliminated from the first equation using the second and third equations. Here, and below, the top symbol in $\pm$ or $\mp$ corresponds to stable, and the bottom symbol to unstable, stratification. If we divide each of the above equations by F^2, differentiate them with respect to x, and use the definition of F, then we obtain

$$x = \tfrac{1}{2}F^{-3/2}(4 \pm 2F)^{1/4}; \qquad f = \frac{4F^{5/2}(4 \pm 2F)^{5/4}}{12 \pm 5F};$$
$$f_T = f_S = \frac{32F^{5/2}(4 \pm 2F)^{1/4}}{12 \pm 5F}. \tag{5.34}$$

These expressions give a parametric representation of the functions $f(x)$, $f_T(x)$ and $f_S(x)$. These functions are the solutions of (5.33) for $x > x_0$, where $x_0 = 1/2F_0^{-3/2}(2F_0 + 4)^{1/4}$ and F_0 is determined from any one of the equations

$$2F_0^2\Gamma_u \mp F_0 = 1; \qquad F_0^2\Gamma_T = 1; \qquad F_0^2\Gamma_S = 1. \tag{5.35}$$

The equations in (5.35) are derived from (5.33) at $x = x_0$ and are equivalent to the macroscopic equations (5.30). Note that (5.27) also yields

$$E_{wT}(k) = \pm\gamma_0\left(\frac{\alpha_T\epsilon_T}{\gamma_0}\right)^{1/2}\left(\frac{\epsilon}{\gamma_0}\right)^{1/6}L^{5/3}f_{wT}(kL);$$
$$E_{wS}(k) = \pm\gamma_0\left(\frac{\alpha_S\epsilon_S}{\gamma_0}\right)^{1/2}\left(\frac{\epsilon}{\gamma_0}\right)^{1/6}L^{5/3}f_{wS}(kL); \tag{5.36}$$
$$f_{wT}(x) = f_{wS}(x) = [\tfrac{1}{8}f(x)f_T(x)]^{1/2},$$

so that the heat (salt) flux spectrum is proportional to the geometric mean of the energy and temperature (salinity) spectra.

Let us now use (5.34) to investigate the asymptotic behavior of turbulence spectra at the far end of the inertial-convective range and in the buoyancy range. At the far end of the inertial-convective range the values of k are high; hence F is low. Therefore, (5.34) gives

$$f(x) = \frac{2^{5/3}}{3}x^{-5/3}\left(1 \pm \frac{5}{12}\cdot 2^{-1/3}x^{-2/3} + \ldots\right);$$
$$f_T(x) = \frac{2\cdot 2^{5/3}}{3}x^{-5/3}\left(1 \mp \frac{1}{12}\cdot 2^{-1/3}x^{-2/3} + \ldots\right); \tag{5.37}$$
$$f_{wT}(x) = \frac{2^{2/3}}{3}x^{-5/3}\left(1 \pm \frac{1}{6}\cdot 2^{-1/3}x^{-2/3} + \ldots\right).$$

Substituting the main terms of these asymptotic expansions into the expressions for $E(k)$ and $E_T(k)$ in (5.31), we obtain the 5/3-power laws (5.7) and (5.12), with numerical coefficients $C_1 = (2^{5/3}/3)\cdot(\gamma_0^{-2/3})$ and $B_1 = 2C_1/\alpha_T$. Also, from (5.36) we obtain for $E_{wT}(k)$

$$E_{wT}(k) = \pm\frac{2^{2/3}}{3}\gamma_0^{1/3}\alpha_T^{1/2}\epsilon_T^{1/2}\epsilon^{1/6}k^{-5/3}. \tag{5.38}$$

In conditions of stable stratification, the spectra $E(k)$ and $E_{wT}(k)$ increase faster and the spectrum $E_T(k)$ increases slower than the 5/3-power law with decreasing wavenumber, while in conditions of unstable stratification, $E(k)$ and $|E_{wT}(k)|$ increase slower and $E_T(k)$ faster than they do by the 5/3-power law.

Within the buoyancy range the values of k are small. Hence, in conditions of stable stratification, the values of F are large. Thus, (5.34) yields

$$f(x) = \frac{2}{5} \cdot 2^{3/5} x^{-11/5} \left(1 + \frac{6}{5} \cdot 2^{3/5} x^{4/5} + \ldots\right);$$

$$f_T(x) = \frac{16}{5} \cdot 2^{1/5} x^{-7/5} \left(1 - \frac{6}{5} \cdot 2^{3/5} x^{4/5} + \ldots\right); \tag{5.39}$$

$$f_{wT}(x) = \frac{2}{5} \cdot 2^{2/5} x^{-9/5} (1 + \ldots).$$

Substituting the main terms of these asymptotic expansions into the expressions for $E(k)$ and $E_T(k)$ in (5.31) and into the expression for $E_{wT}(k)$ in (5.36), we obtain

$$E(k) = \frac{2}{5} \cdot 2^{3/5} \left| \pm \alpha g \left(\frac{\alpha_T \epsilon_T}{\gamma_0}\right)^{1/2} \pm \beta g \left(\frac{\alpha_S \epsilon_S}{\gamma_0}\right)^{1/2} \right|^{4/5} k^{-11/5};$$

$$E_T(k) = \frac{16}{5} \cdot 2^{1/5} \left| \pm \alpha g \left(\frac{\alpha_T \epsilon_T}{\gamma_0}\right)^{1/2} \pm \beta g \left(\frac{\alpha_S \epsilon_S}{\gamma_0}\right)^{1/2} \right|^{2/5} \times$$

$$\times \frac{\epsilon_T}{\alpha_T \gamma_0} k^{-7/5}; \tag{5.40}$$

$$E_{wT}(k) = \frac{2}{5} \cdot 2^{2/5} \left| \pm \alpha g \left(\frac{\alpha_T \epsilon_T}{\gamma_0}\right)^{1/2} \pm \beta g \left(\frac{\alpha_S \epsilon_S}{\gamma_0}\right)^{1/2} \right|^{1/5} \times$$

$$\times \gamma_0 \left(\frac{\alpha_T \epsilon_T}{\gamma_0}\right)^{1/2} k^{-9/5}.$$

These expressions specify (5.19)–(5.20) for turbulence in salt water: for $\epsilon_S = 0$, (5.40) is transformed into (5.19)–(5.20), with concrete numerical coefficients.

Finally, in conditions of unstable stratification we have $F \to 2$ as $k \to 0$. The functions $f(x)$, $f_T(x)$, and $f_{wT}(x)$ tend towards zero as x^5, x, and x^3, respectively. Since they also approach zero as $x \to \infty$, it is clear that, for some positive value of x, these functions reach their maxima. These are real, provided that the corresponding x values exceed x_0.

We can now describe the small-scale parts of the spectra in fully developed turbulence in the following way. The range around the maximum in the kinetic energy spectrum, where most of the energy is concentrated, is followed by the buoyancy range with a 11/5-power law. Then comes the inertial range with a 5/3-power law, followed by the dissipation range, with a complex exponential law of spectral collapse. In the temperature and salinity spectra the range around the maximum is followed by the buoyancy range with a 7/5-power law. Then comes the inertial-convective range, with a 5/3-power law.

After that we find the viscous-convective range, with a minus-one power law and, finally, the viscous-diffusive range, with an exponential spectral collapse. Depending on the buoyancy scale, the buoyancy range can shift either to the range where most of the energy is located and disappear in it or, vice versa, to small scales, where it then occupies the position of the inertial-convective range. The length of the inertial-convective range varies depending on the Reynolds number: it is long at large Re, and vanishes at small Re. Thus, various shapes of turbulence spectra are found at various buoyancy scales and Reynolds numbers.

6. THE SMALL-SCALE STRUCTURE OF TURBULENCE

In this section, special attention will be paid to the fact that the turbulence energy dissipation rate ϵ that is involved in the structure functions and the spectra of fully-developed turbulence and in other expressions that follow from Kolmogorov's similarity hypotheses is not, strictly speaking, a fixed parameter, but is a random field, $\epsilon(\mathbf{x}, t)$. For incompressible fluids, ϵ is related to the velocity field $\mathbf{u}(\mathbf{x}, t)$ by Stokes' expression

$$\epsilon = \frac{\nu}{2} \sum_{\alpha,\beta} \left(\frac{\partial u_\alpha}{\partial x_\beta} + \frac{\partial u_\beta}{\partial x_\alpha} \right)^2 . \tag{6.1}$$

Hence, it fluctuates along with the velocity field. These fluctuations may depend on the properties of the large-scale motion and, above all, on the Reynolds number of the large-scale motion. Since the statistical properties of the field $\epsilon(\mathbf{x}, t)$ will most likely affect the probability distributions of the small-scale turbulence, these probability distributions are not, in fact, universal but depend on Re and on some other characteristics of the large-scale motion. This fact was first stressed by Landau. See the comments on p. 157 in Landau and Lifshitz (1953), mentioned also in the first (1944) edition of the book.

In accordance with this, Kolmogorov's similarity hypotheses for the statistical characteristics of the small-scale structure of turbulence must be made specific, in the sense that they can refer only to *conditional* statistical characteristics, calculated with fixed parameters that determine these characteristics. This was proposed by Kolmogorov (1962a, b) and Obukhov (1962a, b). A simplified version is the following. The small-scale statistical characteristics of turbulence are determined by the velocity at some finite set of time–space points $M_1, \ldots, M_n$. The *mean value* ϵ_G of the field $\epsilon(\mathbf{x}, t)$ is now assumed to be fixed in a certain small time–space region G that contains all the points $M_1, \ldots, M_n$. For fixed ϵ_G the conditional statistical characteristics are calculated by the Kolmogorov similarity hypotheses, while the unconditional characteristics can be obtained from the conditional ones by averaging over the probability distribution for ϵ_G. In this case, the particular form of the region G seems to be of no importance. When he investigated the two-point statistical characteristics for observation points at a distance r, Obukhov used for the region G a sphere of radius $r/2$ with poles at the observation points, so that the quantity

$$\epsilon_r = \epsilon_r(\mathbf{x}_0, t) = \frac{6}{\pi r^3} \int_{|\mathbf{r}'| \leq r/2} \epsilon(\mathbf{x}_0 + \mathbf{r}', t)\, d\mathbf{r}' \tag{6.2}$$

served as ϵ_G.

To develop this idea, Kolmogorov considered the finite-dimensional probability distribution for the relative velocities $\mathbf{v}(\mathbf{r}_k, \tau_k) = \mathbf{u}(\mathbf{x}_0 + \mathbf{r}_k, t_0 + \tau_k) - \mathbf{u}(\mathbf{x}_0, t_0)$ between points at distances $r_k \leqslant r \ll L$ and with time differences $|\tau_k| \leqslant \epsilon_r^{-1/3} r^{2/3}$. He made the similarity hypotheses specific in the following way: *provided that* $\mathrm{Re}_r = \epsilon_r^{1/3} r^{4/3}/\nu$ *is fixed, then the conditional probability distribution for the dimensionless relative velocities*

$$\mathbf{w}(r^{-1}\mathbf{r}_k, \epsilon_r^{1/3} r^{-2/3}\tau_k) = \epsilon_r^{-1/3} r^{-1/3}\mathbf{v}(\mathbf{r}_k, \tau_k) \tag{6.3}$$

is completely determined by this value of Re_r. *Also, if* $\mathrm{Re}_r \gg 1$, *then the probability distribution is independent of* Re_r, *i.e., it is universal*. This specification of the hypotheses shows, in particular, that the conditional averages of the squared velocity differences $v_j(\mathbf{r}, 0)$, i.e., the conditional spatial structure functions of the velocity field, are products of $\epsilon_r^{2/3} r^{2/3}$ and functions of $\mathrm{Re}_r = \epsilon_r^{1/3} r^{4/3}/\nu$ for $r \ll L$. When $\mathrm{Re}_r \to \infty$, they tend towards constants (at this limit the conditional structure functions differ from $\epsilon_r^{2/3} r^{2/3}$ only by numeric factors, i.e., they obey the 2/3-power law). In a similar way, the conditional spatial spectrum of the field $\mathbf{v}(\mathbf{r}, 0)$ at a fixed $\epsilon_{1/k}$, where $1/k \ll L$, is:

$$E(k \mid \epsilon_{1/k}) = \epsilon_{1/k}^{2/3} k^{-5/3} \varphi(\nu \epsilon_{1/k}^{-1/3} k^{4/3}). \tag{6.4}$$

For small arguments, the function φ approaches a constant so that the conditional spectrum obeys the 5/3-power law asymptotically. The similarity hypotheses for the conditional statistical characteristics of the small-scale components of the scalar field T can be specified in a similar manner, provided that the random quantities ϵ_r and ϵ_{Tr} are fixed.

To determine the unconditional statistical characteristics of small-scale turbulence and, above all, the unconditional structure functions and spectra, it is necessary to average the corresponding conditional characteristics over the probability distribution of ϵ_r. It is, therefore, necessary to know the probability distribution for the field $\epsilon(\mathbf{x}, t)$. Numerous experimental measurements, reviewed by Monin and Yaglom (1975, §25.3), show that these probability distributions differ greatly from Gaussian distributions.

According to (6.1), the field ϵ is expressed in terms of velocity derivatives. In particular, for locally isotropic turbulence, $\bar{\epsilon} = (15/2)\nu\overline{(\partial w/\partial x)^2}$. Measurements show that the probability distributions for the nth derivatives, e.g., of the longitudinal velocity $u(x)$, have a positive kurtosis, which rapidly increases with n and Re. In a similar way, the probability distributions for spectral components of the function $u(\mathbf{x}_0, t)$ measured at the output of a narrow bandpass filter with a bandwidth δf around the mean frequency f_m have a kurtosis that increases with f_m for fixed $\delta f/f_m$ and with Re for fixed δf and f_m. These and other data show that the high-frequency velocity fluctuations are close to zero over comparatively long time intervals (or large distances), which are intermittently interrupted by short ones (or thin spatial layers), in which the fluctuations have large amplitudes. With increasing Re this intermittency increases and the range of scales in which it is fairly prominent expands. A simplified model of a random quantity with this kind of intermittency will be a quantity with a finite probability p of being equal to zero and a probability $1 - p$ of having values other than zero, with a Gaussian distribution. Its kurtosis then is equal to $3p(1-p)^{-1}$; it is large when $(1-p)$ is small.

To construct a more detailed model of random fields with small-scale intermittency, let us take $\epsilon(\mathbf{x}, t)$ to be any local non-negative hydrodynamic characteristic that is determined only by small-scale turbulence components. Thus, it can represent not only the turbulent energy dissipation rate, but also the squared product or the absolute value of any spatial derivative of the velocity field, the temperature field, or some other scalar field; the rate of destruction of temperature inhomogeneities; the squared vorticity; the squared difference of either velocity or temperature at two closely spaced points; or the product of two, or several, similar quantities raised to a certain power. Now, let $V''' \subset V'' \subset V'$ be similar one-, two-, or three-dimensional spaces with scales $l''' < l'' < l'$, ϵ''', ϵ'', ϵ' be the mean values of $\epsilon(\mathbf{x}, t)$ over these spaces. Assume that the field $\epsilon(\mathbf{x}, t)$ obeys the following kind of *self-similarity* (Novikov, 1962, 1971; Gurvich and Yaglom, 1967): *if the Reynolds number is sufficiently great,* $l' \ll L$ *and* $l'' \gg \eta'' = (\nu^3/\epsilon'')^{1/4}$, *then the probability distribution of* ϵ''/ϵ' *depends only on the scale ratio* l''/l; *moreover, if* $l''' \gg \eta''' = (\nu^3/\epsilon''')^{1/4}$, *then the ratios* ϵ'''/ϵ'' *and* ϵ''/ϵ' *are statistically independent.*

Let us now consider a sequence of inter-embedded and similar spaces $V_0 \supset V_1 \supset \cdots \supset V_N$, in which the scale ratio of subsequent spaces is the same: $l_1/l_0 = l_2/l_1 = \cdots = l_N/l_{N-1} = \lambda < 1$. In this case, $l_0 \simeq L$ and l_N is so small that the fluctuations of ϵ inside V_N can be neglected. Let ϵ_j be the mean value of the field $\epsilon(\mathbf{x}, t)$ over the space V_j. Introduce the random quantities $e_j = e_j(\epsilon_{j-1}) = \epsilon_j/\epsilon_{j-1}$, taken at a fixed value of the denominator ϵ_{j-1}. Here, ϵ_0 can be identified with the mathematical expectation $\bar{\epsilon}$ of the field $\epsilon(\mathbf{x}, t)$. Also, all quantities e_j with indices j, such that $L \gg l_{j-1} > l_j \gg \eta_j = (\nu^3/\epsilon_j)^{1/4}$, prove to be independent, equally distributed random quantities. Therefore, $\epsilon_n = \bar{\epsilon} e_1 e_2 \ldots e_n$, which yields

$$\ln \epsilon_n = \ln \bar{\epsilon} + \sum_{j=1}^{n} \ln e_j. \tag{6.5}$$

For fully developed turbulence at large values of Re, the self-similar range is sufficiently wide for the sum at the right-hand side of (6.5) to involve many independent, equally distributed terms $\ln e_j$. Therefore, by the central limit theorem of probability theory, this sum has an approximately normal distribution, with a mathematical expectation m_n and a variance σ_n^2 that are given by:

$$m_n = \ln \bar{\epsilon} + A_1(\mathbf{x}, t) + nm; \qquad \sigma_n^2 = A(\mathbf{x}, t) + n\mu. \tag{6.6}$$

Here, m and μ are the mathematical expectation and the variance, respectively, of the equally distributed random quantities $\ln e_j$. The terms A_1 and A depend on the properties of the large-scale motion. Simplified models for the spatial intermittency of velocity fields and turbulent energy dissipation have been proposed by Novikov and Stuart (1964) and Novikov (1965, 1966). These models assume that the variance of equally distributed random quantities $\ln e_j$ is infinite, so that the random quantity $\ln \epsilon_r$ is distributed by one of the so-called 'stable non-Gaussian' distributions. Let us consider a fixed distance $r \ll L$, with n chosen so that $l_n \simeq r$. In this case, $n \simeq \ln (L/r)$ and (6.6) takes the form

$$\begin{aligned} m_r &= \overline{\ln \epsilon_r} = \ln \bar{\epsilon} + A_1(\mathbf{x}, t) + m \ln \frac{L}{r}; \\ \sigma_r^2 &= \overline{(\ln \epsilon_r - m_r)^2} = A(\mathbf{x}, t) + \mu \ln \frac{L}{r}. \end{aligned} \tag{6.7}$$

The second relation in (6.7) was first derived by Kolmogorov (1941, 1962a) for the dissipation rate of the turbulent kinetic energy. Note that $\bar{\epsilon}_r = \bar{\epsilon}$ at any r; hence the relations $m = -\frac{1}{2}\mu$ and $A_1 = -\frac{1}{2}A$ can be readily obtained for the lognormal random quantity ϵ_r. For this random quantity, the moments of arbitrary order are

$$\overline{(\epsilon_r)^q} = (\bar{\epsilon})^q \exp[\tfrac{1}{2}q(q-1)\sigma_r^2] = B_q(\mathbf{x}, t)(\bar{\epsilon})^q \left(\frac{L}{r}\right)^{[\mu q(q-1)]/2}. \tag{6.8}$$

Here the coefficient $B_q(\mathbf{x}, t)$ can depend on the peculiarities of the large-scale motion. Kolmogorov (1962a, b) used this expression for the turbulent energy dissipation rate. For the structure functions of the velocity field in the inertial range this yields the refined expression:

$$\begin{aligned} D_n(r) &= \overline{|u(\mathbf{x}+\mathbf{r}, t) - u(\mathbf{x}, t)|^n} = C_n\overline{(\epsilon_r)^{n/3}}r^{n/3} \\ &= C_n'(\bar{\epsilon})^{n/3}r^{n/3}\left(\frac{L}{r}\right)^{\mu n(n-3)/18} \end{aligned} \tag{6.9}$$

The presence of a multiplier that involves a power of L/r and the possible dependence of the coefficient C_n' on the large-scale motion are refinements of the conventional Kolmogorov similarity theory, which neglects the fluctuations of the turbulent energy dissipation rate. In particular, for $n = 2$, (6.9) yields $D_2(r) \simeq r^{(2/3)-(\mu/9)}$, so that instead of the conventional 2/3-power law we have a refined law, with the exponent $(2/3) - (\mu/9)$. The parameter μ, in principle, can be estimated from the structure functions of the velocity field with

$$\mu = \frac{18}{n(n-3)}\left\{\frac{n}{3} - \frac{\ln[D_n(r)/D_n(r_1)]}{\ln(r/r_1)}\right\}, \tag{6.10}$$

which follows from (6.9).

Expressions similar to (6.9) and (6.10) can also be written for the joint structure functions of the velocity and temperature fields, $D_{mn} = \overline{(\Delta_r u)^m(\Delta_r T)^n}$. Expressions of this kind were reported by Monin (1969) at the International Symposium on Oceanic Turbulence (1968) in Vancouver; they were subsequently detailed by Korchashkin (1970) and Van Atta (1971, 1973, 1974). We shall return to this later.

Let us now choose the spaces V_j that are used in the derivation of (6.5) as line segments along the X-axis. Instead of (6.2) we then have $\epsilon_r = (1/r)\int_0^r \epsilon(x)\,dx$. Hence,

$$\overline{\epsilon_r^2} = \frac{2}{r^2}\int_0^r dr_1 \int_0^{r_1} B_{\epsilon\epsilon}(r')\,dr',$$

where $B_{\epsilon\epsilon}(r) = \overline{\epsilon(x)\epsilon(x+r)}$ is the correlation function of the random field $\epsilon(x)$. Hence,

$$B_{\epsilon\epsilon}(r) = \frac{1}{2}\frac{d^2}{dr^2}(r^2\overline{\epsilon_r^2})$$

and (6.8), with $q = 2$, yields

$$B_{\epsilon\epsilon}(r) = B'(\mathbf{x}, t)(\bar{\epsilon})^2\left(\frac{L}{r}\right)^{\mu}. \tag{6.11}$$

The multiplier $(L/r)^\mu$ is large, so that the quantity $B_{\epsilon\epsilon}(r)$ is also large compared with $(\bar{\epsilon})^2$. As a result,

$$b_{\epsilon\epsilon}(r) = \overline{[\epsilon(x) - \bar{\epsilon}]\,[\epsilon(x+r) - \bar{\epsilon}]} = B_{\epsilon\epsilon}(r) - (\bar{\epsilon})^2 \simeq B_{\epsilon\epsilon}(r),$$

and the one-dimensional spectrum, which is defined by $E_{\epsilon\epsilon}(k) = (1/\pi)\int_0^\infty b_{\epsilon\epsilon}(r)\cos kr\,\mathrm{d}r$, becomes

$$E_{\epsilon\epsilon}(k) = B''(\bar{\epsilon})^2 L(kL)^{-1+\mu} \tag{6.12}$$

in the self-similar range. Here, $\mu < 1$ and, using (6.11),

$$B'' = \frac{B'}{\pi}\int_0^\infty x^{-\mu}\cos x\,\mathrm{d}x = \frac{B'\Gamma(\mu)}{2}\cos\frac{\pi\mu}{2}.$$

Thus, the parameter μ can be estimated by the spectrum $\epsilon(x)$ in the self-similar range. Spectral measurements of the slightly smoothed field of $(\partial w/\partial x)^2$ carried out by Gurvich and Zubkovsky (1963, 1965) gave $\mu \simeq 0.4$, while numerous spectral measurements for the fields $(\partial u/\partial x)^2$ and $(\partial T/\partial x)^2$ carried out by other workers resulted in $\mu = 0.3$–0.5. Similar estimates of μ were obtained by Kholmiansky (1970, 1972, 1973) and other workers by measuring the dependence of the third and fourth moments of ϵ_r on r, according to (6.8). At the same time, the exponents of the higher moments, $q = 5$ and 6, proved to be smaller than those obtained by (6.8). This problem will be discussed below.

The model with a discrete division of the spaces $V_0 \supset V_1 \supset \cdots \supset V_N$ discussed here, which results in a lognormal distribution of ϵ_r, was calculated by Yaglom (1966) and Gurvich and Yaglom (1967). Later, Novikov (1969, 1971) developed a more general approach which demands no discrete division. This approach employs division coefficients given by

$$e_{r,l}(x, x') = \epsilon_r(x')/\epsilon_l(x), \quad r < l \tag{6.13}$$

for values of r and l taken from the self-similar range and for $|x' - x| \leqslant (l - r)/2$. Due to the scale similarity, the probability distribution of $e_{r,l}$ depends only on l/r, and for $l > \rho > r$ the division coefficients $e_{r,\rho}$ and $e_{\rho,l}$ are statistically independent. The equality $e_{r,l} = e_{r,\rho}e_{\rho,l}$ then yields, for the moment $a_q(l/r) = \overline{(e_{r,l})^q}$, the functional relation $a_q(l/r) = a_q(l/\rho)a_q(\rho/r)$. The continuous solutions of this relation are the functions

$$\overline{(e_{r,l})^q} = \left(\frac{l}{r}\right)^{\mu_q}. \tag{6.14}$$

Here, $\overline{e_{r,l}} = 1$, i.e., $\mu_1 = 0$. To calculate the second moment ($q = 2$), let us consider n division coefficients

$$e_j = e_{l/n,l}\left(x, x - \frac{n-1-2j}{2n}\,l\right), \quad j = 0, 1, \ldots, n-1.$$

which are non-negative, equally distributed random quantities. In this case, the definition of the division coefficients (6.13) yields $\Sigma_{j=0}^{n-1} e_j = n$. Squaring the expression for e_j and then averaging it, we can reduce the result to the form

$$\overline{(e_j)^2} = a_2(n) = n - \frac{1}{n}\sum_{j\neq k}\overline{e_j e_k} < n,$$

i.e., $\mu_2 < 1$. This expression and (6.14) for $q = 2$ give $e_{r,l} \leqslant l/r$. Hence we obtain, for $q > 2$,

$$\overline{(e_{r,l})^q} = \overline{(e_{r,l})^{q-2}(e_{r,l})^2} \leqslant \left(\frac{l}{r}\right)^{q-2} \overline{(e_{r,l})^2} = \left(\frac{l}{r}\right)^{q-2+\mu_2},$$

i.e.,

$$\mu_q \leqslant q - 2 + \mu_2, \quad \text{for } q > 2. \tag{6.15}$$

Due to the equality $e_{r,l} = e_{r,\rho}e_{\rho,l}$ and the statistical independence of $e_{r,\rho}$ and $e_{\rho,l}$, the characteristic function $\varphi(\theta, l/r) = \exp(i\theta \ln e_{r,l})$ of the random quantity $\ln e_{r,l}$ at $r < \rho < l$ obeys the functional relation $\varphi(\theta, l/r) = \varphi(\theta, l/\rho)\varphi(\theta, \rho/r)$. Hence,

$$\varphi\left(\theta, \frac{l}{r}\right) = \left(\frac{l}{r}\right)^{-\alpha(\theta)}. \tag{6.16}$$

To ensure the normalizability of the characteristic function, $\varphi(0, l/r) = 1$, we must ensure $\alpha(0) = 0$. Also, since $\alpha(-iq) = -\mu_q$, we have $\alpha(-i) = 0$ and $0 < -\alpha(-2i) = \mu_2 < 1$. Equation (6.16) leads to the fact that the cumulants

$$s_q\left(\frac{l}{r}\right) = (-1)^q \left(\frac{d^q \ln \varphi}{d\theta^q}\right)_{\theta=0}$$

of the random quantity $\ln e_{r,l}$ are proportional to $\ln (l/r)$, and hence the centered and normalized random quantity

$$\zeta_{r,l} = \left[\ln e_{r,l} - s_1\left(\frac{r}{l}\right)\right] \left[s_2\left(\frac{r}{l}\right)\right]^{-1/2}$$

has its first cumulant equal to zero and its second cumulant equal to one. All the others are proportional to some negative exponent of $\ln (l/r)$ and hence tend towards zero for $l/r \to \infty$. It can be proved that the probability distribution of $\zeta_{r,l}$ then tends to a normalized Gaussian distribution. The division coefficients $e_{r,l}$ are, therefore, asymptotically lognormal for $l/r \to \infty$ in the self-similar range. In a similar manner, one can prove that ϵ_r is lognormal, as was established above within the framework of a particular scheme of discrete division of space.

Kraichnan (1974) stated that the basic idea of the Kolmogorov–Obukhov theory (1962), which postulates a self-similar turbulent cascade process in which the intermittency systematically increases with decreasing scale, does not demand that the mean energy dissipation rate is lognormal inside small spaces, but also admits other possibilities. In this respect, the corresponding experimental data increase in importance. The statistical characteristics of the division coefficients calculated by Kholmiansky (1973) and Van Atta and Yen (1975) from wind velocity fluctuations in the atmospheric surface layer have confirmed the realization of the self-similarity requirement at scales exceeding the Kolmogorov microscale η.

Empirical probability distributions for non-negative small-scale characteristics of the type $\epsilon(\mathbf{x}, t)$ were plotted by Gurvich (1966, 1967). See also Gurvich and Yaglom (1967), Kholmiansky (1970, 1972), and Monin and Yaglom (1975). These empirical functions can be approximated fairly well by lognormal distributions but only with pronounced deviations at small values of the random quantity under consideration. The empirical probabilities exceed those of the lognormal distribution, since small values are measured with an unavoidable instrumental error. For example, Figure 6.1 gives plots of the random

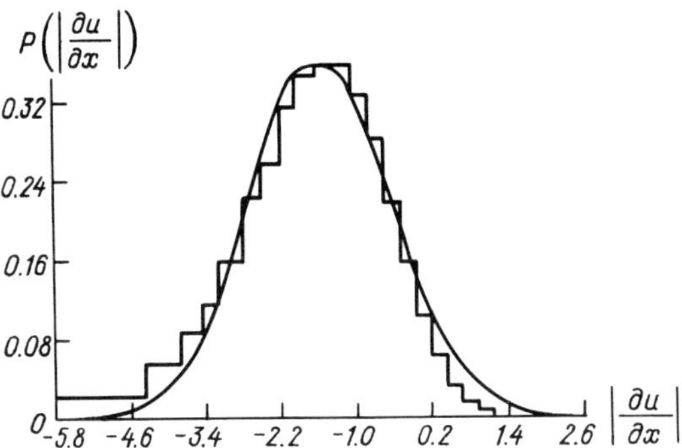

Fig. 6.1. The probability density of ln $|\partial u/\partial x|$: observed (stepwise curve) and approximated by the best, in least-square terms, normal curve (smooth curve) (Stewart *et al.*, 1970).

values of ln $|\partial u/\partial x|$ obtained by Stewart *et al.* (1970), with the best (in the sense of the least square deviation) normal curve approximating this histogram. Note that the probability density of the lognormal distribution with a variance of σ^2 and a mean value of $-\sigma^2/2$, predicted for ϵ_r by the calculations given above, is small at low values of ϵ_r (at $\epsilon_r = 0$ it equals zero) and has a pronounced maximum $p_{\max} = (2\pi\sigma^2)^{-1/2} e^{\sigma^2}$ at $\epsilon_r = x = e^{-3\sigma^2}/2$ ($x \to 0$ and $p_{\max} \to \infty$ at $\sigma^2 \to \infty$). In the range $\epsilon_r < x$ this function is compensated for by instrumental errors, which results in overestimated empirical probabilities of small values of ϵ_r.

Note also that the empirical probabilities of high values of ϵ are underestimated compared compared with the lognormal distribution (see Figure 6.1 and similar plots). This can result from insufficient sampling to obtain the histogram: extremely high values of ϵ are scarce and, in the case of a small sample, even absent. As a result, the right-hand side of the histogram is statistically unstable, i.e., it varies noticeably from sample to sample. However, the data available seem to demonstrate that the underestimation of empirical probabilities for high values of ϵ is of a systematic character. Hence, it should be interpreted in some other way. Stewart *et al.* (1970) attributed this deviation from the lognormal distribution to the fact that the latter requires a high value of n in (6.5) or a high value of ln l/r in the Novikov scheme (Novikov, 1969, 1971). In other words, not only the Reynolds number Re_r, but also its logarithm ln Re_r must be large. If ln Re_r and the number of terms in (6.5) are insufficiently large, as in the case of oceanic measurements, the sum in (6.5) takes on high values with a smaller probability than that predicted by the normal distribution.

Due to the underestimated empirical probabilities of large values, the empirical high-order moments of ϵ appear to be appreciably lower than the moments of the lognormal distribution. This difference increases with the order of the moments. In line with Novikov's inequality, the exponents μ_q in the expressions (6.14) for the moments of the division coefficients in (6.14) increase no faster than a linear function of q. In the case of the moments (6.8) of the lognormal distribution, however, μ_q proves to be a quadratic function of q. Note that, for such a rapid increase in q, the set of all the moments does not unambiguously determine the probability distribution. Also, in the presence of dynamic equations the initial values of all the moments do not unambiguously determine their future values, even in case when the dynamic problem has a unique solution (see Orszag, 1970). On the other hand, Novikov's inequality (6.15) ensures an unambiguous determination of the probability distribution by its moments (6.14). This difference, however, is not surprising, since for $\ln (l/r) \to \infty$ the moments of the division coefficients do not approach those of any limiting distribution, but increase indefinitely. The finite limit, i.e., the normal distribution, features only the probability distribution of the centered and normalized logarithm of the division coefficient.

To illustrate these statements, Novikov (1971) considered a statistical model with a division coefficient $e_{r,2r}$ in the range $0 \leqslant y \leqslant 2$, with a constant probability density $p(y) = \frac{1}{2}$. For this model

$$\mu_q = \log_2 \left(\frac{1}{2} \int_0^2 y^q \, dy \right) = q - \log_2 (q+1). \tag{6.17}$$

In particular, $\mu_1 = 0$, $\mu_2 \simeq 0.41$, $\mu_{2/3} \simeq -0.07$, $\mu_{4/3} \simeq 0.11$, $\mu_3 = 1$, and $\mu_4 \simeq 1.68$. The characteristic function for $e_{r,2r}$ is $\varphi(\theta, 2) = 2^{i\theta}(1 + i\theta)^{-1}$, so that $\alpha(\theta) = -\log_2 \varphi(\theta, 2) = -i\theta + \log_2 (1 + i\theta)$. The lognormal approximation is obtained by replacing the function $\alpha(\theta)$ by the first two terms of its Taylor series:

$$\tilde{\alpha}(\theta) = \frac{1 - \ln 2}{\ln 2} i\theta + \frac{1}{2 \ln 2} \theta^2 .$$

This yields the moments

$$\tilde{\mu}_q = -\tilde{\alpha}(-iq) = -\frac{1 - \ln 2}{\ln 2} q + \frac{1}{2 \ln 2} q^2 ,$$

which differ appreciably from the real moments (6.17). This difference increases with q: in particular, $\tilde{\mu}_1 \simeq 0.27$, $\tilde{\mu}_2 \simeq 1$, $\tilde{\mu}_3 \simeq 5$, and $\tilde{\mu}_4 \simeq 10$. With $\alpha(\theta)$ known, one can obtain the characteristic function

$$\varphi\left(\theta, \frac{l}{r}\right) = \left(\frac{l}{r}\right)^{-\alpha(\theta)} = \frac{\exp\left(i\theta \ln \frac{l}{r}\right)}{(1 + i\theta) \log_2 \frac{l}{r}}$$

for the logarithm of any division coefficient $e_{r,l}$.

The Fourier transform in θ of this function yields the probability density $p_1(z, l/r)$ for $\ln e_{r,l}$. Hence, one obtains for $e_{r,l}$ the probability density $p(y, l/r) = y^{-1} p_1(\ln y, l/r)$, which differs from zero only for $0 \leqslant y \leqslant l/r$. In this range it is equal to

$$\left[\frac{1}{r} \Gamma\left(\log_2 \frac{l}{r}\right)\right]^{-1} \left(\ln \frac{l}{ry}\right)^{\log_2 [(l/r) - 1]} .$$

This function deviates from the lognormal one in the same sense as the experimental data; at $l/r > 2$ and $y = 0$ it tends towards infinity, while the lognormal function becomes zero. In this approach neither the moments nor the probability density approach the corresponding characteristics of the lognormal distribution.

Thus, the lognormal distribution overestimates the probability of high values of ϵ. It also overestimates the higher moments of this random quantity. However, it can be used to describe the central part of the probability distribution of ϵ and to calculate the low-order moments. Hence, an empirical verification of (6.9), which follows from the lognormal approximation, is of interest. This verification was carried out with data on wind velocity fluctuations in the atmospheric boundary layer over the sea (Van Atta and Chen, 1970; Van Atta and Park, 1972). Although the estimates of the high-order structure functions were insufficiently reliable, the data agreed with (6.9) for $n = 4$, 5, 6, 7 when μ was taken as equal to 0.5. For $n = 8$ or 9 one must choose lower values of μ.

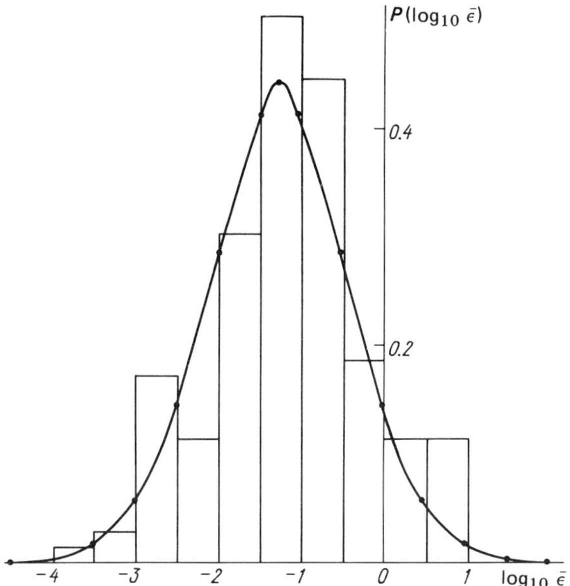

Fig. 6.2. Histogram of $\log_{10} \epsilon$ as approximated by a normal distribution (Liubimtsev, 1976).

Similar calculations were carried out (Vasilenko *et al.*, 1975) with data on velocity fluctuations in the upper ocean. Beliayev *et al.* (1974) reported spectra of the longitudinal velocity (u') obtained during the ninth expedition of the 'Akademik Kurchatov' in 1971 and the seventh expedition of the 'Dmitriy Mendeleyev' in 1972. These spectra were used to determine the corresponding values of ϵ. Figures 6.2 and 6.3 (see Liubimtsev, 1976) show a histogram and an empirical integral probability distribution for $\log_{10} \epsilon$. These are approximated fairly well by a normal distribution with a mathematical expectation of -1.15 and a standard deviation of 0.82. Thereafter, the series $u_k = u(k\Delta t)$, where $\Delta t = 1/600$ s and $k = 0, 1, 2, \ldots, 2047$, was used to calculate the structure functions $D_p(r)$ of various orders p for $r = r_j = jU\Delta t$, with $j = 0, 1, 2, \ldots, 49$. Here, U was the

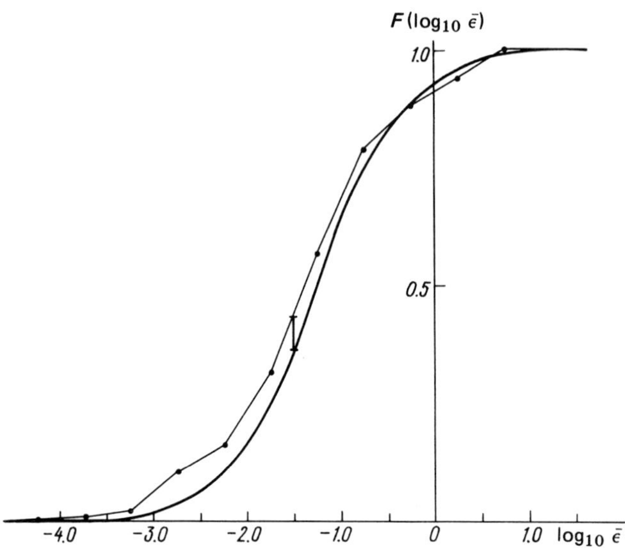

Fig. 6.3. The statistical distribution of $\log_{10} \bar{\epsilon}$ approximated by a normal distribution (Liubimtsev, 1976).

towing velocity of the turbulence meter. Although there was substantial scatter, double-logarithmic plots of the structure functions could be approximated by straight lines. The slopes of the straight lines enabled the determination of the exponents $\mu_{p/3}$. The exponents obtained in this way from data reported elsewhere (Kholmiansky, 1972, 1973; Stewart *et al.*, 1970; Van Atta and Chen, 1970; Van Atta and Park, 1972) are presented in Figure 6.4 as functions of p. The theoretical curve 1 represents the Novikov expression (6.17), curve 2 represents (6.9) with $\mu = 0.51$. Figure 6.4 shows that the exponents $\mu_{p/3}$ first follow curve 2 at small values of p. Later they drop below that curve and, finally, at high values of p, the exponents also drop below curve 1. However, the reliability of the data decreases with increasing p.

We conclude this section with a brief look at the effect of ϵ-fluctuations on the shape of the turbulence spectrum in the dissipation range. We employ the lognormal distribution of ϵ as a first approximation. Let us discuss a simple discontinuous model of the spectrum, with $E(k) = C_1 \epsilon^{2/3} k^{-5/3}$ at $k < 1/(b\eta)$ and $E(k) = 0$ at $k > 1/(b\eta)$. Here, $\eta = (\nu^3/\epsilon)^{1/4}$. The averaging of this spectrum over all values of ϵ yields a non-zero spectrum at all k, so that the inequality $k < 1/(b\eta)$ has a non-zero probability at all k. According to Ellison (1967), the averaged spectrum has the form

$$E(k) = \tfrac{1}{2} C_1 (\bar{\epsilon})^{2/3} k^{-5/3} \mathrm{e}^{-\sigma^2/9} \left\{ 1 - \Phi \left[\frac{2^{2/3}}{\sigma} \left(\ln bk\tilde{\eta} - \frac{\sigma^2}{24} \right) \right] \right\}, \qquad (6.18)$$

where $\tilde{\eta} = (\nu^3/\bar{\epsilon})^{1/4}$. At small values of $k\tilde{\eta}$ this spectrum differs from the initial one by the term $\mathrm{e}^{-\sigma^2/9}$, while at large values of $k\tilde{\eta}$ it is approximately equal to

$$A \exp \left[- \frac{8}{\sigma^2} (\ln k\tilde{\eta})^2 \right].$$

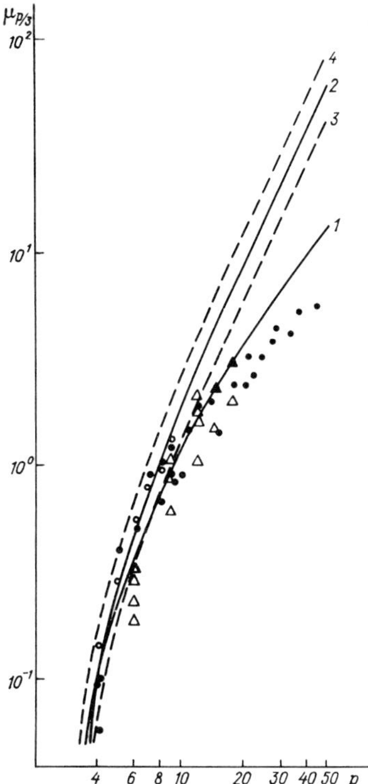

Fig. 6.4. Plots of $\mu_{p/3}$ vs p according to data from different investigators. The theoretical curve 1 is calculated with $\mu_{p/3} = p/3 - \log_2[(p+3)/3]$; the curves 2, 3, 4 with $\mu_{p/3} = (1/18)p(p-3)\mu$ for $\mu = 0.51, 0.35$ and 0.70, respectively (Vasilenko *et al.*, 1975).

Here, A is a slowly varying function of k, i.e., it decreases slowly, yet at a greater rate than any negative power of k. According to Keller and Yaglom (1970) the average spectrum always decreases, provided that the initial spectrum decreases faster than $\exp[-C(\ln k\tilde{\eta})^2]$. The actual rate of decrease of the initial spectrum affects only the lower limit of the applicability of this average spectrum. The averaging procedure does not affect the final rate of decrease of the spectrum in any other way. The deviation of high values of ϵ from the lognormal distribution causes a somewhat faster decrease at fairly high values of $k\tilde{\eta}$ than in Ellison's model.

CHAPTER II

Small-Scale Turbulence

7. INSTRUMENTS FOR THE MEASUREMENT OF SMALL-SCALE TURBULENCE

7.1. Experimental Techniques

Studies of small-scale turbulence in the ocean require devices that have sensitive elements with a very high space–time resolution. In the measurement of turbulence within the dissipation range this resolution must be of the order of millimeters and hundreds of Hertz. To construct equipment of this type is a substantial problem, which accounts for the fact that the first prototypes of small-scale oceanic turbulence meters were built only recently.

Bowden and Fairbairn (1952, 1956) created a device based on electromagnetic principles of velocity measurement. The sensor of the device was about 10 cm in dimension, with a passband up to 1 Hz. Therefore, it could measure only comparatively large-scale velocity variations. Attempts were made to construct equipment with a higher resolution at the end of the nineteen-fifties in the Department of Marine and Mainland Water Physics of the Moscow State University, under the direction of Kolesnikov (1958, 1963). These facilities, called turbulence meters, were equipped with sensors for velocity, temperature, and electrical conductivity fluctuations. They were tested repeatedly in various parts of the oceans. Somewhat later, a sensitive device of a thermo-anemometric type, with a passband up to 500 Hz, was constructed in Canada (Grant *et al.*, 1962). Temperature sensors with a time lag of about 0.1 s were employed in submarine measurements by Liberman (1951). Temperature fluctuation sensors with a short time lag were used by English (1953), Kontoboitseva (1962), Nemchenko (1962a, b), and by other investigators.

In the following years, progress in the construction of devices to investigate small-scale ocean turbulence was made mainly in Canada, the U.S.A., and the U.S.S.R. In Canada, the studies initiated by Grant, Stewart, and Moilliet were continued by Nasmyth (1972, 1973) who constructed a towed system equipped with thermo-anemometric velocity and temperature sensors, with a passband of up to 1000 Hz and a spatial resolution of up to 2–3 mm. The spatial resolution of the electrical conductivity sensors amounted to several centimeters. The system towed behind the vessel was also supplied with thermistor sensors, to measure low-frequency temperature variations in the vicinity of the fluctuation sensors, and a mean flow velocity meter.

In the U.S.A., it was Gibson's group that developed water turbulence meters of a short time lag. One of their devices was equipped with thermo-anemometric sensors for the measurement of temperature fluctuations with a sensitive film of 0.5 mm (Williams and Gibson, 1974). In another modification of their turbulence meter, tested in 1974 during the expedition of the Soviet research ship 'Dmitriy Mendeleyev' (Gibson *et al.*, 1975), the hot-film sensors for the measurement of velocity fluctuations and the cold-film sensors for temperature fluctuations were mounted on two arms, with a spacing of 1.5 m. The sensors were quartz-coated and enabled a spatial resolution of 1 mm. The upper arm was also equipped with a propeller flow meter with magnetically controlled contacts and with a vibrotronic pressure meter. The maximum immersion depth of the turbulence meter was 250 m.

In the Institute of Marine Hydrophysics of the Ukrainian Academy of Sciences, turbulence meters were modified recently by increasing the number of measurement channels and by automating the collection and processing of data (Kolesnikov, 1970; Kolesnikov *et al.*, 1966, 1969; 1972, Kushnir and Paramonov, 1972). One version of the autonomous deep-sea automated turbulence meter (AGAT) constructed in this Institute ensures 0.05 cm s^{-1} sensitivity in the velocity channel and 5×10^{-4} °C sensitivity in the temperature channel. The All-Union Institute of Physical-Technological and Radio-Technical Measurements has developed a turbulence meter based on optical and acoustic principles (Beliayev *et al.*, 1975; Trokhan and Stefanov, 1974). This device has three ultrasonic emitters, which occupy three sides of a square, and two lasers, which emit light beams in the plane perpendicular to the direction of ultrasonic wave propagation (Figure 7.1). Alternatively emitted ultrasonic pulses modulate the light beams in the measurement zone; the light signals are then transformed by photomultipliers and go to the phase detector inputs. The pulse duration at the detector's output is proportional to the phase difference of the input signals which makes it possible, by way of analog processing, to estimate two velocity components (perpendicular to the laser beam directions) and the speed of sound in the water. The measurement length of the device, i.e., the distance between the two laser beams, is 5 mm, and the frequency ranges from 0.1 to 1000 Hz. The measurement error in ocean conditions is some 0.1%.

Great progress in constructing low-inertia ocean turbulence meters has been made recently in the Shirshov Institute of Oceanology of the U.S.S.R. Academy of Sciences. The meters constructed were, as a rule, also equipped with sensors for measuring the mean values and slow variations (with a frequency of less than 1 Hz) of the field variables. This changed the meters into complex systems, which enabled not only turbulent fluctuation characteristics to be investigated, but also enabled the association of them with the background processes. The 'Tunets' system developed in the Special Design Bureau of Oceanological Technology of the Institute of Oceanology of the U.S.S.R. Academy of Sciences is an example of such a complex measuring system.

The 'Tunets' system is designed to study small-scale fluctuations of the various field variables. It can simultaneously measure and record on magnetic tape a number of variables: velocity fluctuations and mean flow, temperature and electrical conductivity of water, speed of sound in the ocean, and some other parameters. Depending on the problem, the system can be modified to incorporate various combinations that involve up to 12 sensors at any one time. The towed version of the system is also supplied with up to ten additional temperature thermistors, whose data are recorded either on magnetic tapes or by pen recorders. For instance, one of the modifications intended for the ocean environment employs a thermo-anemometric sensor for velocity fluctuations, three hydroresistive sensors for flow velocity fluctuations, three hydroresistive sensors for

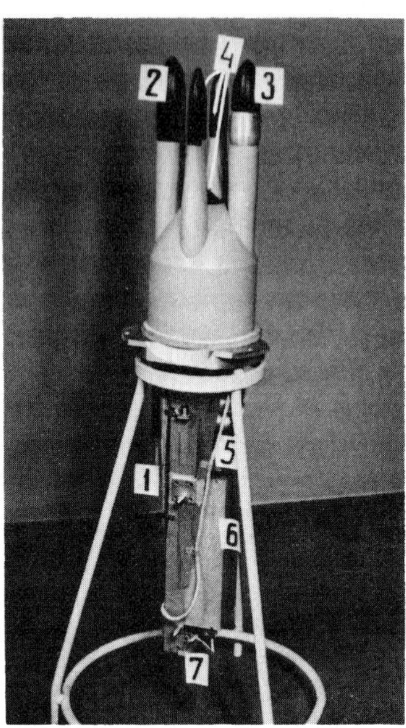

Fig. 7.1. General view of the optical-acoustic turbulence meter designed at VNIIFTRI (All-Union Institute of Physical-Technological and Radio-Technical Measurement) (Beliayev *et al.*, 1975). (1) OKG – 13 laser, (2) fairing unit with transmitter optics, (3) fairing unit with receiver optics, (4) piezoceramic transmitters, (5) electronic unit, (6) high-voltage units, (7) joint.

electrical conductivity fluctuations, five cold-film (thermo-resistive) sensors of temperature fluctuations, a three-component sensor for vibrations, three thermistor sensors for mean temperature, a sensor for the velocity of the unit relative to the medium, a sensor for the speed of sound, and a depth meter (see Figure 7.2).

After initial processing, the signals from these sensors are recorded on magnetic tape in analog or digital form. These signals can be displayed visually on either oscillographs or pen recorders. The running time signal, produced by the system time marker, is recorded automatically.

A thin platinum film that coats a quartz glass substrate serves as the sensitive element of the thermo-anemometric velocity fluctuation sensor (VFS) (Vorobiov *et al.*, 1973). This instrument employs a constant current that heats a sensitive element. The temperature difference between the sensitive element and the surrounding water can be as great as 10–20°C. A velocity change of the flow around the sensitive elements results in a variation of the heat exchange and thus in a variation of the temperature and electrical conductivity of the element. The latter is registered by the electronic circuitry of the instrument. The operating frequency range of the detector is 1–250 Hz, its sensitivity is 0.6 V per 1 m s^{-1} change in the flow velocity, the root-mean-square noise level in the operating frequency band reduced to flow velocity units is 0.12 mm s^{-2}, and a 1°C change in the ambient temperature is equivalent to a 20 cm s^{-1} change in the water velocity around the sensor. The sensor is designed to operate in the velocity range 1–5 m s^{-1} (relative to water) and at a cable length of up to 700 m.

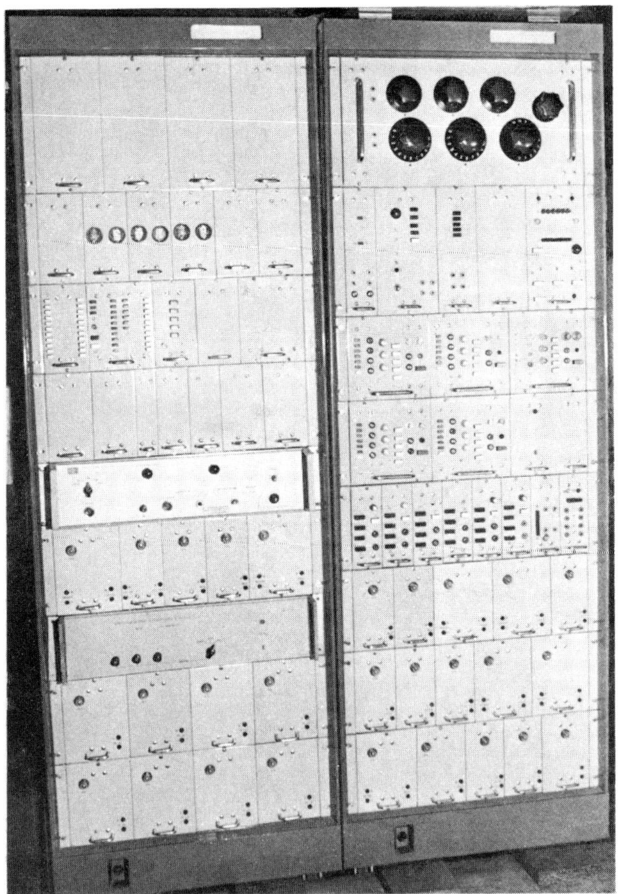

Fig. 7.2. Positions of the 'Tunets' system. ('Tunets' is Russian for tuna fish.)

In the hydroresistive velocity fluctuation sensor (HVFS) the sensitive element is the volume of water between the central and circular electrodes of the sensor. The electrodes are mounted on a hemispherical surface. The central electrode has the form of a dot, 0.5 mm in diameter, and the circular electrode is located along the circumference of a hemisphere, 9 mm in diameter (Figure 7.3). The volume of the heated water is of a rather complex shape. However, with reasonable accuracy, its effective volume can be assumed to be 3 mm^3. The output signal of the sensor contains both the mean flow and the velocity fluctuations, their levels differing appreciably. The fluctuating signal is taken from the operational amplifier of the sensor, which cuts off the mean signal and amplifies the fluctuations. The operating range is 1–1000 Hz for the fluctuating signal. The sensitivity of the HVFS (when heated to 20°C) is 0.15 V m^{-1} s^{-1} for the mean velocity signal, and 90 V m^{-1} s^{-1} for the fluctuation signal. The root-mean-square noise level, reduced to velocity units, is 0.5 mm s^{-1} within the operating frequency range. The towing (probing) velocity of the sensor relative to the water is usually within the range

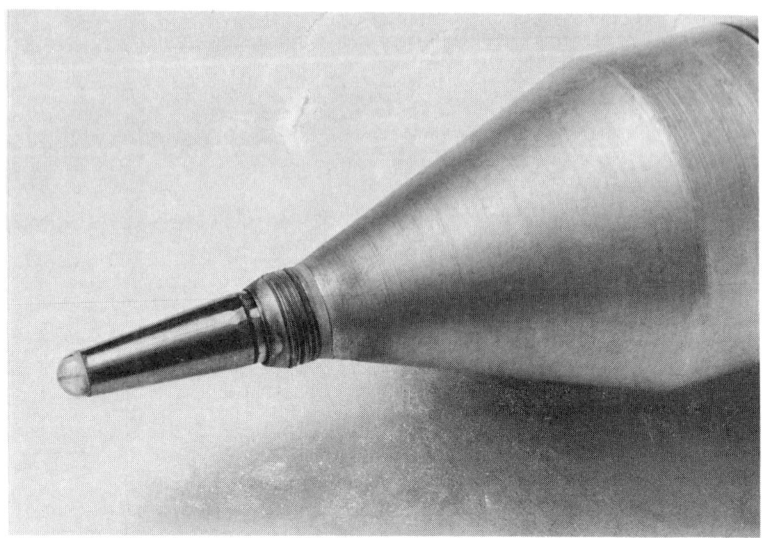

Fig. 7.3. Hydro-resistive velocity fluctuation sensor.

1–10 m s^{-1}. The frequency range of an HVFS is wider than that of a VFS, but an HVFS is somewhat affected by temperature and electrical conductivity variations in the environment. However, the reliability of the HVFS sensors makes them indispensable for long-term and repeated measurements of velocity fluctuations in the ocean.

The hydroresistive electrical conductivity sensor (HECS) is designed in the same way as the VFS. However, it has no heat regulator and the power of its heat generator is low. The sensor divides the signal into constant (varying slowly) and fluctuation components, the operating band (as with the HVFS) being 1–1000 Hz relative to the fluctuation signal. It has a sensitivity of 0.083 V with respect to the mean electrical conductivity (with the electrical conductivity varying by one millisiemens per millimeter) and 1.66 V mho^{-1} cm^{-1} with respect to the fluctuation signal. The root-mean-square level of its own noise, reduced to electrical conductivity units, is 3×10^{-4} mho cm^{-1} over the operating frequency band, the minimum tolerable towing (probing) velocity being 1 m s^{-1}. The HECS can be obtained in two versions, with a 'micro' or a 'macro' electrode. In the latter version, the effective water volume employed as a sensitive element reaches 1 cm^3. The sensor readings, however, are practically unaffected by water heating processes, even at low towing velocities.

The temperature fluctuation sensor (TFS) employs sensitive units of the same type as those of the VFS. However, the heating current passing through the sensitive element is extremely low, so that the temperature of the element is not raised. Changes in the environmental temperature result in variations of the resistance, which are detected by the sensor circuitry. The sensitivity is 8 V per 1°C, the operating frequency band is 1–250 Hz, and the noise level is 0.0008°C.

To measure the constant and low-frequency components of temperature, 'Tunets' employs frequency-modulated mean temperature meters (FMTM). The sensitive element

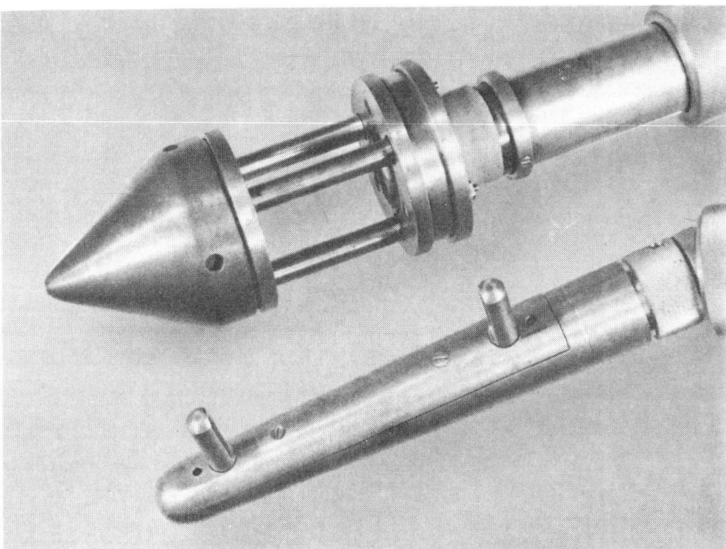

Fig. 7.4. Speed of sound sensors.

in the meter is a MT-54 thermistor which ensures a time constant of no more than 0.3 s. The changes in the thermistor resistance induced by the environmental temperature lead to frequency changes in the signal generator of the sensor. Over a temperature range of –2 to 30°C the frequency varies from 5 to 11 kHz. This makes possible a temperature resolution of about 0.05°C.

The speed of sound sensor (SSS) is based on the principle of the so-called sing-around. The generator emits pulses with a frequency which depends on the speed of sound in water. The pulse goes from the emitter to a reflector and back again (the emitter is also a receiver) and then to the amplifier input (Figure 7.4). The time interval between pulses is determined by the time it takes sound to make the double transit. Since the path length is known and constant, the frequency is directly associated with the speed of sound in water. The SSS transit length is 5 cm, the mutual conductance is 10 Hz of signal frequency variation for a change of 1 m s^{-1} in the speed of sound. The instrumental error is 0.01% of the value measured.

The acoustic sensor of the 'Tunets' system can be used to measure the velocity of the system relative to the water. In this sensor, also based on the 'sing-around' principle, the distance between the two signal emitter-receivers is 10 cm. If the functions of the emitter and receiver change in a cyclic way, then the pulse frequencies will differ by magnitudes proportional to the component of the flow velocity parallel to the emitter–receiver direction, since the velocity of the sound pulse either increases or decreases due to the flow. The mutual conductance of the sensor is 0.1426 Hz for a change in velocity of 1 m s^{-1}. The measurement error is 5% of the upper measurement limit, which is equal to 3 m s^{-1}. The detector resolution corresponding to this flow velocity is 7 cm s^{-1}.

The 'Tunets' system also employs combinations of sensitive elements, in particular a meter with a two-film sensor, of which one film is included in the VFS scheme and the other in the TFS scheme. In this case, the electrode has the form of two unconnected half-rings. To measure mean temperatures, the towing cable of the system can be equipped with additional MT-54 thermistors, having a measurement accuracy of 0.05°C. These thermistors provide information about the temperature field in the vicinity of the measured fluctuation characteristics. As will be shown below, this is of paramount importance to the interpretation of the data obtained. The towed body includes a depth meter based on a commercial pressure gauge of the DDV type and a three-component accelerometer for capsule vibrations. The capsule velocity is not only measured by an acoustic detector, but also by an electromechanical meter based on a cup-shaped propeller placed in a cylinder. There is a magnetic relay in the propeller body that begins to operate as soon as an eight-tab steel sprocket located on the propeller axis begins to rotate. In this situation, the pulse frequency is proportional to the oncoming flow velocity. The electromechanical meter is characterized by a mutual conductance of 34 Hz m^{-1} s^{-1} and measurement errors not exceeding 3% (according to calibration tests on hydrostands).

Apart from the 'Tunets' system, the Institute of Oceanology also employs simpler turbulence probes, designed to operate as towed probes or to carry out probing from an anchored or drifting ship, or even a ship under way. The universality of the apparatus proves to be highly useful for investigations of the microstructure of the field variables in the ocean (Figure 7.5). The turbulence probe 'Grif', developed in the Atlantic department of the Institute of Oceanology, is equipped with a hydro-resistive velocity fluctuation

Fig. 7.5. Probing low-inertia meter developed by the Special Design Bureau of Oceanological Technology of the P. P. Shirshov Institute of Oceanology (U.S.S.R. Academy of Sciences).

sensor, an electrical conductivity sensor of the capacitive type, with a sensitive unit in the form of tubular ceramic condensers, temperature sensors of the thermistor type, an electromechanical relative velocity sensor, and a depth (pressure) meter of the vibrotronic type. The information collected by these sensors is converted into a digital code, which is recorded using a digital tape recorder. The analog signal can be also stored on self-recorders and oscillographs. The microstructure probe designed in the Special Design Bureau of Oceanological Technology of the Institute of Oceanology is equipped with hydro-resistive sensors for velocity fluctuations and electrical conductivity, and also with frequency meters for mean temperature, speed of sound and depth. The probe is immersed in the water with the help of a triple cable which serves to transmit the signals from the sensors and to switch on the power supply unit. When a single cable is used, the set of sensors has to be limited to those measuring electrical conductivity, mean temperature, and depth. Another probe designed in the Atlantic Department of the Institute of Oceanology employs cylindrical electrodes coated with condenser ceramics as electrical conductivity sensors. The noise level of this 'Sigma' probe is equivalent to temperature fluctuations of about 10^{-4}°C, and the spatial averaging of the electrical conductivity fluctuations amounts to several cubic centimeters (Beliayev *et al.*, 1974c). To eliminate disturbances induced by probe displacements due to rolling of the ship, the device was equipped with probes sliding freely along the cable and with completely autonomous measuring units. The transmission of information from a freely sliding probe was accomplished by an inductive method through the cable line and the information was recorded by a portable tape recorder inserted into the probe. The number of sensors varied, depending on the experimental goals.

The autonomous measuring device 'hydroplane' employs the principle of a plane-shaped body that glides down to a depth of about 300 m (Figure 7.6). Once this depth is reached, the ballast is jettisoned by signals from the pressure sensor, the plane acquires positive buoyancy, and the hydroplane starts rising to the surface. The trajectory of the hydroplane can be set by a programming device that governs the control units. The sensors are mounted on long probes up front; this ensures that measurements are carried out in undisturbed water. The information is stored on portable tape recorders or self-recorders. Submarines and remotely controlled probes of various kinds hold much promise as smooth sensor carriers (Ozmidov *et al.*, 1980). Such carriers guarantee a low noise level and offer the possibility to measure turbulence in such extreme conditions as storms or under the Arctic ice.

Noise control in ocean microstructure measurements is a complex problem in itself. Disturbances induced by the rolling of the ship lie within the basic energy range of small-scale turbulence. Therefore, investigations of turbulence generation mechanisms necessitate the elimination of these disturbances. If it is impossible to completely remove such disturbances, then it is desirable to register the inherent rolling of the ship (or rather the carrier oscillations) with accelerometers. These oscillations can subsequently be filtered out from the records obtained. If vibrational disturbances cannot be completely eliminated, they must also be recorded by vibration detectors to make it possible to filter them out and to judge the quality of the data obtained. Noise with the frequencies of alternating currents (e.g., 50 Hz) that is often observed in records can be readily filtered out using narrow-band filters or simply 'cut off' from the spectra of the signals studied.

Before they were used in the field, velocity fluctuation sensors were tested on a special test stand with a subsurface flow of water. The following sensor parameters were determined: output voltage versus flow velocity, sensitivity to velocity fluctuations at mean flow velocities corresponding to towing and probing speeds, amplitude–frequency

Fig. 7.6. General view of the hydroplane before being lowered from the research vessel 'Dmitriy Mendeleyev'.

characteristics, and intrinsic noise level (Palevich *et al.*, 1973). Temperature and electrical conductivity sensors were tested on a stand with controlled temperature conditions (Nabatov *et al.*, 1974).

In the last few years, the basic method of measuring turbulence in the ocean has been that of towing a turbulence meter behind a ship (Ozmidov, 1973; Paka, 1972, 1974). For this purpose, special towing lines, with deepening and fairing units attached to reduce the hydrodynamic resistance of the line, were employed (Chistiakov, 1973). This technique enables vast ocean areas to be investigated and extensive data on statistical processes to be collected (Figure 7.7). A comparatively high towing velocity (up to 10–12 knots) makes it possible to employ the hypothesis of frozen turbulence to re-calculate temporal turbulence characteristics into spatial ones. However, fast towing conditions do not allow investigations at great depths (in the expeditions of the Institute of Oceanology the devices have been submerged to a maximum depth of 250–270 m). Also, frequencies below several tenths of a Hertz have to be discarded, since at these frequencies the records are greatly distorted by the rolling of the ship. Probing devices that are connected to the ship by a cable have the same drawbacks (Korchashkin and Ozmidov, 1973), but they make depths of 1500–2000 m, and even deeper, attainable. When a great length of cable is paid out, the rolling effect of the ship is considerably reduced by the slack in the cable. Probing devices suffer substantially from high-frequency vibrations arising in the towing line. As mentioned above, freely sliding probes and autonomous measuring devices eliminate these disadvantages.

Because the horizontal variability of hydrophysical field variables is much smaller than its vertical counterpart, measurements from towed probes produce records with homogeneous (stationary) portions. These can be employed to determine the statistical characteristics of the variable investigated. At the same time, long-term records obtained

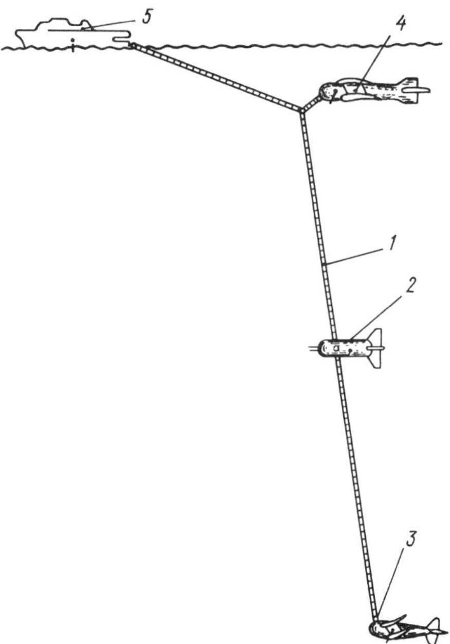

Fig. 7.7. Schematic representation of towing a turbulence meter (2). (5) – ship, (1) – towing line, (3) – deepening unit, (4) – stabilizing unit.

by towing make it possible to investigate the phenomenon of intermittency, i.e., of sharp changes of the fluctuation level in space. Measurements made by probing are quite valuable since they enable the simultaneous recording of the vertical fine structure and the high-frequency fluctuations of the hydrophysical field variables. Information of this kind deals with the relationship between the characteristics of the high-frequency fluctuations and the fine-structure parameters. Therefore, it enables the investigation of the processes of turbulence generation and damping, as well as high-frequency internal waves and finestructure inhomogeneities in their interrelation and interdependence.

Interesting information concerning the space–time structure of turbulence can be obtained by towing with a periodic change in depth. This method, first employed by Canadian investigators (Nasmyth, 1972) enables one to obtain vertical profiles of the turbulence characteristics in the layer scanned by the device. To control the motion of the device, use is made of a special winch which is operated either manually or automatically by a program with set trajectory parameters (frequency and amplitude).

The existence of vertical fine-structure with fairly long lifetimes (minimum 1 h) is evidence for the presence of low-intensity turbulence, since the turbulence is unable to destroy the interfaces between the layers of fluid quickly. Moreover, such turbulence must have small Reynolds numbers. Indeed, the characteristic scale of turbulence in quasi-stationary layers of fluid will be the width of these layers (i.e., some tens of centimeters or a few meters) and the characteristic velocity will be the velocity gradients across the layers (about 1 cm s^{-1}). In this case, the Reynolds number, approximately

10^3-10^4, shows that the turbulence in the interlayers is not fully developed. As a result, there cannot be a pronounced universal inertial range in the turbulence spectrum. The statistical characteristics of this type of turbulence depend upon the peculiarities of the layer in which it occurs, rather than on the general depth at which this layer is located. However, since the statistical characteristics of the vertical fine-structure are related to the large-scale peculiarities of the region under study, it would be natural to expect the parameters of the small-scale turbulence to be dependent, on average, on the general hydrometeorological situation in the region.

This brief description of small-scale turbulence in the ocean acounts for the necessity to combine fluctuation measurements with the recording of both local and average background conditions (Ozmidov, 1973). If these data are neglected, then the turbulence characteristics become much less valuable and turn into a set of results that shows no relation between the phenomenon studied and the determining parameters. This conclusion formed the basis of planning small-scale ocean turbulence investigations in the expeditions of the Institute of Oceanology of the U.S.S.R. Academy of Sciences (Ozmidov, 1971a, b, 1971b, 1974b, 1978a, 1979).

As shown above, the basic hydrodynamic parameters that characterize the processes of turbulence generation and damping are: the gradients of mean velocity and mean water density, and the rates of turbulent energy and temperature (density) inhomogeneity dissipation. These parameters are supplemented by the physical constants of the medium: the coefficients of molecular viscosity and heat conductivity (diffusion), and a buoyancy parameter. As a result, it is expedient to carry out small-scale turbulence studies in ocean regions (polygons) with typical, regional values of these parameters. Large variations of these parameters can be ensured only when working in ocean regions that differ substantially with respect to meteorological and hydrological conditions, in particular, in zones with positive and negative heat fluxes through the ocean surface. In the former case (when the water is heated) the stratification is stable, in the latter case (when the water is cooled) convective mixing and the destruction of density gradients can occur. In addition to these two extreme cases there may be intermediate situations with various vertical distributions of density and velocity. Various combinations of background conditions resulted in different cruise routes for expeditions devoted to ocean turbulence studies (Figure 7.8). Some of the routes were meridionally extended, which allowed investigations under both summer and winter conditions, the polygons being located partly in zones with strong flows and partly in zones with comparatively moderate dynamic conditions.

In these expeditions, the large-scale background conditions were measured using standard equipment and specially devised facilities. The former included flow meters of the BPV type and commerical bathymeters to detect the large-scale components of the vertical profiles of water velocity and density. For this purpose, the polygons were, as a rule, equipped with buoy stations furnished with strings of propellers in order to carry out a hydrological survey. In this situation the number of bathymeters in series was increased. In addition to the propellers, the buoy stations were equipped with photothermographs that detected the water temperature at a few vertical points at 5–10 min intervals. Flow velocities were detected by the propellers at 5–10 min intervals. Thus, the buoy station measurements enabled investigations of velocity and temperature fields within time scales ranging from tens of minutes to several days (standard operating

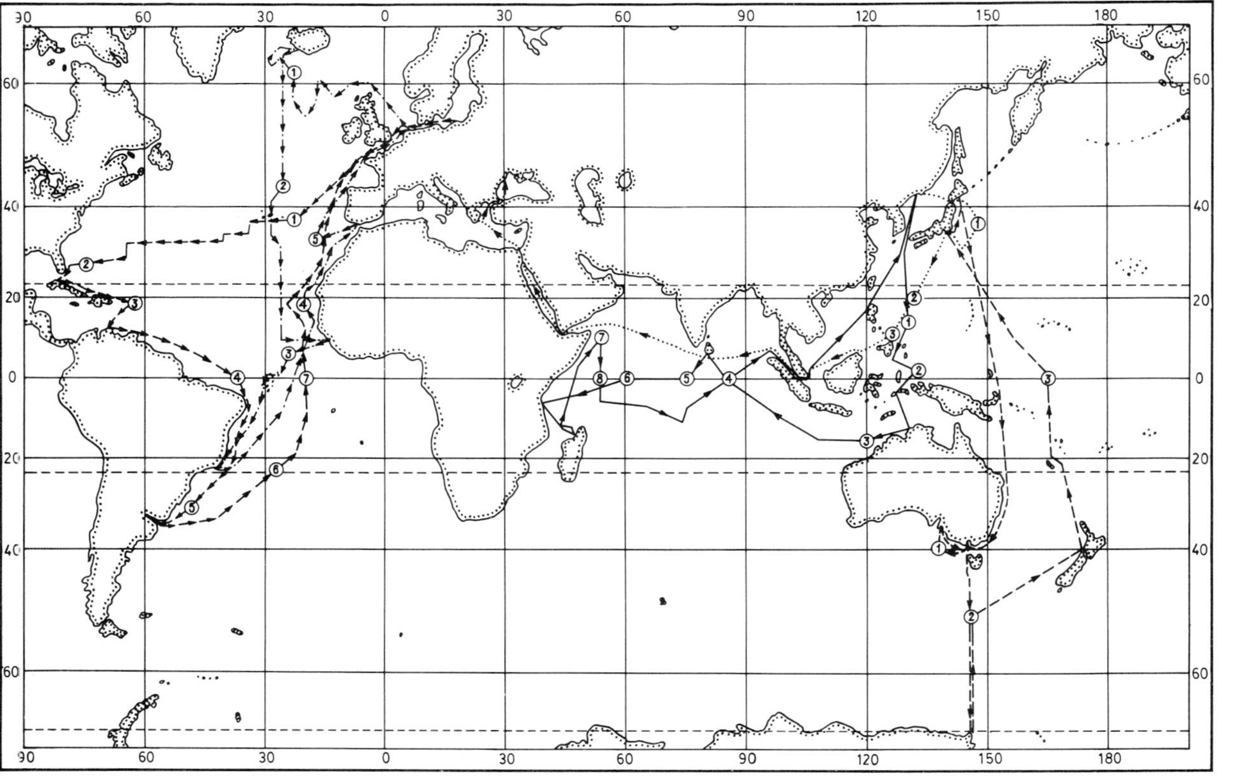

Fig. 7.8. Scheme of some routes of scientific research ships of the P. P. Shirshov Institute of Oceanology, U.S.S.R. Academy of Sciences, for studying turbulence and the fine structure of hydrophysical fields in the ocean. The numbers refer to polygons.

time on a polygon). As a rule, each polygon had two or three buoy stations, which also ensured the detection of horizontal gradients in the velocity and temperature fields in the areas studied (Figure 7.9). To detect more subtle details of the background conditions, non-standard meters were used. Among these it is necessary to mention a thermotrawl developed at the Institute of Oceanology by a group headed by V. Paka. This device is a towed system incorporating a string of thermo-resistive sensors (with an inertia of about 1 s). Thermotrawl measurements from a ship under way produce a space–time profile of the temperature field in the upper ocean (to depths of 200–250 m). To accomplish this, the information from the sensors moving along a given horizontal plane is put into a computer which runs a special program to translate the sensor readings into isotherm shifts. These are presented graphically by means of plotters. To calculate purely spatial statistical characteristics of the temperature field (primarily internal waves) by thermotrawl data, it is necessary to take series of profiles in various planes, preferably at different towing velocities, which makes it possible to eliminate distortions of the internal waves due to the Doppler effect.

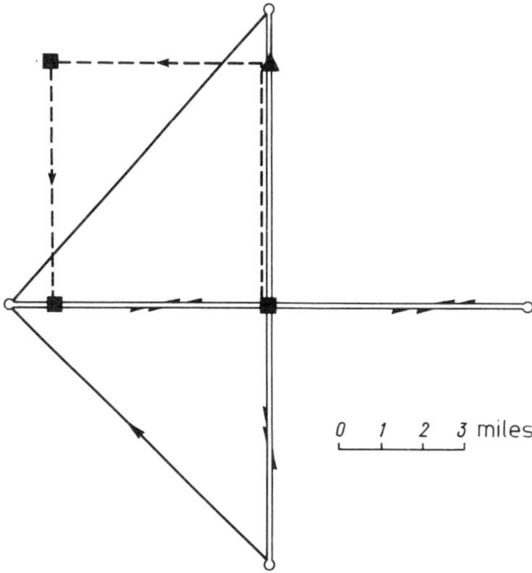

Fig. 7.9. Operation scheme for one of the polygons. Solid line – thermotrawl towing, dashed line – turbulence meters towing, squares and triangles – locations of buoy stations and sounding using probing devices.

Probing devices with a resolution of about 1 m were also widely employed in the polygons to gain information relating to the fine structure of the hydrophysical fields. The most convenient was the 'Aïst' probe devised at the Institute of Oceanology (Figure 7.10). This probe made available information concerning the temperature and the electrical conductivity of water at discrete points along the direction of motion of the descending device (Borkovsky *et al.*, 1972). The information obtained with the sensors of this probe was cabled to the ship and either recorded in code on punch tapes or entered directly into a computer to obtain profiles of water salinity and density.

Information concerning the fine structure of the velocity field was obtained by an acoustic probe designed at the Pacific Institute of Oceanology. This cross-beam probe (Volkov *et al.*, 1974; Shevtsov and Volkov, 1973) involves an emitter that scatters ultrasound from seawater inhomogeneities and small inclusions carried by the flow (suspension, plankton, etc.). On both sides of the emitter are two symmetrically-mounted sound receivers oriented such that the axes of their directional patterns cross the emitter axis at the same point, making an angle of 40°. This was the method employed to establish the area to be investigated. The scattered energy arrives at the receivers and is transformed into electrical signals which travel to amplitude sensors through two separate channels. Thereafter, the frequency differences that result from the frequency shift due to the Doppler effect of sound scattered by moving inhomogeneities are separated.[1] After frequency-analog transformation, the signals are recorded on self- or tape recorders. The absolute velocity detection error does not exceed 1 cm s^{-1}. The flow direction is traced with a sensor that measures the phase of the electrical signal induced in a coil by its rotation in the Earth's magnetic field. The relative position of the coil and the body of the device is controlled by a photosensor which produces a short pulse at a definite position.

Fig. 7.10. The 'Aïst' probe.

Detailed information concerning the structure and time evolution of the temperature field was obtained by a radiometric buoy complex designed by the Institute of Oceanology and the Moscow Institute of Physical Engineering. The complex consisted of a set of about 10 sensors mounted along a cable running from a buoy station. The buoy station included electronic facilities, power supply units, and a transmitting station. On board the ship there were a receiving station, decoding units, and digital tape recorders to record the information obtained. The information from the sensors was recorded at 1–3 s intervals, and the operating range was about 10 miles.

Investigations within each polygon were carried out in the following order. Firstly, exploratory measurements of the temperature profile were carried out by a probing unit or a bathythermograph. If the vertical temperature distribution satisfied the necessary requirements, then the region was selected as a polygon for the measurement of flow velocities, the installation of radio buoys and buoy stations equipped with propellers and photothermographs, the measurement of hydrological parameters in series, the measurement of parameters using the 'Aïst' probe, and the observation of sea roughness and meteorological conditions. The roughness of the sea was observed visually, by means of floating beacons supplied with wave recorders, and in a number of cases from the ship by optical-acoustic wave recorders. The complex of meteorological observations included standard measurements as well as special ones; in particular, measurements of the heat, momentum, and moisture fluxes through the ocean surface. The meteorological situation prior to the observation time was also taken into consideration (Shishkov, 1974).

A series of background measurements was followed by towing and probing carried out with high-frequency or combined measurement systems. The towing was usually carried out along a square having sides 5–10 miles long. The probings were carried out either throughout the bulk of the ocean (1500–2000 m depth) or in layers that were the most interesting from a specific viewpoint (e.g., those layers characterized by high density or velocity gradients). In the latter case, frequent probings were employed.

This range of devices supplied investigators with extensive information,[2] which presented quite a problem from the viewpoint of data processing and analysis (Ozmidov *et al.*, 1974). Most of the information was recorded on magnetic tapes, either in analog or in a code format convenient for computer input (Beliayev, 1973; Vorobiov and Palevich, 1974; Koshliakov and Sorohtin, 1972). However, some oceanological devices (e.g., the BVP propeller) display the information in a form that is not suitable for computer input. This makes data processing extremely time- and labor-consuming. Statistical processing of the information was performed by marine computers of the 'Minsk' type. For this purpose, a program library, including a fast Fourier transform program, was established. However, even the extremely time-saving fast Fourier transform algorithm introduced into commercial computer programs did not allow real-time processing of the information. Indeed, even the possibilities of such a powerful computer as the IBM 7090 are limited to information from a single sensor with a transmission band of about 1 kHz. This limitation in the data collection and processing system can be overcome by creating specialized processors for spectral analysis. The operating speed of specialized computers, compared with commercial ones, can be substantially increased by a fixed succession of operations, by simultaneous operations with the real and imaginary parts of complex numbers, and by more efficient use of computer storage.

Specialized processors allow more useful experiments to be carried out, since the possibility of statistical processing of the information during the course of the measurements enables one, if necessary, to vary experimental conditions. A specialized processor of this type based on commercial minicomputers is now under development at the

Institute of Oceanology as one component of a new instrument complex intended for studies of small-scale turbulence and the accompanying background processes.

8. STATISTICAL CHARACTERISTICS OF TURBULENCE

Any hydrophysical characteristic in the ocean (flow velocity component, water temperature and salinity, content of oxygen or any other chemical element, diffraction coefficient, etc.) forms a certain four-dimensional field $u(x, y, z, t)$, where x, y, z are Cartesian coordinates with the origin on the ocean surface with the z-axis directed downwards, and t represents time. In general, the field $u(x, y, z, t)$ is random, nonisotropic and nonstationary. It is expedient to start investigating this field by determining the limits within which the space and time scales of the field occur. The horizontal scale of the ocean itself can be employed as the maximum length scale L_{max}, while the minimum scale can be estimated from the theory of locally isotropic turbulence: $L_{min} = \sqrt[4]{\nu^3/\epsilon}$. Reasonable estimates of the turbulent energy dissipation rate ϵ fall in the range 10^{-1}–10^{-5} cm^2 s^{-3}, and the kinematic viscosity $\nu \simeq 0.01$ cm^2 s^{-1}. Therefore, L_{min} is of the order of 1 mm–1 cm. The length-scale range of oceanological field variations thus involves nine to ten orders of magnitude.

The minimum time scale of the field can be estimated by the theory of locally isotropic turbulence: $T_{min} = (\nu/\epsilon)^{1/2}$. This yields $T_{min} = 1$–100 s at the same ϵ-values. Determining T_{max} is harder, since the time scale of the large-scale motion has no fixed value. Indeed, the hydrological conditions in the ocean are known to vary annually; the field variations exhibit not only cyclic changes over several years, but are also conditioned by climatic changes, to say nothing of the evolution of the World Ocean during the course of geological time. In studying ocean turbulence it would be reasonable to confine ourselves to the annual changes of background conditions. We set $T_{max} = 3 \times 10^7$ s. In this case, the time scale ranges over 7 to 8 orders of magnitude.

To describe the random field $u(x, y, z, t)$ in detail it is necessary, as stated in §2, to know either all the multi-dimensional probability densities $P_{M_1 \ldots M_n}(u_{j_1}, \ldots, u_{j_n})$ or the corresponding characteristic functions $\varphi_{M_1 \ldots M_n}(\theta_{j_1}, \ldots, \theta_{j_n})$. Generally speaking, experimental investigations of the field should be carried out with sensitive devices operating for sufficiently long periods of time at all types of finite sets of points in the ocean. It is quite obvious that such experiments are impossible, which necessitates looking for a compromise. The problem would become much simpler if the field could be assumed to be homogeneous and stationary. In this case, the field would be fairly completely described by its spectrum $S(\mathbf{k}, \omega)$, where $\mathbf{k}$ and ω are the wavenumber and frequency, respectively. However, in real fields not only purely random fluctuations but also 'trend' components arise, which violate homogeneity and stationarity. These are associated with the configuration of the ocean and with external forces. Therefore, although more complex, the sum of a number of deterministic components $u_i(x, y, z, t)$ and some, presumably homogeneous and stationary, stochastic 'additions', defined by their spectra $S_i(\mathbf{k}, \omega)$, constitute a more realistic model of the field $u(x, y, z, t)$. In this case, a given regular field component $\overline{u_i}(x, y, z, t)$ can be roughly assumed to be the main energy supply for the random components that are its small-scale neighbors (Ozmidov, 1966).

This model of oceanological fields facilitates and, to a certain extent, optimizes experimental studies in the ocean. Indeed, with a superpositional picture of the field, experimental investigations can be carried out separately in different space–time windows. If possible, these should include not less than one deterministic component and a random range of scales next to it. In this case, the duration of observations must be chosen on the basis of the periodicity of the regular component, and the sampling time on the basis of the high-frequency limit of the range of random components. Hence, and particularly important, random field components can be studied in a localized typical region, i.e., in a polygon.

If an experiment is carried out in a comparatively high-frequency window, then the observation time can be short. For instance, when studying small-scale turbulence, which has internal waves and fine-structure inhomogeneities as energy-providing background processes, the observations can last for several days. On the other hand, to investigate large-scale ('horizontal') turbulence generated by Rossby waves, horizontal shear in oceanic flow systems, and large-size baroclinic formations, the observations carried out in the polygon can take several months (e.g., the 'Polygon-70' experiment which will be briefly discussed in Chapter 3).

When choosing the sampling duration, it is desirable to know the characteristics of the high-frequency components. This enables one to predict and to allow for possible distortions induced by discretization noise in the process under study. As already known from the Kotelnikov theorem, these distortions are not observed if the sampling time $\Delta\tau$ or Δx relates to the frequency (or wavenumber) of the highest-frequency components of the signal f_n, as $\Delta\tau \leqslant 1/(2f_n)$. In order to investigate turbulence fluctuations up to the viscous range of the spectrum, it is therefore essential, according to the above estimates, to have a spatial resolution as small as millimeters, or even fractions of a millimeter, and a time resolution (when the sensor is fixed relative to the surrounding medium) as small as seconds, or even fractions of a second.

Carrying out measurements while towing or probing, one can only obtain one-dimensional profiles of the field investigated. These are random functions of a spatial coordinate and of time. At sufficiently high towing and probing speeds the frozen turbulence hypothesis can be employed for the high-frequency components. The realizations obtained are then regarded as purely spatial. Multidimensional distribution laws can be derived for these realizations at all points along the track. It is also possible to confine oneself to either one-dimensional distribution laws or estimates of the statistical moments of this or that order. Once the regular, non-stationary components have been removed from the record, it is desirable to calculate the correlation and spectral functions by employing ergodicity, i.e., by assuming that averaging over an ensemble can be substituted for by averaging over space (or time). Multichannel measuring systems, which give synchronous information concerning the fields under investigation at two or more points (or tracks), make it possible to also obtain the joint characteristics of two (or several) random functions. The same characteristics can also be calculated when measuring different fields located practically at the same point in the flow.

When information goes from the sensors to the magnetic storage media, care should be taken not to distort the signal and to avoid information losses due to the limited dynamic range of the media. Thus, for instance, the dynamic ranges of the tape recorders employed in the special expeditions of the Institute of Oceanology were 42 and 66 dB for the analog and the digital equipment, respectively. The capabilities of recording units can be estimated if certain assumptions concerning the spectral composition of the recorded signal are made. It would be natural to approximate the spectral density $E(f)$, as a power law of the type $E(f) \simeq f^{-\alpha}$. The mean signal amplitude then equals the square root of $E(f)$, while the instantaneous signal amplitude can substantially differ from this value. Therefore, this possibility must be taken into account when estimating the frequency bandwidth in which the signal has to be recorded without distortion. The frequency limits of the band are denoted as f_{lower} and f_{upper}. If we assume that the momentary signal amplitudes in the band Δf can vary from $\sqrt{E(f)\Delta f/n}$ to $n\sqrt{E(f)\Delta f}$, where n is a given number, then we obtain $N = n^2(f_{\text{upper}}/f_{\text{lower}})^{\alpha/2}$, where the dynamic range of the recording device, N, is expressed in an equivalent number of discrete signal levels. In our case, N = 126 for the analog signal and N = 2048 for the digital signal. Figure 8.1 presents (on a semilogarithmic scale) the $f_{\text{upper}}/f_{\text{lower}}$ nomograms for n = 2 and

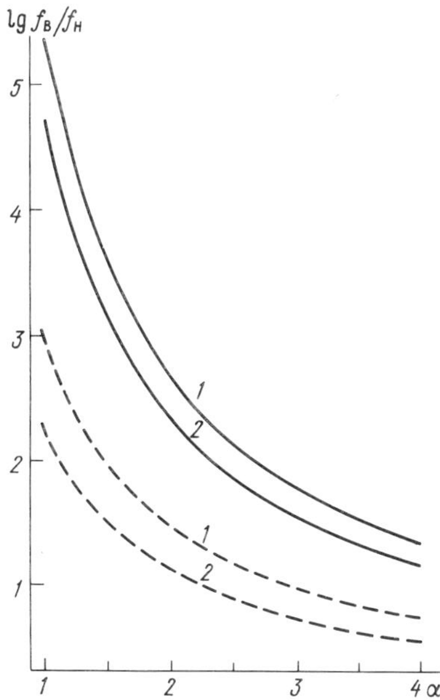

Fig. 8.1. Nomogram of lg f_u/f_l for turbulence signals recorded by analog (dashed lines) and digital (solid lines) tape recorders. (Figure 1): n = 2, (Figure 2): n = 3.

n = 3, with the exponent α varying from 1 to 4. For α = 2 and n = 2–3, the possible frequency-limit ratio is seen from the nomogram to be 14–32 for the analog tape recorder and 250–510 for the digital one. Thus, when recording signals on analog tape recorders, we have to divide the signal into separate frequency bands (overlapping bands are preferred) in order to 'pack in' all the information on the magnetic tape without introducing distortion. In Institute of Oceanology expeditions, the signal was, as a rule, divided into two channels, one of which was used for the filtered high-frequency portion. The signal was divided by filters with a mutual conductance drop of 32 dB per octave. Statistical analysis was then carried out for each signal separately, with a subsequent 'splice' of the results on the frequency (wavenumber) scale.

The sampling interval of each of the filtered signals was related to its maximum frequency f_{upper}. This frequency, in turn, was determined by the spectral composition of the process under study, the channel characteristics and the filter parameters. In most cases, the sampling interval was chosen with a certain margin compared with the theoretical value predicted by the Kotelnikov theorem; it equalled $\Delta t = \frac{1}{3} f_{\text{upper}}$. To verify the operation of the channel that provides the input for the computer, a sinusoidal signal with frequency f_0 was used. If the duration of the signal transmission is limited, then its spectral density is known to be different from a δ-function. It has the shape of the curve in Figure 8.2. The dimensionless frequency $2(f - f_0)T_{\text{max}}$ is plotted along the X-axis, where T_{max} is the width of the correlation window and A is the sine-wave amplitude. Only if the calculated spectral density of the sine wave that passed through the entire input channel coincided with the theoretical one would the choice of all channel parameters and channel operation be considered reliable and suitable for the experimental conditions.

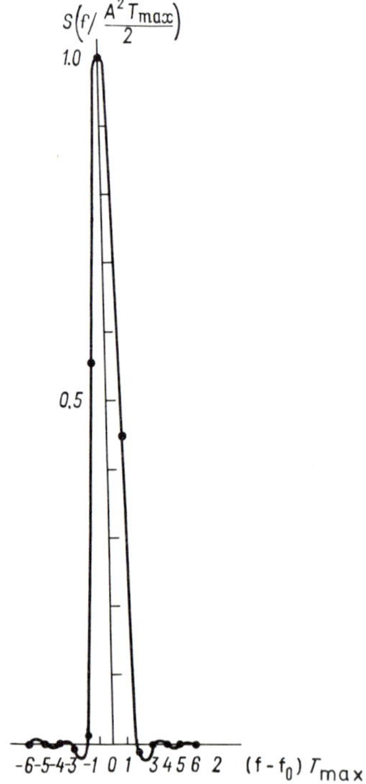

Fig. 8.2. Theoretical spectral density of a time-limited sine wave (solid line) and calculated spectral density of a sinusoidal signal (dots).

To calibrate the input channel, the tape recorder input was fed with calibration signals, i.e., sine waves with different frequencies and amplitudes, as well as constant voltages of different magnitudes. With a harmonic signal introduced, the computer scale unit Δ was determined by

$$\Delta = \frac{2\sqrt{2} u_{\text{inlet}}}{a_{\text{max}} - a_{\text{min}}}, \tag{8.1}$$

where u_{inlet} is the effective voltage of the calibration signal at the tape-recorder input, and a_{min} and a_{max} are the minimum and maximum values of the sine wave introduced into the computer in digital form. The mean computer scale unit is evaluated for each input channel for a series of input channel frequencies and amplitudes. The calibration is performed at the beginning and at the end of each information input session.

Statistical processing of the data is carried out with the following programs. For each series of values of x one calculates the mean value $\bar{x}$, the mean-square deviation σ, the skewness S and the kurtosis K. In the case of discrete series with N numbers, these quantities are determined as follows:

$$\bar{x} = \frac{1}{N} \sum_{t=1}^{N} x_t; \tag{8.2}$$

$$\sigma = \left[\frac{1}{N-1} \sum_{t=1}^{N} (x_t - \bar{x})^2 \right]^{1/2}; \tag{8.3}$$

$$S = \frac{m_3}{m_2^{3/2}}; \tag{8.4}$$

$$K = \frac{m_4}{m_2^2} - 3, \tag{8.5}$$

where

$$m_r = \frac{1}{N} \sum_{t=1}^{N} (x_t - \bar{x})^r, \quad r = 2, 3, 4.$$

Along with the moments of the distribution, it is necessary to calculate the probability density of the values. For a given number of classes n (n is an input parameter of the program) within the limits $(-as, +as)$, where a is a given multiplier, one determines the probability density $P_j^e = N_j/Nh$, $j = 1, 2, \ldots, n$. Here, N_j is the number of points in the jth class and $h = 2as/n$. The integral distribution function P_r^e is estimated as $P_r^e = \Sigma_{j=1}^{r} P_j^e$, where $r = 1, 2, \ldots, n$. The theoretical probability density P^t is calculated within the same ranges and with the same number of classes, for instance, by the formula

$$P^t(x) = \frac{1}{\sqrt{2\pi}\sigma} \exp\left(-\frac{x^2}{2\sigma^2}\right),$$

where σ^2 is the variance of the process. The quantity P^t is calculated for the points $x_r = -ax + (r - \frac{1}{2})h$, where $r = 1, 2, \ldots, n$. The theoretical probability density is compared with the experimental one using one of the statistical criteria, e.g.,

$$\chi^2 = Nh \sum_{j=1}^{n} \frac{(P_j^e - P_j^t)^2}{P_j^t},$$

with united neighboring intervals, where $P_j^t Nh < 8$.

The structure function of the process can be constructed either before or after the initial series is filtered. The calculations are carried out using the expression

$$D_k = \frac{1}{N-k} \sum_{t=1}^{N-k} (x_{t+k} - x_t)^2. \tag{8.6}$$

Filtering is carried out, e.g., by employing a cosine-filter, given by

$$\tilde{x}_t = \frac{1}{l} \sum_{i=t-l/2}^{t+l/2} \left[1 + \cos \frac{2\pi(t-i)}{l}\right] x_i, \quad t = l/2 + 1, \ldots N - l/2, \tag{8.7}$$

where l is the smoothing interval. After the smoothing procedure, the series length shortens to $\tilde{N} = (N - l)$ numbers. The high-frequency filtering of the series is performed by the simple procedure of subtracting the smoothed quantities from the initial ones by the formula $\tilde{\tilde{x}}_t = x_t - \tilde{x}_t$.

The correlation function of the process is given by

$$B_{x,k} = \frac{1}{N-k} \sum_{t=1}^{N-k} (x_t - \bar{x})(x_{t+k} - \bar{x}), \quad 0 \leqslant k \leqslant L - 1, \tag{8.8}$$

where $\bar{x} = (1/N)\Sigma_{t=1}^{N} x_t$ is calculated by the filtered series. Here and below, the tilde is omitted for simplicity. The correlation function $B_{x,k}$ can be normalized by dividing it by $B_{x,0}$.

The spectral density of the process is given by

$$E_{x,i} = 2\Delta t \left[B_{x,0} + 2 \sum_{k=1}^{L-1} B_{x,k} w_k \cos \frac{\pi k i}{F} \right], \quad 0 \leqslant i \leqslant F, \tag{8.9}$$

where Δt is the discrete time step of the series and w_k is the correlation window of the calculations.

As a rule, the following correlation windows are used:

$$\text{(a)} \quad w_k = \begin{cases} 1, k \leqslant L, \\ 0, k > L. \end{cases}$$

$$\text{(b)} \quad w_k = \begin{cases} 1 - \dfrac{k}{L}, k \leqslant L, \\ 0, k > L. \end{cases}$$

$$\text{(c)} \quad w_k = \begin{cases} \dfrac{1}{2}\left(1 + \cos \dfrac{\pi k}{L}\right), k \leqslant L, \\ 0, k > L. \end{cases} \tag{8.10}$$

$$\text{(d)} \quad w_k = \begin{cases} 1 - 6\left(\dfrac{k}{L}\right)^2 + 6\left(\dfrac{k}{L}\right)^3, k \leqslant \dfrac{L}{2}, \\ 2\left(1 - \dfrac{k}{L}\right)^3, \dfrac{L}{2} < k \leqslant L, \\ 0, k > L. \end{cases}$$

The correlation window (8.10c), referred to as the Tukey window (Blackman and Tukey, 1958) is most often employed to calculate turbulence characteristics in the ocean. If this window is used, then the spectral density of the process is calculated by

$$E(f_p) = 2\,\Delta t \sum_{r=0}^{m} \delta_r B_r \left(1 + \cos\frac{r\pi}{m}\right) \cos\frac{pr\pi}{m}, \tag{8.11}$$

where

$$\delta_r = \begin{cases} \frac{1}{2} & \text{at } r = 0 \text{ and } r = m, \\ 1 & \text{at } 0 < r < m. \end{cases}$$

Also, p ranges from 0 to m, the corresponding frequency being $f_p = p/2m\,\Delta t$). Figure 8.3 depicts the Tukey correlation window and the corresponding spectral window

$$W(f) = T_m \left(\frac{\sin 2\pi f T_m}{2\pi f T_m}\right) \left(\frac{1}{1 - (2fT_m)^2}\right). \tag{8.12}$$

In the spectral analysis by the Tukey method it is of great importance to choose correctly the truncation points of the correlation function m, because their position determines the frequency bandwidth b of the spectral window (Jenkins and Watts, 1971, Vol. 1; 1972, Vol. 2). In the program employed in the Institute of Oceanology $b = 1.33/(m\,\Delta t)$. The divergence of the spectral estimates reduces with increasing equivalent number of degrees of freedom. For the Tukey window, this is

$$\nu = 2.66\frac{N}{m}. \tag{8.13}$$

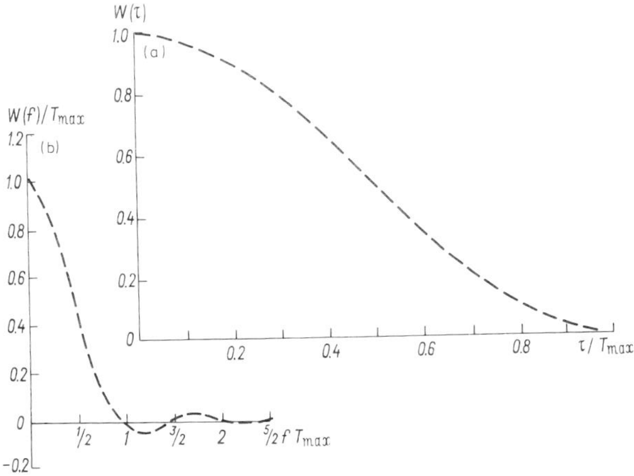

Fig. 8.3. Correlation window (a) and spectral window (b) of the Tukey filter.

Figure 8.4 shows the $100(1-\alpha)\%$ significant limits of spectral estimates for $\alpha = 0.01$, 0.05, and 0.20 as functions of the number of degrees of freedom ν. To increase ν and to reduce the reliability interval of the spectral estimates, it is necessary to choose a small time lag m compared with the length N of the series analyzed. In that case, however, the frequency bandwidth of the spectral window may prove to be too large and the spectral estimate will be substantially shifted if the spectrum under study is not that of white noise. Therefore, it is impossible to obtain a non-shifted spectral estimate with negligible deviation. In each particular case one, therefore, has to establish certain compromise values of m. Since the exact shape of the spectrum is, as a rule, unknown before the calculations are performed, it would be desirable to estimate the spectrum for a number of spectral window widths. When defining the truncation points, the correlation functions must be taken into account and special attention has to be paid to the time lags at which they become negligible. This procedure of estimating the spectrum at a number of values of m is referred to as the 'spectral window contraction method'. Note that, at a given value of m, only components with $\Delta f \leqslant 1.33/(m\,\Delta t)$ can be resolved in the spectrum obtained. If Δf is assigned, then m is obtained by this expression for a known time step Δt. Equation (8.13) is then used to determine the total length N of the series that ensures the required Δf-value.

To carry out relative statistical analysis of two series, one must calculate the correlation functions

$$B_{xy,k} = \frac{1}{N}\sum_{t=1}^{N-k}(x_t-\bar{x})(y_{t+k}-\bar{y}), \quad 0 \leqslant k \leqslant L-1; \tag{8.14}$$

$$B_{xy,-k} = \frac{1}{N}\sum_{t=1}^{N-k}(x_{t+k}-\bar{x})(y_t-\bar{y}), \quad 0 \leqslant k \leqslant L-1; \tag{8.15}$$

the normalized correlation functions

$$b_{xy,k} = \frac{B_{xy,k}}{\sqrt{B_{x,0}B_{y,0}}}; \qquad b_{xy,-k} = \frac{B_{xy,-k}}{\sqrt{B_{x,0}B_{y,0}}}; \tag{8.16}$$

the even and odd parts of the correlation function

$$l_{xy,k} = \tfrac{1}{2}(B_{xy,k}+B_{xy,-k}), \quad 0 \leqslant k \leqslant L-1; \tag{8.17}$$

$$q_{xy,k} = \tfrac{1}{2}(B_{xy,k}-B_{xy,-k}), \quad 0 \leqslant k \leqslant L-1; \tag{8.18}$$

the cospectrum

$$L_{xy,i} = 2\,\Delta t\left[l_{xy,0} + 2\sum_{k=1}^{L-1} l_{xy,k}w\cos\frac{\pi ik}{F}\right], \quad 0 \leqslant i \leqslant F; \tag{8.19}$$

the quadrature spectrum

$$Q_{xy,i} = 4\,\Delta t\sum_{k=1}^{L-1} q_{xy,k}w_k\sin\frac{\pi ik}{F}, \quad 1 \leqslant i \leqslant F-1; \tag{8.20}$$

$$Q_{xy,0} = Q_{xy,F} = 0; \tag{8.21}$$

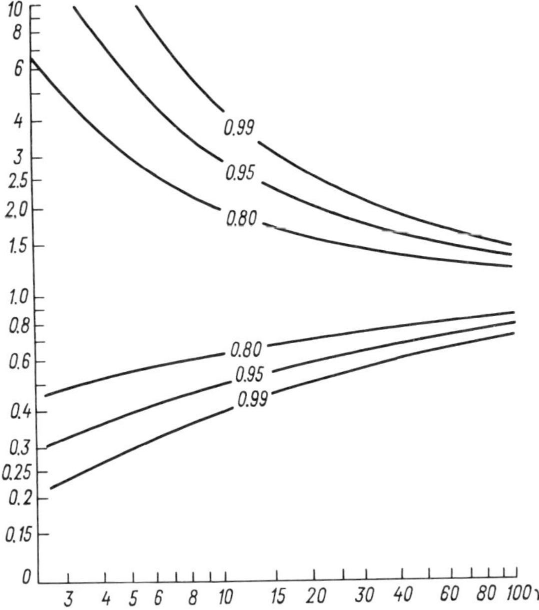

Fig. 8.4. Reliability intervals of spectral estimates for various numbers of degrees of freedom ν.

the joint amplitude spectrum

$$A_{xy,i} = \sqrt{L_{xy,i}^2 + Q_{xy,i}^2}, \quad 0 \leqslant i \leqslant F; \tag{8.22}$$

the phase spectrum

$$\varphi_{xy,i} = \arctan\left[-\frac{Q_{xy,i}}{L_{xy,i}}\right], \quad 0 \leqslant i \leqslant F; \tag{8.23}$$

and the squared coherence spectrum

$$K_{xy,i}^2 = \frac{A_{xy,i}^2}{E_{x,y}E_{y,i}}, \quad 0 \leqslant i \leqslant F. \tag{8.24}$$

When analyzing the two series, the final choice of calculation parameters can be also made by the spectral window contraction method. It is the coherence spectrum $K(f)$ that is most sensitive to the choice of parameters. In this case, the $100(1-\alpha)\%$ significant limit for $\arctan|K(f)|$ is determined by

$$\arctan|K(f)| \pm \eta\left(1-\frac{\alpha}{2}\right)\frac{1}{\sqrt{\nu}}, \tag{8.25}$$

where $\eta[1-(\alpha/2)]$ designates the $100[1-(\alpha/2)]\%$ quantile of the normal distribution. If the reliability intervals are chosen, one can determine the necessary number of degrees of freedom from (8.25), and m and N from (8.13). The use of $\arctan|K(f)|$ in (8.25), instead of the coherence coefficient, is justified by the fact that the reliability interval of this function is constant throughout the entire frequency range.

If the $\tan \varphi(f)$ distribution is approximated by a normal law, then the $100(1 - \alpha)\%$ reliability intervals can be approximated as

$$\tan \varphi(f) \pm \eta \left(1 - \frac{\alpha}{2}\right) \sqrt{\sec^4 \varphi(f) \frac{1}{\nu} \left(\frac{1}{K^2(f)} - 1\right)}. \tag{8.26}$$

Figure 8.5 demonstrates the 95% reliability intervals for various numbers of degrees of freedom ν of a random phase shift estimate and for various values of the coherence spectrum.

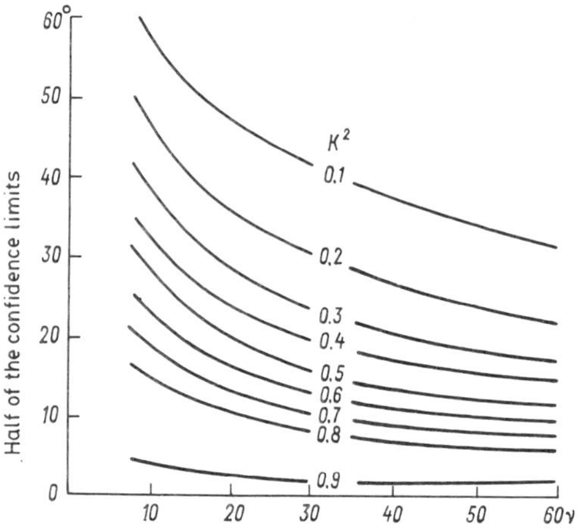

Fig. 8.5. 95%-reliability intervals for the phase spectrum, with various coherence spectra and different numbers of degrees of freedom ν.

When vector fields have to be analyzed, the longitudinal B_{LL} and transverse B_{NN} correlation functions can be estimated as

$$B_{LL}(\mathbf{r}) = \frac{1}{T} \int_0^T u_L(\mathbf{x}) u_L(\mathbf{x} + \mathbf{r})\, \mathrm{d}t; \tag{8.27}$$

$$B_{NN}(\mathbf{r}) = \frac{1}{T} \int_0^T u_N(\mathbf{x}) u_N(\mathbf{x} + \mathbf{r})\, \mathrm{d}t, \tag{8.28}$$

where the quantities designated by L and N in the right-hand sides are the vector components parallel and normal to the separation vector $\mathbf{r}$, respectively.

The longitudinal and transverse spectra of the vector field $\mathbf{u}(\mathbf{x})$ are given by the expressions

$$F_1(k) = \frac{1}{\pi} \int_0^\infty \cos kr\, B_{LL}(r)\, \mathrm{d}r; \tag{8.29}$$

$$F_2(k) = \frac{1}{\pi} \int_0^\infty \cos kr\, B_{NN}(r)\, \mathrm{d}r. \tag{8.30}$$

The kinetic energy spectrum of the vector field is

$$E(k) = \frac{1}{\pi} \int_0^\infty (kr \sin kr - k^2 r^2 \cos kr) B_{LL}(r) \, \mathrm{d}r \tag{8.31}$$

for a solenoidal field, and

$$E(k) = \frac{1}{\pi} \int_0^\infty (kr \sin kr - k^2 r^2 \cos kr) B_{NN}(r) \, \mathrm{d}r \tag{8.32}$$

for a potential field.

The two-point triple moments of the vector field $\mathbf{u}(\mathbf{x})$ are often of interest. They are given by

$$B_{LL,L}(\mathbf{r}) = \frac{1}{T} \int_0^T u_L^2(\mathbf{x}) u_L(\mathbf{x} + \mathbf{r}) \, \mathrm{d}t; \tag{8.33}$$

$$B_{NN,L}(\mathbf{r}) = \frac{1}{T} \int_0^T u_N^2(\mathbf{x}) u_L(\mathbf{x} + \mathbf{r}) \, \mathrm{d}t; \tag{8.34}$$

$$B_{LN,N}(\mathbf{r}) = \frac{1}{T} \int_0^T u_L(\mathbf{x}) u_N(\mathbf{x}) u_N(\mathbf{x} + \mathbf{r}) \, \mathrm{d}t. \tag{8.35}$$

The two-point quadruple moments are

$$B_{LL,LL}(\mathbf{r}) = \frac{1}{T} \int_0^T u_L^2(\mathbf{x}) u_L^2(\mathbf{x} + \mathbf{r}) \, \mathrm{d}t; \tag{8.36}$$

$$B_{LLL,L}(\mathbf{r}) = \frac{1}{T} \int_0^T u_L^3(\mathbf{x}) u_L(\mathbf{x} + \mathbf{r}) \, \mathrm{d}t, \tag{8.37}$$

and, in a similar way,

$$\begin{matrix} B_{LL,NN}, & B_{LN,LN}, & B_{NN,NN}, & B_{NN,MM}, & B_{NM,NM}, \\ B_{LLN,N}, & B_{LNN,L}, & B_{NNN,N}, & B_{NNM,M}, & \end{matrix} \tag{8.38}$$

where M designates the direction normal to L and N.

The three-point triple moments of a vector field can be obtained by

$$B_{ijl}(\mathbf{r}, \mathbf{r}') = \frac{1}{T} \int_0^T u_i(\mathbf{x}) u_j(\mathbf{x} + \mathbf{r}) u_l(\mathbf{x} + \mathbf{r}') \, \mathrm{d}t. \tag{8.39}$$

In the case of a scalar field, this reads

$$B(\mathbf{r}, \mathbf{r}') = \frac{1}{T} \int_0^T \vartheta(\mathbf{x}) \vartheta(\mathbf{x} + \mathbf{r}) \vartheta(\mathbf{x} + \mathbf{r}') \, \mathrm{d}t. \tag{8.40}$$

In a joint analysis of the vector field $\mathbf{u}(\mathbf{x})$ and the scalar field $\vartheta(\mathbf{x})$, two-point triple moments of the type

$$B_{L\vartheta,\vartheta}(\mathbf{r}) = \frac{1}{T}\int_0^T u_i(\mathbf{x})\vartheta(\mathbf{x})\vartheta(\mathbf{x}+\mathbf{r})\,\mathrm{d}t \tag{8.41}$$

might be of particular interest. This also holds for the corresponding spectral functions

$$F_{L\vartheta,\vartheta}(k) = \frac{1}{2\pi^2}\int_0^\infty \left\{-\frac{\cos kr}{kr} - \frac{\sin kr}{(kr)^2}\right\} B_{L\vartheta,\vartheta}(r)\,\mathrm{d}r, \tag{8.42}$$

and for three-point moments of the type

$$B_{j\vartheta\vartheta}(\mathbf{r},\mathbf{r}') = \frac{1}{T}\int_0^T u_j(\mathbf{x})\vartheta(\mathbf{x}+\mathbf{r})\vartheta(\mathbf{x}+\mathbf{r}')\,\mathrm{d}t \tag{8.43}$$

and the corresponding spectral functions.

If required, more complex characteristics can be defined by analogy with these expressions.

A very important characteristic of two series of observations is the two-dimensional joint probability density of their values. If the series $\{x_i\}$ contains M samples, the series $\{y_i\}$ N samples and, for the sake of clarity, $M \leqslant N$, one first determines the mean values $\bar{x}$ and $\bar{y}$, the root-mean-square deviations S_x and S_y, and the covariance S_{xy} by the expressions

$$\bar{x} = \frac{1}{M}\sum_{i=1}^{M} x_i; \qquad \bar{y} = \frac{1}{M}\sum_{i=1}^{M} y_i; \tag{8.44}$$

$$S_x^2 = \frac{1}{M-1}\sum_{i=1}^{M}(x_i-\bar{x})^2; \qquad S_y^2 = \frac{1}{M-1}\sum_{i=1}^{M}(y_i-\bar{y})^2; \tag{8.45}$$

$$S_x = \sqrt{S_x^2}; \qquad S_y = \sqrt{S_y^2};$$

$$S_{xy} = \frac{1}{M-1}\sum_{i=1}^{M}(x_i-\bar{x})(y_i-\bar{y}). \tag{8.46}$$

The series are then centered and normalized, i.e., reduced to the form

$$x_i' = \frac{x_i-\bar{x}}{S_x}; \qquad y_i' = \frac{y_i-\bar{y}}{S_y}. \tag{8.47}$$

The series $\{x_i'\}$ and $\{y_i'\}$ are used to determine their amplitude ranges: (a, b) for the series $\{x_i'\}$ and (c, d) for $\{y_i'\}$. As a calculation input parameter, one selects the number of subdivisions k_x and k_y in the ranges of variation of $\{x_i'\}$ and $\{y_i'\}$, i.e., $[-nS_x, +nS_x]$

and $[-nS_y, +nS_y]$, where n is another input parameter. Thus, the width of each subdivision used to classify the samples in the series is given by the expressions

$$W_x = \frac{2nS_x}{k_x}; \qquad W_y = \frac{2nS_y}{k_y}. \tag{8.48}$$

The joint probability density is determined by

$$\hat{P}(x, y) = \frac{R_{x',y'}}{RW_xW_y}, \tag{8.49}$$

where $R_{x',y'}$ is the number of pairs of samples that simultaneously fall within a given interval and R is the number of samples that lie within the entire range considered. The same limits of variability can be used to calculate a theoretical joint probability density, e.g., a Gaussian one. This makes it possible to compare the empirical and the theoretical distributions by one of the agreement criteria, e.g., the χ^2-test.

The so-called 'sliding' variances and structure functions (Beliayev and Liubimtsev, 1977) prove to be extremely useful for investigating the intermittency of ocean turbulence. The former are calculated from the series by selecting successive or overlapping segments of length n:

$$(S^2)^n_{(j-1)p} = \frac{1}{n-1} \sum_{i=1}^{n} [x_{(j-1)p+i} - \bar{x}_{(j-1)p}]^2, \tag{8.50}$$

where $j = 1, 2, \ldots,$ is the integer closest to N/n. The overlap parameter p varies from 1 to n.

The sliding structure function is calculated using the expression

$$D^n_{(j-1)p,k} = \frac{1}{n-k} \sum_{i=1}^{n-k} [x_{(j-1)p+i+k} - \bar{x}_{(j-1)p+i}]^2, \tag{8.51}$$

where $j = 1, 2, \ldots,$ is the integer closest to the number N/n, the time lag k is always less than n, and the overlap parameter p varies from 1 to n.

The spectral functions can be calculated using the fast Fourier transform method (FFT) instead of the Tukey method. The FFT method allows a significant reduction of the computer time necessary for processing long observational time series. Computations of spectra by the Tukey method and by FFT require computer time proportional to N^2 and $N \log_2 N$, respectively, where N is the number of samples in the series. The FFT method is an iterative calculation of the Fourier coefficients of the initial series with the Fourier coefficients of the odd and even elements. The discrete Fourier transform of the series x_n $(n = 0, 1, \ldots, N-1)$ is determined by

$$S_k = \frac{1}{N} \sum_{n=0}^{N-1} x_n \exp\left(-2\pi i k \frac{n}{N}\right), \tag{8.52}$$

where $k = 0, 1, \ldots, N-1$. Grouping the terms in (8.52) that contain only the odd and even samples of the initial series, one obtains

$$S_k = \frac{1}{N} \sum_{n=0}^{(N/2)-1} x_{2n} \exp\left(-2\pi ik \frac{2n}{N}\right) + \frac{1}{N} \sum_{n=1}^{(N/2)-1} x_{2n+1} \times$$

$$\times \exp[-2\pi ik(2n+1)/N] = \frac{1}{2(N/2)} \sum_{n=0}^{(N/2)-1} x_{2n} \exp\left(-2\pi ik \frac{n}{(N/2)}\right) +$$

$$+ \exp\left(-\frac{2\pi ik}{N}\right) \frac{1}{2(N/2)} \sum_{n=0}^{(n/2)-1} x_{2n-1} \exp\left(-\frac{2\pi ikn}{N/2}\right). \tag{8.53}$$

From (8.52) we obtain for $k = 0, 1, \ldots, \frac{1}{2}N - 1$

$$S_k = \tfrac{1}{2} A_k + \tfrac{1}{2} \exp\left(-\frac{2\pi ik}{N}\right) B_k. \tag{8.54}$$

Here, A_k and B_k are the Fourier coefficients of the two series that are derived from the initial series by selecting odd and even samples, respectively. They contain $\frac{1}{2}N$ samples each. Since

$$A_{k+(N/2)} = A_k \quad \text{and} \quad B_{k+(N/2)} = B_k, \tag{8.55}$$

(8.54) yields

$$S_{k+(N/2)} = \tfrac{1}{2} A_k - \tfrac{1}{2} \exp(-2\pi ik/N) B_k. \tag{8.56}$$

The determination of the Fourier coefficients of the initial series with N samples is reduced to that of the coefficients of two series, each containing $N/2$ samples. The coefficients A_k and B_k can be calculated in a similar way, i.e., by subdividing each of the series obtained into two subseries containing $N/4$ samples. If $N = 2^\gamma$, where γ is an integer, then, employing the same procedure, one finally obtains one-sample series whose Fourier coefficients are naturally equal to the sample values themselves. When computed using the FFT method, the series is subdivided into lengths containing 2^γ samples. These segments can overlap. The Fourier coefficients for each segment are found by the expressions

$$a_k = \frac{1}{N} \sum_{n=0}^{N-1} x_n \cos(-2\pi kn/N), \tag{8.57}$$

$$b_k = \frac{1}{N} \sum_{n=0}^{N-1} x_n \sin(-2\pi kn/N), \tag{8.58}$$

where $k = 0, 1, \ldots (N/2) - 1$. The values of the kth histogram with the corresponding frequency $f_k = k/(N\,\Delta t)$ are

$$|S_k|^2 = a_k^2 + b_k^2 \tag{8.59}$$

For a wide range of random processes, $|S_k|^2$ is approximately an independent random quantity at large values of N. For $k \geqslant 1$, $|S_k|^2$ is distributed as x^2 with two degrees of freedom (Bartlett, 1955). In this case, the variance $|S_k|^2$ does not tend towards zero at $N \to \infty$ and the histogram is not a valid spectral estimate. The variance of the spectral estimate can be reduced by averaging the histogram over all of the r segments and by smoothing it over a certain frequency range q. In the case of non-overlapping segments, the number of degrees of freedom of the spectral estimates can be approximately found by the formula $\nu = 2rq$. The accuracy of the spectral estimates can be increased by multiplying the initial samples of the series by an appropriate weighting function. This procedure is equivalent to employing a spectral window that corresponds to this function.

The analysis of broad-band processes faces certain computational difficulties that are associated with the large number of initial samples. In order to estimate these spectra reliably, one may average the spectral density of the process over a limited frequency range (Kholmiansky, 1971), with frequency ranges forming a geometric progression of the type $1/2^{\alpha}$. Acceptable values of the averaging parameter α are determined by the inequality

$$\frac{2^{\alpha[(N/2)-1]} - 1}{2^{\alpha} - 1} < 100,$$

subject to the limited number of frequency ranges permitted by the computer program.

9. VELOCITY FLUCTUATIONS

9.1. Root-Mean-Square Values

The root-mean-square values of velocity fluctuations in the ocean depend strongly on the range of space or time scales over which the results are averaged. This dependence is associated with the complicated spectral composition of the velocity field in the ocean. Indeed, when the field is characterized only by short-period oscillations, averaging of the results over a period larger than the maximum fluctuation period will not result in changes of the mean velocity $\bar{u}_i$ and standard deviation σ_i. If the velocity field is characterized by various components, then an increase in the period (scale) of averaging will result in an increase in σ_i. This dependence reflects the non-negative energy spectrum of any velocity component. The integral of the spectrum is the variance estimate σ_i^2 within the averaging range. This explains why any experimental value of σ_i in the ocean must be accompanied by that of the scale range in which it is determined; otherwise the σ_i-data become useless.

Bowden and Fairbairn (1952, 1956) seem to be the first to have obtained σ_i for small-scale velocity fluctuations in the ocean. They employed a device (mentioned in §7) with a transmission band up to 1 Hz; the measurements were carried out in strong tidal currents in the coastal waters of England. For calculations of the standard deviations of velocity fluctuations, the averaging periods were approximately 10 min. At these time scales σ_i varied from 2.4 to 5.2 cm s^{-1} for the horizontal velocity components and from 1.2 to 2.0 cm s^{-1} for the vertical component. The fluctuation velocities, therefore, were sufficiently high and showed appreciable anisotropy within these scales and in this type of flow (the tidal flow velocities were 24–50 cm s^{-1}).

Velocity fluctuations at higher frequencies were measured by Kolesnikov (1959, 1960) under the Arctic ice and in Lake Baikal, employing model turbulence meters. For a case in which the surface water layer was entrained at a velocity of 5 cm s^{-1} by a drifting block of ice, the velocity fluctuations did not exceed 5–15% of this value and decayed rapidly with depth. In Lake Baikal the velocity fluctuations in the layer under the ice were close to 0.3 cm s^{-1}. The vertical velocity fluctuations proved to be somewhat smaller than the horizontal ones; no information was, however, reported concerning the range of scales pertaining to the data. Similar data taken with turbulence meters in some other ocean regions are available from Panteleyev (1960) and Speranskaya (1960). Approximately at the same time Inoue (1960), Bowden (1962) and Bowden and Howe (1963) published data on velocity component fluctuations obtained with electromagnetic sensors. In the tidal flow of the Mersey river estuary, at an averaging period of 5 min, the root-mean-square velocity fluctuations did not exceed 4% of the mean flow velocity (94–172 cm s^{-1}) in the surface layer (approximately 4 m below the water surface) and 7% in the bottom layers (0.5–2 m above the bottom), where the mean flow velocity was 41–85 cm s^{-1}. employing a similar device, Bowden (1962) observed a 7% fluctuation level for the transverse component of the horizontal velocity and a 14% level for the longitudinal component at 50–175 cm above the bottom in Red Wharf Bay, the averaging period being 10 min.

In 1959–63, Canadian investigators carried out a series of measurements employing low-inertia devices (Grant and Moilliet, 1962; Grant *et al.*, 1959, 1960, 1962, 1963). In these publications, special attention was paid to the calculation of the spectral characteristics of turbulence (these will be discussed later). Note, however, that the order of magnitude of the fluctuating components was, on average, about 1 cm s^{-1} according to these measurements, over scales ranging from the viscous range to about 1 m.

In various Institute of Oceanology expeditions, velocity fluctuations were measured in the tropical and equatorial latitudes of the Atlantic Ocean, in the Indian and Pacific Oceans (Ozmidov, 1971a,b, 1973a, 1974b, 1978a, 1979), in the West Wind Drift (Ozmidov, 1974b), in the Japan Current (Beliayev and Gezentsvei, 1977), in the Gulf Stream (Karabasheva *et al.*, 1975), in the Norwegian and Mediterranean Seas (Pozdynin, 1976), and in the Baltic Sea (Lozovatsky *et al.*, 1977). The bandwidths of the devices employed were, as a rule, from fractions of a Hertz up to 100 or several hundred Hertz. An extreme value of the root-mean-square velocity fluctuations was detected at the lower limit of the upper mixed layer (at 43 m depth) in the system of equatorial currents in one of the polygons in the Indian Ocean. This value amounted to several centimeters per second (Beliayev *et al.*, 1975a).[3] Most likely, such a high level of small-scale turbulence will be seldom observed in the ocean. Typical σ_u values range from several millimeters per second to 1 cm s^{-1}. For example, in tropical latitudes of the Atlantic Ocean, σ_u was close to 1 cm s^{-1} at depths of 30–140 m (Pozdynin, 1974). However, in the frontal zone of the Norwegian Sea, the turbulence level was somewhat higher. Turbulence measurements in the Gulf Stream showed σ_u to be higher in the central part of the Gulf Stream than in the boundary layer between the Gulf Stream and the Labrador Current. In the region where the Gulf Stream merges with the North Atlantic Drift, the root-mean-square amplitude of the velocity fluctuations reached 0.3–0.5 cm s^{-1} (Karabasheva *et al.*, 1975). It is interesting to note that the velocity fluctuation level in the Gulf Stream varied negligibly with depth down to 1300 m, while most ocean regions seem to exhibit values of σ_u that vary with depth, as well as turbulence with a stratified vertical structure. Vertical distributions of σ_u obtained by towing a measuring probe in three polygons in the Atlantic Ocean were reported by Beliayev *et al.* (1973a). The measurements demonstrated that σ_u varied in different ways. In one of the polygons, the smallest amplitudes of velocity fluctuations were observed in the 30–50 m layer (measurements were only carried out below this level to avoid disturbances induced by the ship), while at the 87 m level the amplitudes increased. Note that the water in the vicinity of this level was practically homogeneous with respect to density, while below and above it quite noticeable vertical density gradients were discovered. In the second polygon, the turbulence intensity decreased with increasing depth. Measurements on the third polygon showed a nearly constant level of velocity fluctuations, close to 1 cm s^{-1}.

More details of the structure of the vertical distributions of turbulent velocity fluctuations were determined by probing devices. Figure 9.1 presents vertical profiles of the

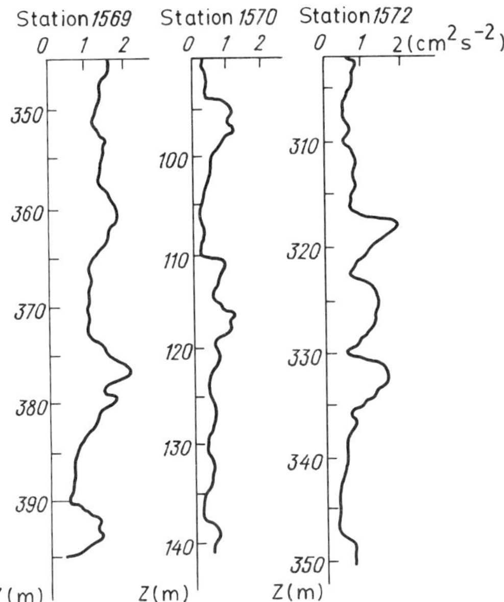

Fig. 9.1. Vertical variance profiles of velocity fluctuations measured at three polygons in the Gulf of Tunis (Pozdynin, 1976).

mean-square velocity fluctuations in different ocean layers, measured in the Gulf of Tunis (Pozdynin, 1976). The profiles are divided into sections with nearly constant values of σ_u^2 in each subdivision. This procedure was carried out while taking into account the resolution of the equipment, estimated by the technique developed by Palevich *et al.* (1973). The subdivisions with approximately constant turbulence level appear to have layer thicknesses from several meters to several tens of meters. The empirical probability density of these thicknesses obeys a logarithmically normal law, as seen in Figure 9.2.

Interesting data concerning the vertical distribution of σ_u throughout shallow seas were obtained by Lozovatsky *et al.* (1977). Figure 9.3 gives the vertical distributions

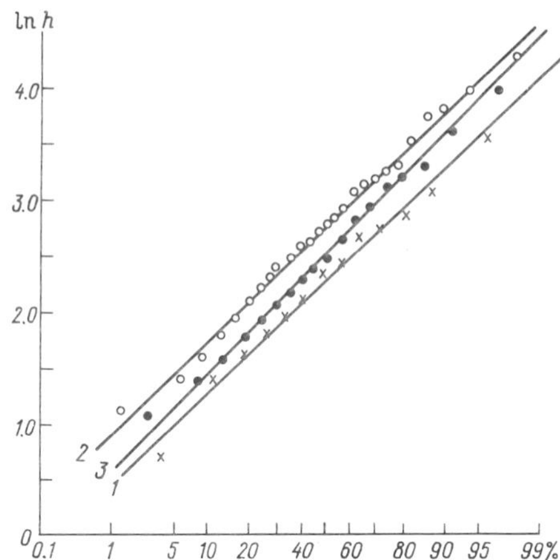

Fig. 9.2. Empirical thickness distributions of layers with the same turbulence level, and straight lines with a log-normal distribution. Data from three polygons (see Figure 9.1).

of σ_u and the mean water temperature from the surface to the bottom of the Bornholm Hollow (52 m deep). From the surface layer down to 5–7 m, the turbulence level was high due to surface disturbances. Below the wind-mixed layer the small-scale velocity fluctuations were negligible and constant down to the lower boundary of the upper homogeneous layer. In the thermocline, the σ_u-fluctuations sharply increased and the turbulence had a pronounced intermittent character. Below the layer of maximum temperature gradient (35–42 m), the turbulence decayed once again, but in the bottom layer (42–52 m) an increase in σ_u was observed. The highly turbulent bottom layer was subdivided into several sublayers with widths of up to 2 m, each characterized by a nearly constant value of σ_u. It is interesting to note that all measurements revealed the 1–2 m-thick layer at the very bottom to be highly turbulent.

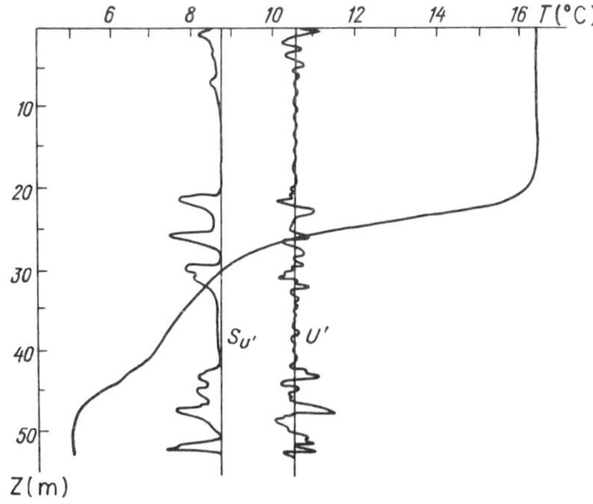

Fig. 9.3. Vertical distributions of mean temperature, velocity fluctuations and their variance (in relative units) from data obtained in the Baltic Sea (Lozovatsky *et al.*, 1977).

The most detailed description of the amplitude range of velocity fluctuations in the ocean can, of course, be obtained with probability distribution laws. However, in order to derive such a law for the entire World Ocean, as well as for different hydrometeorological situations, one needs access to extensive experimental data. The information currently available in the literature gives only approximate distribution laws, which can vary substantially from case to case. One empirical distribution of σ_u was constructed from observational data obtained in the ninth cruise of the 'Akademicik Kurchatov' and in the seventh cruise of the 'Dmitriy Mendeleyev'. Velocity fluctuations of less than 1 cm s^{-1} proved to be most likely, the probability of larger fluctuations was much lower. The histogram can be approximated, for instance, by a hyperbolic distribution law of the type $P(\sigma_u) = 1/(\sigma_u \ln \frac{a}{b})$, where a and b are the probabilities of minimum and maximum values of σ_u. When considering distribution histograms of root-mean-square velocity fluctuations, it is necessary to bear in mind that one usually processes and analyzes those parts of a record in which the effective signal is much larger than the noise level of the measuring device. As a result, the histograms often lack a considerable range of small velocity fluctuations, which might have indicated a pronounced peak at values of σ_u of the order of millimeters per second.

9.2. Correlation Functions and Spectra

Observational data concerning high-frequency velocity fluctuations in the ocean make it possible to calculate not only their mean-square values but also more complex characteristics, particularly correlation functions and their outer (L) and inner (λ_0) scales (English usage: integral scales and microscales). The quantities L and λ_0 are determined from the correlation function $B(\mathbf{r}) = \overline{u(\mathbf{x})u(\mathbf{x} + \mathbf{r})}$ by the expressions

$$L = \frac{1}{B_0} \int_0^\infty B(r)\, \mathrm{d}r, \tag{9.1}$$

and

$$\lambda_0 = \left(- \frac{B(0)}{2B''(0)} \right)^{1/2}. \tag{9.2}$$

In (9.2), the two primes denote the second derivative of $B(r)$ with respect to r. Typical normalized correlation functions of velocity fluctuations are presented in Figure 9.4. That the function $B(r)$ tends towards zero at comparatively small arguments can be attributed largely to the limited frequency band of the measuring channel. Hence, the scale L obtained from these correlation functions appears to range from several tens of centimeters to a meter and to change negligibly from measurement to measurement, provided the integration range is limited to the point on the X-axis at which the value of $B(r)$ drops below 0.05 for the first time. The inner scale λ_0 calculated with (9.2) ranges from 0.8 to 1.9 cm (Pozdynin, 1974).

Detailed information concerning the small-scale structure of turbulence in the ocean was acquired by spectral analysis of the results obtained. The velocity spectra taken in the Discovery Strait (offshore Canada) by Grant *et al.* (1960) showed good agreement with the theory of locally isotropic turbulence (see Figure 9.5). As shown below, such

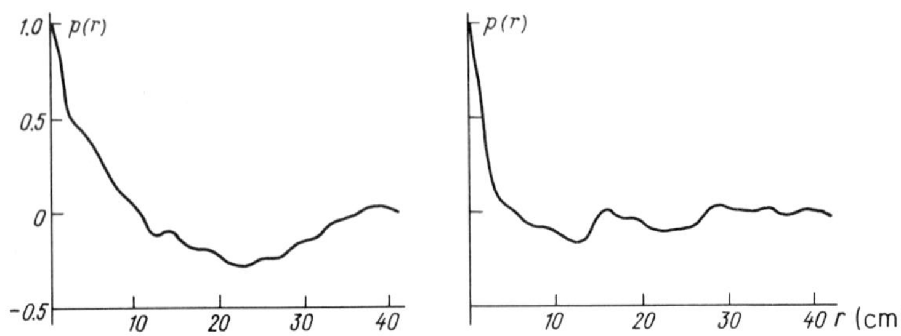

Fig. 9.4. Typical shapes of the normalized correlation functions of velocity fluctuations, from data measured during the ninth cruise of the 'Akademicik Kurchatov' (Beliayev *et al.*, 1973).

agreement is not a universal fact in the ocean. In the case under discussion, it was made possible by the specific properties of the current under study, i.e., by a strong tidal flow with a Reynolds number of $5 \times 10^7 - 3 \times 10^8$, which resulted in fully developed turbulence and in the appearance of a distinct spectral range obeying the laws of locally isotropic turbulence. Under other circumstances, the properties of ocean turbulence and its spectrum can be quite different. It is interesting to note that, according to Grant, Stewart, and Moilliet, the 5/3-power law is valid up to wavenumbers of $k \simeq 1$ cm^{-1}, while for $k > 1$ cm^{-1} the experimental points drop below the 5/3-power law curve because of the effects of molecular viscosity. In keeping with Kolmogorov's similarity hypothesis, the velocity fluctuation spectrum in this range of wavenumbers can be described by a universal dimensionless function of the type

$$\varphi(k\eta) = (\epsilon\nu^5)^{-1/4} E(k), \tag{9.3}$$

where η is the Kolmogorov scale. At small values of $k\eta$ the function φ is proportional to $(k\eta)^{-5/3}$, i.e., the 5/3-power law holds. The function $\varphi(k\eta)$ obtained by Grant, Stewart, and Molliet is illustrated in Figure 9.6. This shows that the experimental points begin to diviate from the 5/3-power law at $k\eta \simeq 1/8$.

Extensive calculations of velocity spectra from observational data obtained during specialized expeditions of the Institute of Oceanology of the U.S.S.R. Academy of Sciences demonstrated a substantial variability in the levels and spectral shapes of small-scale turbulence in the ocean. Thus, for instance, according to data obtained in ten polygons by the 'Dmitriy Mendeleyev' and the 'Akademicik Kurchatov', spectral densities of the velocity fluctuations varied over four orders of magnitude. At wavenumbers between 0.06 and 5 cm^{-1} the slopes of the spectral tangents (the slopes of the logarithmic curves) varied approximately from 1 to 2.2–2.7; often, there was not even a monotonic k-dependence. The ranges, in which the spectral curves followed a 5/3-power law, were not always pronounced and, in a number of cases, were even absent. Sometimes the shapes and levels of the spectra appeared to be substantially different for closely-spaced towing sites, even at the same level (the mean large-scale hydrological conditions at these sites were equal). In polygons, the mean hydrological conditions were apparent only as a shift in the centre of gravity of families of spectra obtained in various polygons.

Figure 9.7 presents data on the spectra of velocity fluctuations obtained by the ships 'Dmitriy Mendeleyev' and 'Akademicik Kurchatov' in a number of polygons in

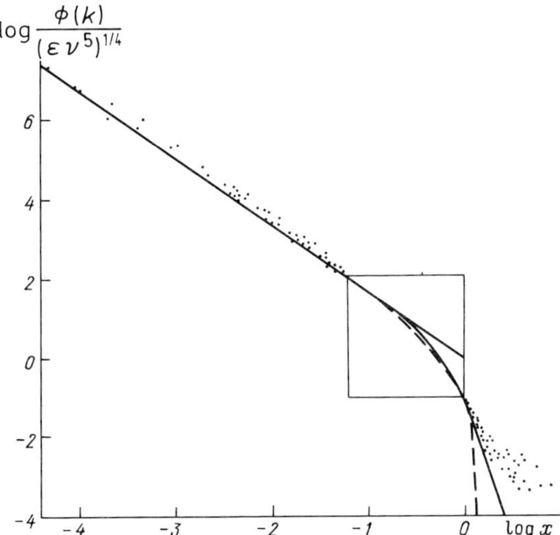

Fig. 9.5. Dimensionless longitudinal spectrum of velocity fluctuations in a tidal flow (Grant *et al.*, 1962).

various regions of the World Ocean and generalized by Beliayev *et al.* (1974b). To simplify the comparison of data from different polygons, each fragment of Figure 9.7 is furnished with a vertical dashed line showing the wavenumber $k = 1\ \mathrm{cm}^{-1}$ and a horizontal dashed line corresponding to the standard spectral density level $E(k) = 10^{-1}\ \mathrm{cm}^3\ \mathrm{s}^{-2}$. Each spectral curve is the average of 5–10 spectra calculated from specific parts of velocity fluctuation records obtained during towing of the measuring device. As seen from Figure 9.7, the results from some polygons exhibit substantial

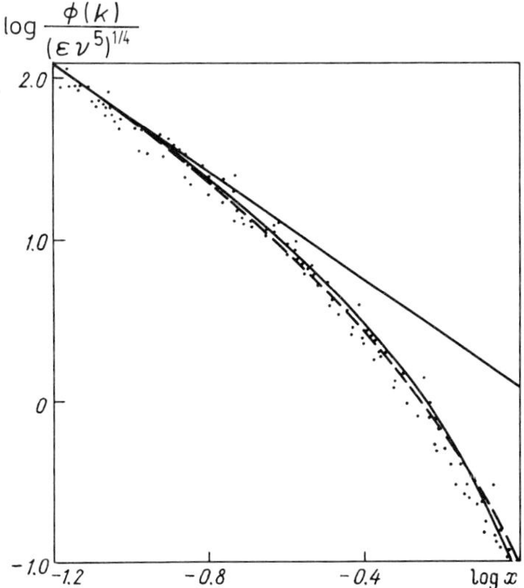

Fig. 9.6. The shape of the spectrum at large wavenumbers (Grant *et al.*, 1962).

variations in $E(k)$, while other polygons are characterized by nearly homogeneous turbulence. In most cases, as has already been mentioned, turbulence levels do not change with depth, but the shapes of the spectra vary from case to case. A comparison of turbulence properties with hydrological conditions reveals a general tendency of velocity fluctuations to decrease in an increasing mean vertical density gradient. Indeed, the most intensive u'-fluctuations were observed in polygons with a very small density stratification, which lacked the layer of density discontinuity that shields deeper water layers from the turbulent energy flux in the upper ocean. The smallest u'-fluctuations were observed in polygons characterized by steep density gradients.

The wide variety of turbulence properties within a fixed wavenumber range in the ocean is, in all probability, indicative of the predominant effects of various forces (buoyancy forces, forces of inertia, viscous forces) on the turbulent vorticity at the scales discussed. To check this assumption and to discover universal spectral shapes, it would be natural to try to compare the empirical spectra obtained in different polygons with theoretical models. The spectra obtained in one of the polygons were, for instance, compared with the universal curve employed by Grant *et al.* (1962). This comparison was carried out using the technique reported by Stewart and Grant (1962) at the best visual coincidence between the experimental points of a spectrum and the universal curve in the middle of the spectral range studied. The results are shown in Figure 9.8, where the experimental spectra are seen to be fairly well described by the universal function, except at small wavenumbers. In this case, the spectra have the shape of typical longitudinal velocity spectra that pertain to not very high Reynolds numbers and, by laboratory measurements, correspond to a flow with $\mathrm{Re} = 3 \times 10^3 - 4 \times 10^4$ (Gibson and Schwartz, 1963).

Quite different results were obtained by approximating a group of spectral curves by a theoretical curve based on another polygon. Here, in the range of small wavenumbers, the experimental curves did not obey a 5/3-power law, which indicated the dominant role of buoyancy forces. Therefore, the universal function $f(x)$ obtained for the spectrum of longitudinal velocity fluctuations in a stratified medium (Monin, 1962) was chosen as an approximating curve. The comparison between experimental curves and the model spectrum was carried out in the following way. Each experimental curve, plotted on a bilogarithmic scale, was approximated by a straight line with a $-5/3$ slope in the high-frequency range. The inertial range of the model spectrum was fitted to the approximating curve. After that, the theoretical curve was shifted along this straight line to make the best visual fit with the results of the experiment (see Figure 9.9). Here $x = kL$, $L = -4\gamma^{-1/2}\alpha^{-3/4}L_*$, L_* is the buoyancy scale, γ is the proportionality coefficient in Heisenberg's hypothesis of spectral eddy viscosity, and α is the ratio between the eddy thermal conductivity and the eddy viscosity. As seen from Figure 9.9, the experimental curves are very close to the model spectrum. In this case, the scale L appears to approximate the size of the fine-structure 'steps' of the density field in the polygon considered.

9.3. Dependence on Local Background Conditions

Small Reynolds numbers and small buoyancy scales for small-scale turbulence in a stratified ocean are substantial reasons for considering the turbulence structure in a given space–time region to be controlled not by the average characteristics of the hydrological

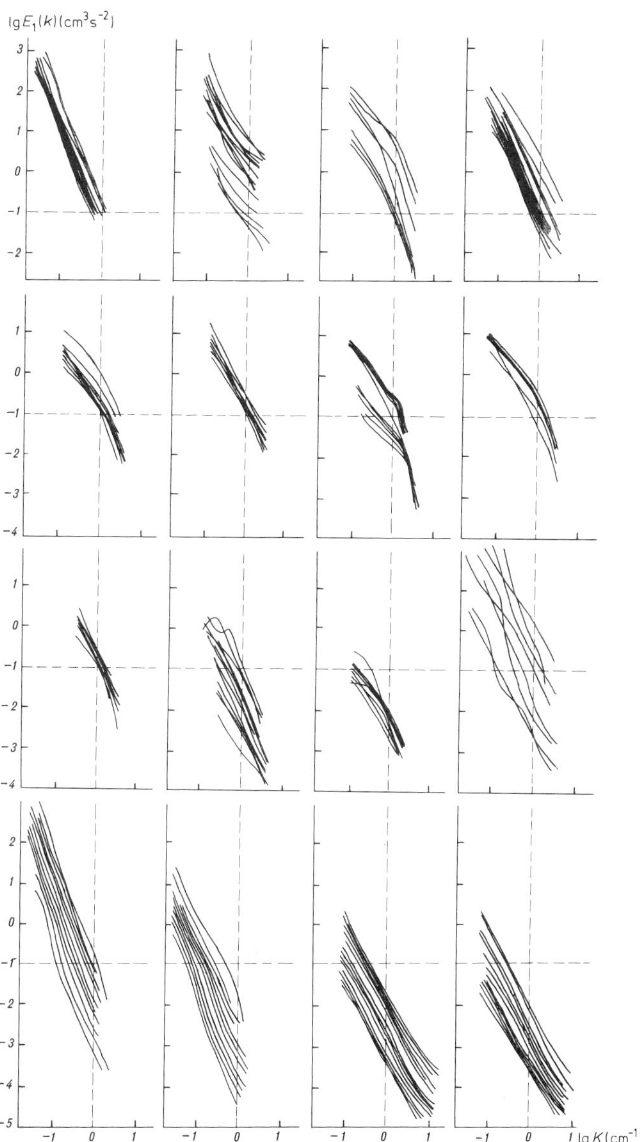

Fig. 9.7. Summarized plots of the velocity fluctuation spectra obtained from measurements taken in various ocean regions on the research vessels 'Dmitriy Mendeleyev' and 'Akademicik Kurchatov'.

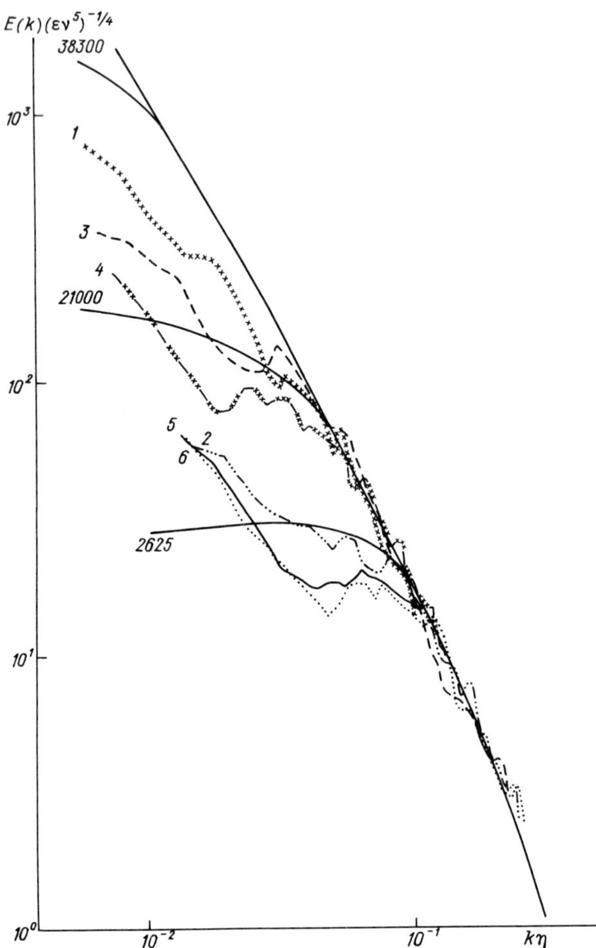

Fig. 9.8. Comparison of a group of dimensionless spectra, with the universal shape of the inertial and viscous ranges. Thin solid lines present turbulence spectra from laboratory measurements at the Reynolds numbers designated (Beliayev *et al.*, 1975).

fields in the polygon, but also by their local values, which depend on the microstructure of the fields. This conclusion followed from an analysis of the first extensive measurements of small-scale turbulence in the ocean (Beliayev and Ozmidov, 1970; Kolmogorov *et al.*, 1971; Ozmidov and Beliayev, 1973) and was later confirmed by combined measurements of turbulent fluctuations and the fine structures of the fields. The microstructure of hydrophysical fields is a typical phenomenon encountered in oceans. As shown by probing measurements, the ocean appears in most cases to be stratified into quasi-homogeneous layers that vary in thickness from tens of meters to decimeters or even centimeters. These layers are subdivided into sublayers with pronounced vertical gradients of the hydrophysical characteristics. The microstructures usually are quite persistent – they have lifetimes of some tens of minutes or hours (see Figure 9.10). Hence, the small-scale

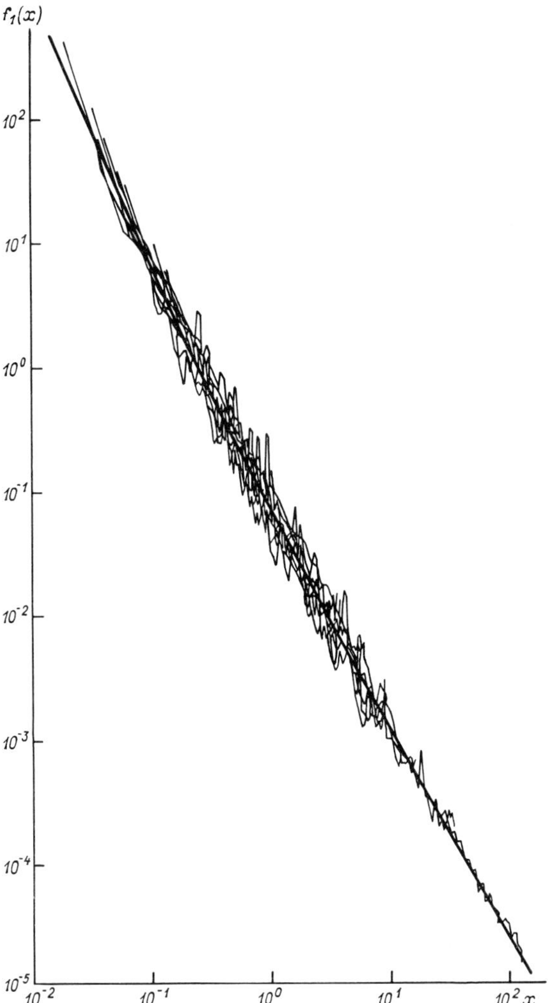

Fig. 9.9. Specimen of a family of spectra, compared with the model spectrum of a stratified flow (Ozmidov *et al.*, 1974).

turbulence in these layers must be determined by the parameters of each layer, rather than by its general depth or the mean characteristics of the bulk ocean. Woods' studies (1968a, b, c) with dye introduced into the surface layer visually showed the relationship between small-scale turbulence and local background conditions (microstructure). The underwater photographs taken by Woods in the Mediterranean clearly exhibit that sometimes a streak of dye, initially in the shape of a sheet or a thread, formed small clouds outlining highly turbulent volumes of water. Simultaneously, thin homogeneous layers, with a similar thickness as the turbulent volumes, were observed in the vertical temperature (and density) profiles.

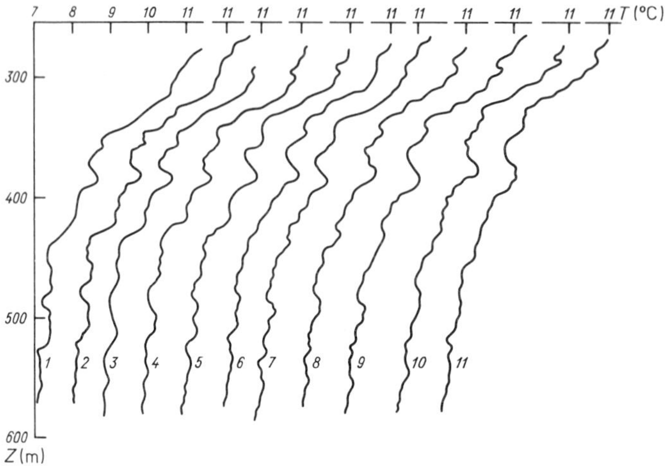

Fig. 9.10. Series of vertical temperature profiles $T_i(z)$ obtained by lowering and lifting a microstructure probing device from a station during the nineteenth cruise of the 'Dmitriy Mendeleyev'.

The qualitative picture of the relationship between the fine structure of the hydrophysical fields and the small-scale turbulence obtained by Woods was quantitatively analysed using complex measuring systems. Such measurements, taken using a towed system equipped with fluctuation and noise sensors, were carried out during the seventh expedition of the 'Dmitriy Mendeleyev' in 1973 in one of the polygons studied, at levels between 43 and 187 m (Beliayev *et al.*, 1975a). These experiments proved that the variance of the velocity fluctuations (over a frequency range 1–150 Hz and with a towing velocity of 2–3 m s^{-1}) varied regularly with the temperature gradient dT/dz in the vicinity of the point at which the fluctuations were measured. The temperature sensors were located next to the velocity fluctuation probes, at a distance of 1 and 6 m; the time constant of the temperature fluctuation sensors (thermistors) was 1 s. In most cases, the thermistors recorded the temperature as a smooth curve. According to the thermistors located at distances of 2 and 6 m down the towing line, the temperature difference ΔT remained practically constant during the course of each measurement. However, at one of the levels (No. 12) the T-records were irregular curves with sharp jumps of 2° and higher, ΔT substantially varying with time. Figure 9.11 presents a plot of temperature recorded with four thermistors in this configuration. A towing line is affected by the rolling of the ship, hence when the thermistors were moved vertically in the temperature gradient layer, they showed temperature variations with the rolling period (8–10 s) and amplitudes proportional to dT/dz. The rolling of the ship did not affect the fluctuation sensors, due to a significant difference in their frequency bands. Clearly, the rolling of the ship did not show up in thermistor records taken in isothermal layers. The horizontal parts of the records in Figure 9.11 correspond to such isothermal layers; the sharp temperature jumps indicate that the sensors crossed boundaries between layers with different temperatures. These gradient fragments also vividly show the temperature oscillations induced by the rolling of the ship. The substantial difference between some curves in Figure 9.11 proves that the vertical temperature profile varies significantly, even in layers only as thick as a few meters. With such a complex local structure of the temperature field it would be natural to suppose that great variability exists in the turbulence characteristics along the measuring track.

Moving averages of the variance of the u'-fluctuations at all measurement levels were obtained by an averaging device with a time constant of about 3 s. At the majority of the levels the local values of σ_u^2 varied between comparatively narrow margins about the mean values. In some cases, however, the curves distinctly exhibited areas with increased velocity fluctuations.

It would be natural to present the interplay of velocity variance and local background conditions as $\sigma_u^2(\mathrm{Ri})$, where Ri is the Richardson number. Calculations of local Richardson numbers in ocean conditions, however, prove to be far from trivial. If one employs data of standard hydrological observations and propeller flow measurements, then the estimates of the derivatives $d\rho/dz$ and du/dz will be quite rough. Therefore, Ri was calculated from the temperature, which was measured with thermistors fitted along the towing line, while the contribution from the salinity to the density field (ordinarily less than the contribution from the temperature) was estimated from the salinity profile averaged over standard hydrological series. The velocity gradient was estimated from the results of acoustic probing of the velocity with as minimum a space–time separation as was possible.

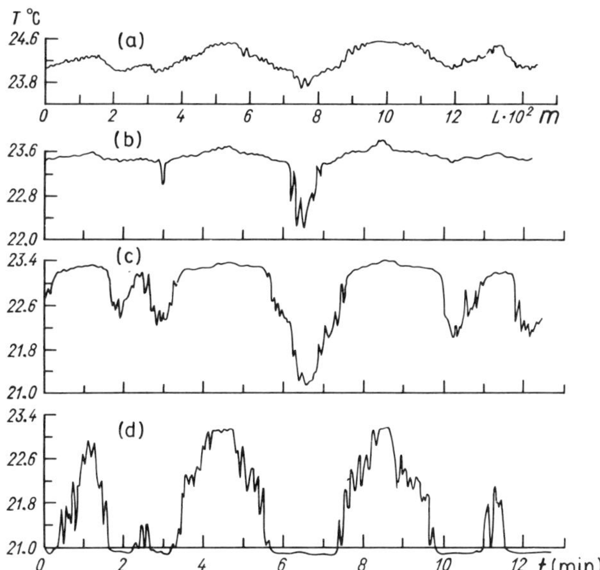

Fig. 9.11. Temperature records derived from measurements taken using a towing line with four thermistors (a–d) which are 5, 1 and 6 m apart, respectively (Beliayev *et al.*, 1975).

When plotting σ_u^2 against Ri, we took into account the difference in the mean velocities of the sensors with respect to the water. This difference resulted (for a fixed frequency band of the device) in a change of the scale range of the fluctuations. To reduce this effect, σ_u^2 was normalized by the square of the towing velocity of the device, $\bar{V}$. When the fluctuation spectrum decreases with increasing wavenumber by a cube law, then the dimensionless ratio σ_u^2/V^2 is independent of the towing velocity. Figure 9.12 shows plots of σ_u^2/V^2 versus Ri. Clearly, the dimensionless quantity σ_u^2/V^2 tends to decrease with increasing Ri. This can be approximated by a hyperbolic dependence of the type $y = ax^{-1}$. The constant a, determined by the least-squares method, appeared to equal 2.2×10^{-4}. Figure 9.12 does not present the points corresponding to the measurement levels 9 and 12 because the temperature gradients and the flow velocity at level 9 turned out to be equal to zero within measurement error, which resulted in an indefinite value of Ri, while level 12 showed a peculiar example of turbulence conditions. At this level, the local σ_u^2-values varied over a wide range. The regions of increased turbulence intensity corresponded to layers with minimum temperature gradients, i.e., the character of the σ_u^2-dependence on Ri at this level was opposite to that depicted in Figure 9.12. The highest turbulence levels corresponded to records characterized not only by high dT/dz-values, but also sharp changes of these gradients along the measuring track. These facts can be correlated with overturning internal waves. These waves induce disturbances in the initial temperature field and thus a step structure with large local values of dT/dz in certain layers. It is interesting to note that a high velocity gradient was recorded at level 12, which could have contributed to the Kelvin–Helmholtz instability in this area. In this case, the largest transverse velocity gradient must be present either at the crests or in the hollows of the internal waves, where dT/dz is also greatest. The local Richardson number can become less than the critical one and the local instability creates a spot having an increased level of turbulence. This turbulence generation mechanism was also confirmed by the fact that the period of turbulence outbursts along the measuring track at level 12 appeared to correspond closely to the length of internal waves observed in the polygon.

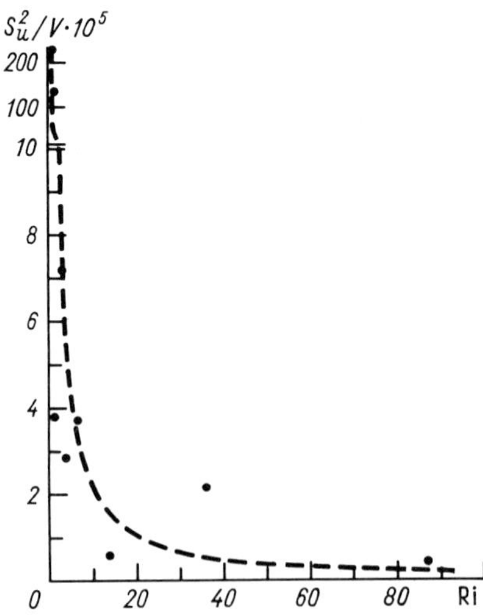

Fig. 9.12. Dependence of the normalized velocity variance on local values of the Reynolds number. The dashed line represents the approximating hyperbola (Beliayev *et al.*, 1975).

The Reynolds number was calculated in the following way. The thickness of the quasi-homogeneous layer estimated from the vertical temperature profile (Figure 9.13) was chosen as the characteristic outer scale of turbulence. The quasi-isothermal layers, varying in thickness from 1 to 35 m and including the measurement level, are marked on the temperature profile. The velocity variations in the corresponding quasi-homogeneous layer, estimated by the method of acoustic probing, were assumed to be characteristic velocity scales. The Reynolds numbers calculated in this way proved to be about 10^4–10^5, i.e., only slightly exceeding the critical Re. These results directly confirm the hypothesis that small Re numbers can occur in the ocean and that undeveloped small-scale turbulence governed by local background conditions does exist.

These conclusions were confirmed by probing with fluctuation and noise sensors. The results of those measurements give a fairly clear relationship between the vertical fine structure and the turbulence fluctuations. For example, by repeated probing of a water layer that included a density discontinuity, Beliayev and Gezentsvei (1977) proved that the Richardson number could fairly often reach the critical value, which resulted in the generation of turbulence spots. The turbulence intensity in these spots depended on the Richardson number. Quite a different quantitative dependence of σ_u^2 on Re was obtained by Pozdynin (1976), who analyzed fluctuation measurements in the Lomonosov Current (ninth cruise of the 'Akademicik Kurchatov', 1971). Measurements were carried out by the method of stepwise probing at 15–20 m intervals, with the ship under way in the direction of the current. The results of these measurements are listed in Table 9.1. The Richardson numbers were calculated from simultaneous measurements with a velocity probe and the 'Aist' probe carried out in the same region during the Atlantic tropical experiment (GATE) in 1974. Such a time separation between fluctuation and 'noise' measurements might, naturally, throw doubt on the feasibility of their comparison. Nevertheless, this procedure seems to be quite acceptable because of the stable vertical structure of the equatorial system of currents and the stationary hydrological conditions in the region. The Richardson-number profile plotted from these measurements is presented in Figure 9.14.

TABLE 9.1

Small-scale velocity fluctuations measured in the Lomonosov current during the ninth cruise of the 'Akademicik Kurchatov' (Pozdynin, 1976)

Depth of measurement (m)	Number of measurements	Root-mean-square amplitude of velocity velocity fluctuations (cm s^{-1})	Root-mean-square error (cm s^{-1})
36	245	1.2	0.06
52	285	1.4	0.06
73	250	0.8	0.04
90	336	1.2	0.05
110	336	1.1	0.04
140	78	0.8	0.07

Note: The root-mean-square error was calculated using the expression for Gaussian processes. The range of space scales of the velocity fluctuations was 2–150 cm.

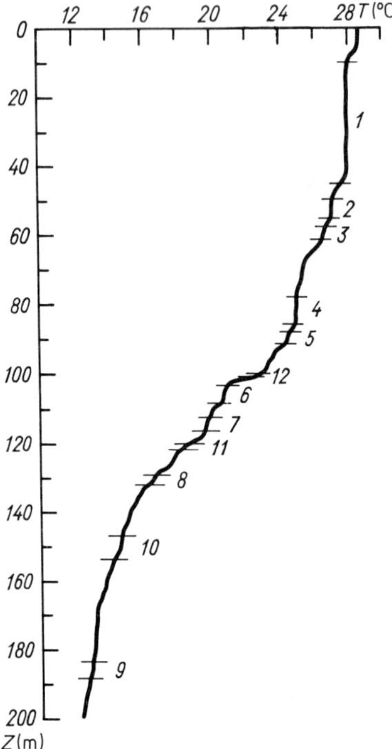

Fig. 9.13. Vertical temperature profile. Quasi-homogeneous layers in which fluctuation measurements were carried out are marked.

The dependence of σ_u^2/V^2 on Ri was also taken as being of the form $y = ax^{-b}$, but in this case the constants appeared to be: $aV = 1.7$, $b = -0.3$, where V is the velocity of the sensor with respect to the water, which is known only approximately ($\simeq 2$ m s^{-1}). This discrepancy with respect to the parameters of the hyperbolic dependence of σ_u^2/V^2 on Ri, determined above, can be explained, on the one hand, by different basic mechanisms of turbulence generation during the measurements and, on the other hand, by insufficiently accurate and asynchronous determination of the Richardson numbers employed in the analysis. It must be remembered that in, the first case, the basic mechanism of turbulence generation was, most likely, the overturning of internal waves, while in the Lomonosov Current the prevailing mechanism seemed to be shear instability of the vertical velocity gradient in plane-parallel flow.

9.4. Spectra of Fluctuation Intensity and Energy Dissipation

Interesting information concerning the properties of small-scale turbulence in the ocean can be gained from the spectra of velocity fluctuation intensity $kE(k)$ and of turbulent energy dissipation $k^2E(k)$, where k is the wavenumber and $E(k)$ the spectral density of velocity fluctuations. The function $kE(k)$ is known to be the contribution of different scales of motion to the total energy of the process. The fact that the function $kE(k)$

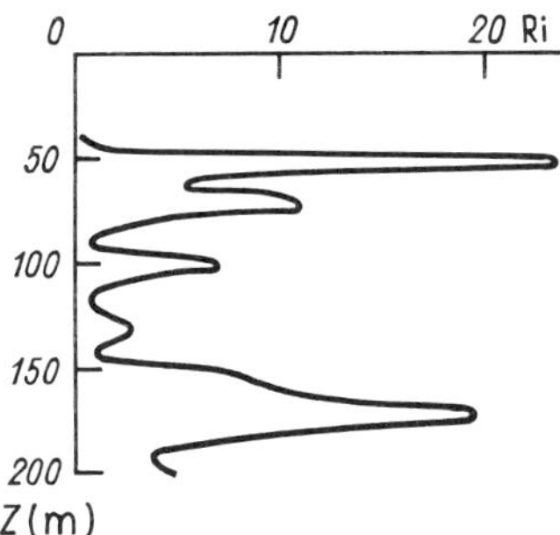

Fig. 9.14. Vertical profile of the Richardson numbers obtained in the Lomonosov Current (Pozdynin, 1976).

has maxima indicates that there exist energy-supplying regions in the wavenumber range under investigation. This allows us to estimate the characteristic scales of the energy-supply processes. At the same time, the positions of the maxima in the dissipation spectrum $k^2E(k)$ characterize the scales of the motions that convert the kinetic energy of the process into thermal energy. Integrating the function $k^2E(k)$ over the entire wavenumber range gives the total dissipation rate of kinetic energy as it is converted into heat, ϵ. If the wavenumber range includes a maximum of $k^2E(k)$, and if the values at the limits of the range are small, then the integration can be reliably carried out over the entire range of k.

The functions $kE(k)$ and $k^2E(k)$ for small-scale velocity fluctuations in the ocean were apparently determined for the first time by Grant *et al.* (1962) from the data mentioned above, which were obtained in a strait with a strong tidal flow. The spectra are given in Figure 9.15. As can be seen, the energy of the velocity fluctuations rapidly and monotonically decreases with increasing k, while the dissipation spectrum has a pronounced maximum at a value of k equal to several cm^{-1}. This behavior of the function $kE(k)$ is quite natural in a tidal flow, since the scale of the energy-supply process in the strait must correspond, e.g., to the width of the strait (1.5 km on average), which is an environment far removed from that represented by the scale range of Figure 9.15.

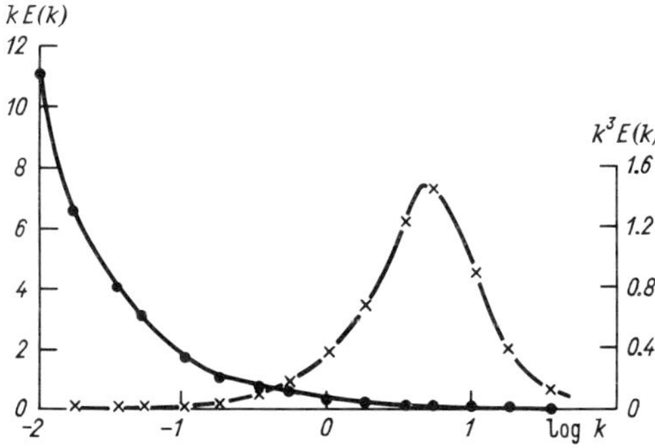

Fig. 9.15. Plots of the spectra of velocity fluctuations and of turbulent energy dissipation (Grant *et al.*, 1962).

Beliayev *et al.* (1974a) also calculated the functions $kE(k)$ and $k^2E(k)$ from data observed in various ocean regions. The calculations were carried out using data from the same records that were employed to determine the function $E(k)$. Note, however, that the functions $kE(k)$ and $k^2E(k)$ in the high-frequency part of the spectrum are more sensitive to disturbances of various kinds, and hence more often exhibit random outbursts negligible in $E(k)$. Nonetheless, the basic shape of the curves can be easily deduced. The functions $kE(k)$ and $k^2E(k)$ were calculated for various measurement levels in three polygons during the ninth cruise of the 'Akademicik Kurchatov' and the seventh expedition of the 'Dmitriy Mendeleyev'. Unlike the $kE(k)$ curve obtained by Grant *et al.*, the curve obtained from these measurements did not always decrease monotonically. In a number of cases it showed pronounced maxima, but in other cases the rate of decrease of $kE(k)$ with increasing k did not change at all. Thus, for example, in one of the polygons studied during the seventh expedition of the 'Dmitriy Mendeleyev', the function $kE(k)$ showed pronounced maxima at wavenumbers approaching 10^{-1} cm^{-1}. Less distinct maxima were discovered at somewhat larger values of k in a polygon studied during the ninth cruise of the 'Akademicik Kurchatov'. At smaller values of k, however, $kE(k)$ increased again. This pointed to a more powerful source of energy in large-scale motions, which were, however, not studied. The values of $kE(k)$ differed, as a rule, substantially at various measurement levels, and without any monotonic dependence on depth. The spectra of the velocity dissipation, analysed for each polygon, had pronounced maxima in the vicinity of $k = 1$ cm^{-1}. The values of these maxima varied from level to level and from measurement to measurement.

9.5. Turbulent Energy Dissipation Rate

As mentioned above, by integrating the function $k^2E(k)$ over the entire wavenumber range one can estimate an important turbulence parameter, namely the turbulent energy dissipation rate ϵ. Recall that ϵ can be determined by the expression

$$\epsilon = \frac{\nu}{2} \sum_{i,j=1}^{3} \overline{\left(\frac{\partial u_i}{\partial x_j} + \frac{\partial u_j}{\partial x_i} \right)^2}, \tag{9.4}$$

where u_i are the velocity components and ν is the kinematic viscosity. For isotropic turbulence, (9.4) yields for the mean value of ϵ:

$$\bar{\epsilon} = 15\nu \overline{\left(\frac{\partial u}{\partial x} \right)^2}, \tag{9.5}$$

where u is the longitudinal component of the flow velocity (along the axis $x = x_1$).

At present, many estimates of ϵ are available, most of them obtained by indirect calculation methods. The first estimates of ϵ were made with the aid of data on the damping of tidal waves. Taylor (1918) analyzed the energy dissipation rate of the tidal flow in the Irish Sea and found that $\epsilon = 8 \times 10^{-2}$ cm^2 s^{-3} (annual average). In the case of stationary wind-driven flows, Shtockman (1946, 1947) found ϵ to lie between 0.6×10^{-2} and 1.8×10^{-2} cm^2 s^{-3} for the tidal flow in the Gulf of Finland, and to approach 1.7×10^{-3} cm^2 s^{-3} for gradient flows off the west coast of the Caspian Sea. Moroshkin (1948, 1951) estimated that ϵ varied from 1.3×10^{-2} to 4.3×10^{-2} cm^2 s^{-3}

for mean conditions in the Baltic Sea, and reached 0.08–0.43 $cm^2 s^{-3}$ under stormy conditions. Taking into account the basic sources of kinetic energy in the ocean (wind, tides, and convection), and assuming that this energy dissipates in the upper 100 m layer, Noss (1957) estimated that ϵ varies from 0.3×10^{-2} to 1×10^{-2} $cm^2 s^{-3}$. Much lower ϵ-values were obtained by Suda (1936) for convection in the waters offshore Japan: $\epsilon = 10^{-4}$–10^{-6} $cm^2 s^{-3}$. Analyzing a group of ϵ-estimates in the ocean, Nan'niti (1964) came to the conclusion that in shallow waters ϵ is approximately 1–10^{-1} $cm^2 s^{-3}$ in the surface and bottom layers, and 10^{-1}–10^{-2} $cm^2 s^{-3}$ in the middle of the flow; in deep water he concluded that ϵ decreases from 10^{-3}–10^{-4} $cm^2 s^{-3}$ in the surface layer to 10^{-7} and even 10^{-8} $cm^2 s^{-3}$ at great depths. A number of estimates of ϵ were made from diffusion experiments in the ocean. Using data on discrete particle diffusion at the surface, Ozmidov (1960) found ϵ to be equal to 2×10^{-3}–6.4×10^{-2} $cm^2 s^{-3}$; the experiments were carried out in calm weather conditions with wind speeds not exceeding 4–5 m s^{-1}. When calculated from data obtained using propeller-type devices, ϵ was commonly between 10^{-2} and 10^{-4} $cm^2 s^{-3}$ (Ozmidov, 1968).

The great discrepancy in these estimates of ϵ can be attributed, first of all, to differences in the types of motion that dissipate the energy. Moreover, in different space–time regions the velocity derivatives (and, consequently, ϵ) can vary substantially, even within homogeneous types of motion, e.g., in surface waves and drift flows. There also exist random ϵ-fluctuations, associated with the turbulent character of flows in the ocean. The dependence of the mean rate of turbulent energy dissipation in the ocean on the type of water motion or, in other words, on the scale of the phenomenon, was discovered by Ozmidov (1960) and was analyzed later in detail by Nan'niti (1964), who plotted ϵ versus the scale l of the process under study. This was approximated by an expression with an exponent between –0.4 and –0.5. For the discrete power supply scheme suggested by Ozmidov (1965b), this plot must be replaced by a stepwise function with characteristic values of ϵ_i for large-, average-, and small-scale processes in the ocean. The analysis of the available ϵ-data (Ozmidov, 1968) made it possible to estimate ϵ as 10^{-1}, 10^{-3}, and 10^{-5} $cm^2 s^{-3}$, respectively, for each of these processes.

Grant *et al.* (1962) were the first investigators to estimate ϵ from direct measurements of small-scale velocity fluctuations in the ocean. They did this by comparing the experimental spectral curves with the theoretical $E(k)$. In total, they carried out 17 comparisons and obtained values of ϵ in the range 0.015–1.020 $cm^2 s^{-3}$, the mean value being 0.268 $cm^2 s^{-3}$. These comparatively large values of ϵ can perhaps be associated with the strong flow in which the measurements were made. Another series of ϵ-estimates was obtained by Stewart and Grant (1962) using the same equipment in a region characterized by weak flows and small wind-driven waves. The results of these measurements, along with the measurement levels and the wave heights, are listed in Table 9.2. The data given in Table 9.2 suggest a certain dependence of ϵ on the wave heights and measurement levels. A quantitative form for the expression of this dependence was suggested by Benilov (1973):

$$\epsilon = \gamma \frac{v^3}{\lambda} \left(\frac{h}{\lambda} \right)^3 e^{-6\pi z/\lambda}, \tag{9.6}$$

where v, h, and λ are the wave velocity, height, and length, respectively, z is the depth, and γ is a dimensionless constant, estimated by Benilov to be approximately one.

TABLE 9.2

ϵ ($cm^2\ s^{-3}$), estimated from observational data by Stewart and Grant (1962)

Measurement level (m)	Wave height (m)					
	0.1	0.2	0.3	0.4	0.5	0.6
1	–	–	–	–	0.042	–
1.5	–	0.015	–	0.023	–	–
2	0.0052	–	0.029	–	0.022	0.045
12	–	–	–	–	0.00025	–
15	0.0011	–	–	–	–	–

According to (9.6), the energy dissipation rate in surface waves must decrease rapidly with depth, which is in fairly good agreement with the data presented in Table 9.2.

Somewhat different values of ϵ estimated from observational data recorded in Discovery Strait ($\epsilon = 5 \times 10^{-1}$–$5 \times 10^{-4}$ $cm^2\ s^{-3}$) were reported by Grant at the Symposium on Ocean Turbulence in Vancouver (see Monin, 1969a). Measurements taken in fast tidal flows (Grant *et al.*, 1968) at depths of 15–24 m yielded values ranging from 17 to 0.52 $cm^2\ s^{-3}$, while in the open ocean (Grant *et al.*, 1968) $\epsilon = 2.5 \times 10^{-2}$–$3.1 \times 10^{-2}$ $cm^2\ s^{-3}$. In the Counter Current in the Pacific Ocean (Williams and Gibson, 1974), ϵ was found to be 8×10^{-2} $cm^2\ s^{-3}$; this estimate was later considered to be too high. The value of ϵ (0.1 $cm^2\ s^{-3}$) obtained on the 'Flip' platform was also overestimated. When corrected, using new calibration data, these estimates decreased to 1.5×10^{-3} $cm^2\ s^{-3}$.[4] Nasmyth (1973) analyzed fluctuation measurements taken off the Canadian coast and came to the conclusion that the mean value of ϵ in the upper 300 m of water in the area that was studied was approximately 10^{-4} $cm^2\ s^{-3}$. Similar results were obtained by Osborn (1974), who studied fluctuations of the horizontal velocity component measured with a free-sliding probe equipped with a thin vane sensor which was deflected by the water.

In the Institute of Oceanology of the U.S.S.R. Academy of Sciences, ϵ was estimated in various ways from fluctuation measurements (Beliayev *et al.*, 1973b). The first method was to calculate ϵ from (9.5) with time derivatives substituted for the spatial ones (following the frozen turbulence hypothesis). These time derivatives were calculated from finite-difference relations with a time step of 1/300 s. The second method consisted of integrating the area under the curve $k^2E(k)$. Yet another method, the one also employed by Grant *et al.*, was to determine the normalization parameters necessary for the comparison of experimental data with the universal spectral curves. Finally, a fourth method was to estimate ϵ from $\epsilon = 15\nu\sigma_u^2/\lambda^2$, where λ is the Taylor microscale.

The quantity ϵ, estimated by all these methods in various regions of the World Ocean proved to be about 1–10^{-3} $cm^2\ s^{-3}$. These comparatively high values were associated, firstly, with the properties of the regions under study (in most cases the measurements were carried out in steep velocity gradients) and, secondly, with the ignoring of records that corresponded to weak signals, close to the background noise level. As a result, in these cases ϵ was estimated mainly for turbulence 'spots'. To calculate mean values of ϵ for a region under investigation, one has to multiply the values obtained by a factor of less than unity that characterizes the intermittency of turbulence spots in the ocean.

Now, let us discuss in detail the first two estimation methods. These are, in essence, identical, since both the differential and the integral expression for ϵ are derived, based on the same assumption of the isotropy of the phenomenon. The spatial derivative in (9.5) was replaced by the finite-difference ratio $\Delta_r u/r$. Here, $\Delta_r u = u(x + r) - u(x)$, where $r = v\,\Delta t$, $x = rn$, $n = 1, 2, 3, \ldots$, and v is the velocity of the sensor with respect to the surrounding water. The time step Δt was chosen by taking into account the upper limit, $f_{\max}$, of the bandwidth of the measuring device and the size d of the sensitive element, i.e., $\Delta t = \frac{1}{2} f_{\mathrm{N}}$, where f_{N} is the Nyquist frequency determined by the relation $f_{\mathrm{N}} = \min(f_{\max};\, v/2d)$. For this time step, only the signal distortions induced by the device itself affect the estimates of ϵ. Indeed, let us present the x_1-component of the velocity fluctuations as the Fourier–Stieltjes integral

$$u(\mathbf{x}) = \int \mathrm{e}^{i\mathbf{k}\cdot\mathbf{x}}\, \mathrm{d}Z_1(\mathbf{k}), \tag{9.7}$$

where $\mathbf{x}$ is the radius vector, $\mathbf{k}$ is the wavenumber vector, $\mathrm{d}Z_1(k)$ is the random amplitude, and the integration is carried out over the entire wavenumber range. Averaging (9.7) over the sensor length d (oriented along the X-axis), and differentiating it with respect to x, we have

$$\begin{aligned} \frac{\partial \bar{u}}{\partial x} &= \frac{\partial}{\partial x}\left[\frac{1}{d}\int_{-d/2}^{d/2} u(x + y, x_2, x_3)\, \mathrm{d}y\right] \\ &= i\int \varphi\left(\frac{kd}{2}\right) k\, \mathrm{e}^{i\mathbf{k}\cdot\mathbf{x}}\, \mathrm{d}Z_1(\mathbf{k}), \end{aligned} \tag{9.8}$$

where

$$\varphi\left(\frac{kd}{2}\right) = \left[\sin\left(\frac{kd}{2}\right)\Big/\left(\frac{kd}{2}\right)\right].$$

Multiplying (9.8) by its complex conjugate, averaging the result over all possible $\mathrm{d}Z_1(\mathbf{k})$-values (assuming the amplitudes $\mathrm{d}Z_1(\mathbf{k})$ to be uncorrelated), and integrating it with respect to k_2 and k_3, we obtain

$$\overline{\left(\frac{\partial \bar{u}}{\partial x}\right)^2} = \int_0^\infty \varphi^2\left(\frac{kd}{2}\right) k^2 E(k)\, \mathrm{d}k. \tag{9.9}$$

A comparison of (9.9) with the dissipation integral shows that the distortion of the signal spectrum by the sensor can be described by the term $\varphi^2(kd/2)$. If we now require that the signal distortion level should be no more than 0.7 of the signal amplitude (0.5 for square quantities), then we obtain an upper limit for the frequency band studied. Indeed, $\varphi^2(kd/2) \geqslant 0.5$ for $kd/2 \leqslant 4\pi/9 \simeq \pi/2$. Taking into account that $k = 2\pi f/v$, we obtain $f \leqslant v/2d$. On the other hand, the substitution of (9.7) into

$$\Delta_r u/r = [u(x + r, x_2, x_3) - u(x, x_2, x_3)]$$

yields an expression that, at $r = d$, coincides with the right-hand side of (9.8). Hence,

$$\overline{\left(\frac{\partial \bar{u}}{\partial x}\right)^2} \equiv \overline{\left(\frac{\Delta_r u}{r}\right)^2}, \qquad (9.10)$$

i.e., no additional error is introduced into ϵ-estimates provided Δt is set equal to d/v at $v < 2df_{max}$. If $v > 2df_{max}$, d in this expression for Δt must be replaced by $v/2f_{max}$.

In the third method of estimating ϵ, one has to use the Kolmogorov hypothesis, which presents the spectral density of velocity fluctuations as

$$E(k) = \epsilon^{1/4} \nu^{5/4} F(\xi), \qquad (9.11)$$

where $F(\xi)$ is a universal function and ξ is a dimensionless wavenumber determined by

$$\xi = \frac{k}{\epsilon^{1/4} \nu^{-3/4}}. \qquad (9.12)$$

On a logarithmic plot, (9.11) and (9.12) can serve as expressions for a coordinate transformation that moves the system along the vector ($\ln \epsilon^{1/4} \nu^{-3/4}$; $\ln \epsilon^{1/4} \nu^{5/4}$). Dividing (9.11) by (9.12), we obtain

$$\frac{E(k)}{\nu^2 k} = \frac{F(\xi)}{\xi}, \qquad (9.13)$$

which implies that any point on the straight line $F = \xi$ must also belong to the straight line $E = \nu^2 k$. The slopes of these two straight lines are identical and equal to +1 on a logarithmic plot. A model spectrum, e.g., the one suggested by Monin (1962), is then plotted logarithmically and the straight line $\ln F = \ln \xi$ is drawn with a point (0, 0) marked on it. The straight line $\ln E(k) = 2 \ln \nu + \ln k$ is then drawn on the logarithmic plot of the empirical spectra, with the values of ν chosen depending on the water temperature and salinity in the polygon. Thereafter, both plots are moved relative to one another in order to attain the best fit of the empirical points to the universal curve. The desired estimate of ϵ is then calculated from the coordinates of the point that coincides with the calibration, based on (9.11) or (9.12).

Extensive calculations using these methods were carried out for ten polygons studied during the ninth expedition of the 'Akademicik Kurchatov' and the seventh expedition of the 'Dmitriy Mendeleyev'. The results are presented in Figure 9.16. The polygons represented in Figure 9.16 are arranged in order of decreasing mean spectral level obtained from the same experimental data. A reduction in the mean dissipation rate from polygon to polygon may be assumed to result from different hydrological conditions. Figure 9.16 shows that the vertical distribution of ϵ in each of the polygons is very non-uniform: in some cases ϵ either decreases with depth (polygons 9.6 and 7.6), or increases (polygons 9.5 and 7.7). Most common, however, are the cases where no distinct dependence of ϵ on depth is observed, and there is an appreciable scatter. This variability is probably associated with the space–time structure of the local fine-structure background conditions in the polygons. Analysis of ϵ-values for polygons with significant variations in their mean vertical density gradient has revealed no distinct dependence of ϵ on $d\rho/dz$. Unfortunately, the idea of a joint analysis of ϵ and Ri values for all the polygons cannot be realized because data on velocity gradients are lacking.

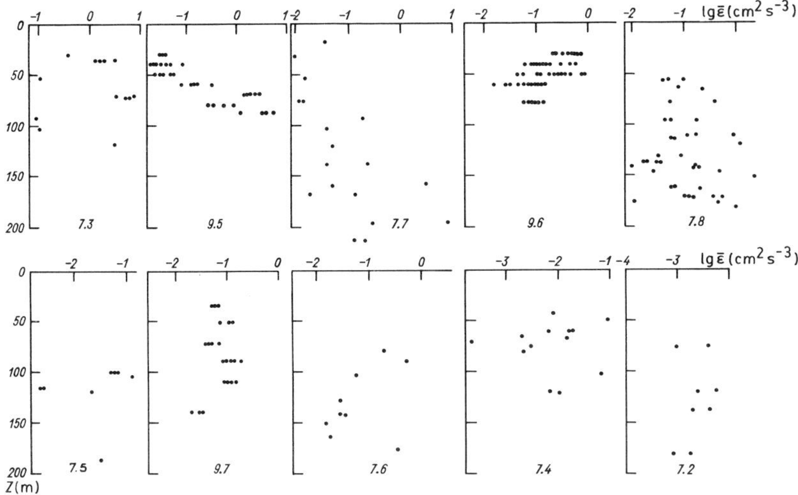

Fig. 9.16. Turbulent energy dissipation rates for polygons studied during the seventh expedition of the 'Dmitriy Mendeleyev' and the ninth expedition of the 'Akademicik Kurchatov'. Figures on the plot fragments denote the expedition and polygon numbers, respectively (Liubimtsev, 1976).

9.6. Climatology of Small-Scale Turbulence

To conclude this section, let us consider the possibility of predicting small-scale turbulence parameters from the mean hydrometeorological conditions prevailing in the region under investigation (i.e., turbulence climatology). As shown above, there exists a deterministic relationship between the statistical characteristics of turbulence and the parameters of the fine-structure fields which serve as background conditions for the evolution of turbulence. The statistical characteristics of the fine structure must, in their turn, be associated with the large-scale processes of power supply and energy distribution in the region under study. Thus, one may assume that there is an oblique two-step relationship between the large-scale hydrometeorological background conditions and the small-scale turbulence. This relationship manifests itself, naturally, only in the dependences between the large-scale background parameters and the statistical characteristics of a variety of small-scale turbulence parameters of a given region. For instance, one may try to find the relationship between the mean hydrometeorological conditions in a polygon and the turbulence intensity distribution (or moments of this distribution) at a certain fixed wavenumber k. Such a relationship between the large-scale properties of the density field in polygons and the mean value (mathematical expectation) of the random quantity E (k = 1 cm^{-1}) has been analyzed qualitatively when Figure 9.7 was discussed. However, a more detailed analysis of this kind and the establishment of a relationship between large-scale properties and the highest moments of the parameter distributions in the turbulence requires extensive statistical data. Unfortunately, the number of measurements taken in each of the polygons was not sufficient for such calculations. Therefore, we used the observational data obtained in three polygons studied during the ninth expedition of the 'Akademicik Kurchatov' and seven polygons studied during the seventh expedition of the 'Dmitriy Mendeleyev' (Beliayev *et al.*, 1974d) as an example. A group of 102 curves of one-dimensional spectral densities of the longitudinal velocity component for these polygons is presented in Figure 9.17. Each curve was obtained by averaging 5–10 individual spectra. The wavenumbers ranged from 4.6×10^{-2} to 5.9 cm^{-1} and the measurement levels varied from 20 to 213 m depth. A histogram of log $E(k_0)$, with k_0 = 1 cm^{-1}, was then plotted for a group of spectra. From this distribution we calculated the mean value m, the variance σ^2, the skewness S, the kurtosis K, and the variances $D(S)$ and $D(K)$ of the quantities S and K. These estimates are listed in Table 9.3.

The proximity of the resulting empirical distribution to the normal law was estimated by the Kolmogorov criterion. This showed a 70% probability of coincidence, i.e., quite a fair agreement. Thus, the distribution law of log $E(k_0)$ for velocity fluctuations at the scale considered appeared to be close to normal. If this law is assumed to be sufficiently

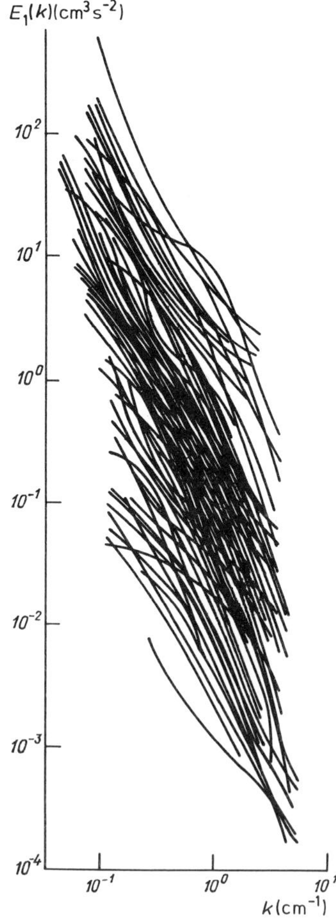

Fig. 9.17. Correlation plot of $E(k)$ obtained from measurement data taken at 10 polygons during the seventh expedition of the 'Dmitriy Mendeleyev' and the ninth expedition of the 'Akademicik Kurchatov'.

universal, then the relationship between the mean conditions and the turbulence parameters in polygons is reduced to that between these conditions and the moments of the distribution.

Relationships of this type can be found not only for the parameters of the distribution laws for spectral energy levels, but also for the characteristics of the shapes of the curves (for instance, their slope when plotted logarithmically). For a universal spectrum, the slope depends on the part of the spectrum it corresponds to; for a non-universal spectrum it depends on the factors governing the spectral shape.

TABLE 9.3

Parameters of the probability distribution of $\log E(k_0)$ for velocity fluctuations, estimated at $k_0 = 1\ \text{cm}^{-1}$

m	δ^2	S	K	$D(S)$	$D(K)$
−0.94	0.64	−0.09	0.01	0.06	0.20

A necessary condition for locally isotropic turbulence at a scale range l in the presence of a free boundary (the ocean surface) is the explicit geometrical relation $h/l > 1$, where h is the depth of the measurement level. This relation holds in all measurements analyzed. Hence, in the upper mixed ocean layers, where the density gradients are shallow and the turbulence level is high, turbulence spectra often exhibit parts that obey the 5/3-power law. In stratified layers, the large-scale boundary of the inertial range can shift towards smaller scales due to the effects of buoyancy forces. If one estimates the buoyancy scale using the relation $L_0 = \epsilon^{1/2} N^{-3/2}$, where N is the Väisälä frequency (Ozmidov, 1965c), then the quantity L_0 proves to vary from fractions of a centimeter to tens of meters at reasonable values of ϵ and N. Thus, in a number of cases, the spectrum in question may belong entirely or partially to the buoyancy range. Molecular viscosity effects can be traced in turbulence spectra starting with eddy sizes of about 8η, where η is the Kolmogorov scale. For $\nu = 10^{-2}\ \text{cm}^2\ \text{s}^{-1}$ and with $\epsilon = 10^{-4}$–$1\ \text{cm}^2\ \text{s}^{-3}$, the quantity η varies from 0.03 to 0.3 cm, and hence the effects of viscosity can manifest themselves in seawater at inhomogeneity sizes of up to 2.4 cm. Variations in the viscosity ν due to temperature and salinity changes can increase this estimate by a factor of 1.3. Therefore, at the scale range considered, different universal turbulence conditions may exist, characterized by various slopes (exponents) $-\alpha$ of the spectral curves (assuming that $E(k) \simeq k^{-\alpha}$). Experimental spectral curves fail, as a rule, to be approximated by a single power-law dependence throughout the entire wavenumber 'window'. Hence, the $\log k$-range from −1 to 0.5 was divided into six equal subranges, in which the spectral slope (in a logarithmic plot) can be considered to be constant. In each of these subranges the slopes α of all the spectral curves depicted in Figure 9.17 were estimated to vary over a wide range, i.e., from 0.5 to 4.5 (Beliayev *et al.*, 1976). The histograms of the α-distributions were plotted for each of the $\log k$-subranges, with the values of α subdivided into classes with ranges as wide as 0.5. Table 9.4 presents the mean value m, the root-mean-square deviation σ, the skewness (S), and the kurtosis (K) of the distributions of α.

The distributions obtained in this way were used to plot two-dimensional probability density distributions of the spectral slopes of $\log E(k)$ (Figure 9.19). The numbers on

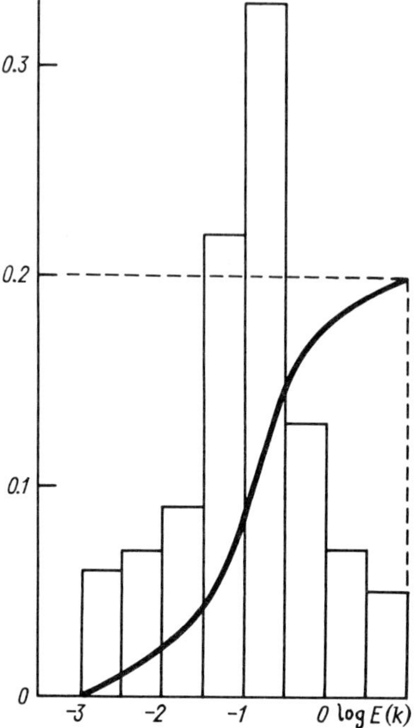

Fig. 9.18. Histogram of $\log E(k_0)$ at $k_0 = 1\ \mathrm{cm}^{-1}$, from data in Figure 9.17. The solid line is the integral distribution curve.

TABLE 9.4
Parameters of the probability distribution of α.

Distribution parameter	Number of subrange and boundary values of $\log k$					
	1	2	3	4	5	6
−1	−0.75	−0.5	−0.25	0	0.25	0.5
m	1.91	1.91	1.83	2.07	1.87	2.29
δ	0.76	0.62	0.79	0.76	0.70	0.85
S	0.23	1.42	0.82	0.26	0.54	0.19
K	−0.78	1.47	0.48	−0.53	0.24	−0.55

the isolines denote the probability density at the corresponding scales of turbulence. When plotting the isolines, we used a rectangular grid, with a step of 0.25 along the $\log k$-axis and a step of 0.5 along the α-axis. The values in each of the subranges were transferred to the centers of the corresponding grid. The isolines were drawn by linear interpolation between neighboring points. Figure 9.19 shows that in subranges 2 and 3 slopes between 1.5 and 2.0 are most frequent, while in subrange 1 the probability density of α has a relative minimum. In subranges 4 to 6 the isolines generally deflect towards larger α-values at smaller scales of turbulence. This behavior can be attributed to the inertial range of turbulence prevailing in subranges 2 and 3, buoyancy effects in

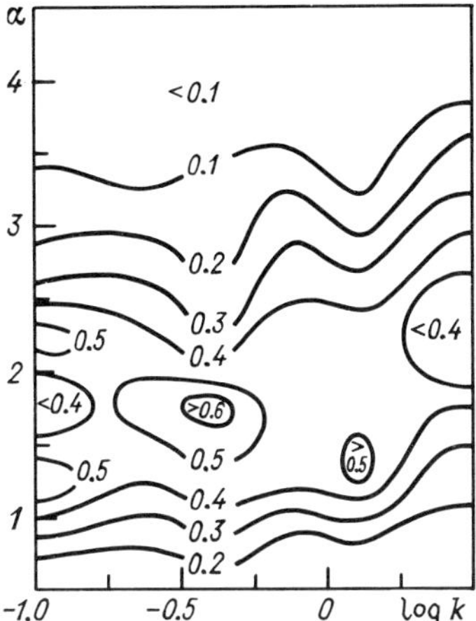

Fig. 9.19. Isolines of the two-dimensional probability density of the slope of velocity fluctuation spectra.

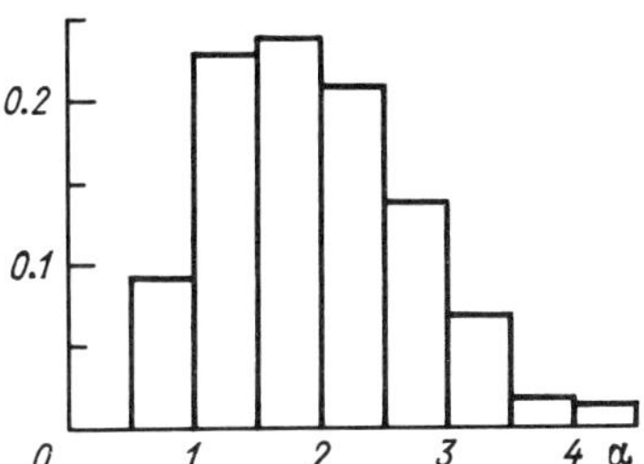

Fig. 9.20. Overall histogram of the probability distribution of the slope of velocity fluctuation spectra (Beliayev *et al.*, 1976).

subrange 1 (with a high probability of $-11/5$ slopes), and viscous effects in subrange 6 (a steeper slope, i.e., increasing values of α). In undeveloped turbulence, the slopes of the velocity spectra can vary over wide ranges and at times can be less than one. According to Figure 9.19, values of α close to 1 are comparatively common throughout almost the whole scale range studied (subranges 1 to 5). This is in agreement with the frequent occurence of low Re for small-scale turbulence in the ocean. Figure 9.20 presents a summary histogram of the distributions of α in all subranges of the k-axis. A characteristic feature of the histogram is its skewness. On average, the most probable value of α is 1.5–2.0. However, often one can observe other spectral slopes. These result, as stated above, from different conditions of the small-scale turbulence in the ocean.

10. TEMPERATURE FLUCTUATIONS

10.1. An Indirect Method of Estimating Temperature Fluctuations

Temperature fluctuations in a turbulent flow, T', are largely secondary compared with the velocity fluctuations. Indeed, if a flow has a gradient of mean temperature $\overline{T}$, movement of the fluid relative to a motionless temperature sensor in a direction not parallel to the isotherms will be interpreted by the sensor as a temperature fluctuation. On the other hand, in a flow without a temperature gradient even intensive turbulent velocity fluctuations cannot create any significant temperature fluctuations. Molecular temperature flucturations (caused by chaotic molecular motion) will induce T variations in flows with a homogeneous temperature and in motionless liquids. The variations induced in this way, however, can be neglected when compared with turbulence fluctuations. The temperature fluctuations in a turbulent flow with a temperature gradient can be determined by the expression

$$T' = (\text{grad}\, \overline{T} \cdot \mathbf{u}')t_0, \tag{10.1}$$

where $\mathbf{u}'$ is the velocity fluctuation vector, and t_0 is the maximum period of velocity fluctuations. The quantity t_0 can be determined in another way, i.e., as the period (characteristic time) that is used to divide turbulence fields into averages and fluctuations. In this case, the bar over T denotes an average of the temperature field over a time scale t_0. The time average can, of course, be replaced with the corresponding spatial average or by a more general probability average (over an ensemble of results). Expression (10.1) shows that the fluctuation T' can depend significantly on the averaging scale. Generally speaking, u_i'-values increase with increasing scale values, but grad $\overline{T}$ can decrease in this situation. This accounts for the fact that there is probably no unambiguous scale dependence for turbulent temperature fluctuations, unlike that for turbulent velocity fluctuations. As seen from (10.1), in order to determine T' one must know grad $\overline{T}$ and $\mathbf{u}'$ (with t_0). Hence, T' in a turbulent flow can be found without direct measurements, i.e., by way of calculation. However, it does not mean that T' need not be measured directly, since, firstly, the determination of grad $\overline{T}$ and $\mathbf{u}'$ in the ocean is even more complex than measurements of T' and, secondly, direct estimates are always preferable to indirect ones.

The most natural scale to use for dividing any hydrophysical field in the ocean into an average value and fluctuations is that at which the minimum spectral density of the field occurs. If the spectrum does not exhibit such a minimum, then the choice of t_0 becomes largely arbitrary, and the procedure of dividing the field into an average and fluctuations depends, in each particular case, on the problem being solved, the bandwidth of the device employed, etc. Therefore, we do not initially give t_0 a value, but first consider the available data concerning grad $\overline{T}$ in the fine structure of temperature in the ocean and the characteristics of the temperature fluctuations at various averaging scales.

10.2. Local Temperature Gradients in the Ocean

Temperature gradient distributions computed from measurements taken using low-inertia probes were plotted by Korchashkin (1976, 1977). The division of the vertical temperature profiles into sections, each with an approximately constant temperature gradient, was carried out by a program with the following division criteria:

$$T_i - \widetilde{T}_i \leqslant \Delta T, \qquad \frac{\text{grad}\, T_i - \text{grad}\, \widetilde{T}_i}{\text{gr}\widetilde{\text{a}}\text{d}\, T_i} \leqslant \Delta\, \text{grad}\, T_i,$$

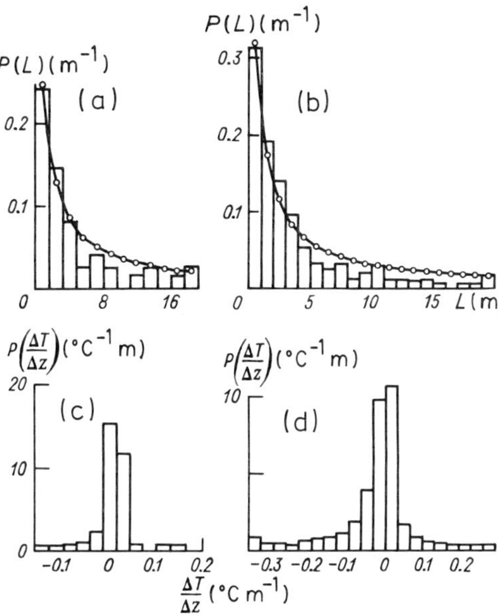

Fig. 10.1. Empirical probability densities of vertical temperature gradients (c, d) and layer thicknesses, with approximating hyperbolic curves (a, b) in Antarctic waters (a, c) and the equatorial region (b, d) of the Pacific Ocean (Korchashkin, 1976).

where ΔT is the absolute error in temperature measurements, Δ grad T_i is the relative error in the determination of the temperature gradient, $\widetilde{T}$ and $\widetilde{\text{grad}}\ T$ are linear (relative to depth) approximating functions calculated by the least squares method (by discrete points), $i = 1, 2, 3, \ldots$ are the sample numbers, and Δ grad T is taken as being equal to 0.5. Figure 10.1 presents the distributions of layer thickness and temperature gradient obtained during the eleventh expedition of the 'Dmitriy Mendeleyev' in Antarctic waters and in equatorial latitudes of the Pacific Ocean. Here, the empirical distribution of layer thicknesses can be approximated by a hyperbolic curve (see Figure 10.1). This law may be considered to be an approximation of the more general lognormal law over a limited range of its argument. The lognormal law, in turn, is apparently quite a general law describing the distributions of the small-scale inhomogeneities of hydrophysical fields in the ocean (fine-structure layer thicknesses, turbulence spots, energy dissipation, etc.). This fact can, apparently, be associated with a general property of the generation mechanisms of these inhomogeneities, which is similar to the mechanism of successive independent crushings of particles (Kolmogorov, 1941a). The temperature gradient distributions, however, are too complex to be described by a few parameters. In this case, the type of law and its parameters seem to depend in an essential way on the large-scale hydrological conditions of the region under investigation. For example, Figure 10.1 reveals essential differences between the vertical temperature gradients in the equatorial Pacific Ocean and in Antarctic waters. The same conclusion can be drawn concerning the mathematical expectations m, the root-mean-square deviations σ, and the third and fourth normalized moments S and K for a number of stations in this region (Table 10.1).

The mean gradients for the equatorial zone are positive, and equal to 0.01 °C m^{-1} in the layer 0–900 m while in Antarctic waters they are negative, and equal to –0.03 or –0.04 °C m^{-1} in the layer 0–430 m. At the same time, the scatter of grad $\overline{T}$-values in the equatorial region is somewhat greater than in the Antarctic. The most significant difference between these two regions is found in the skewness of the grad T-distributions.

TABLE 10.1

Statistical parameters of the probability distributions temperature gradients for a number of stations in the equatorial Pacific Ocean and in Antarctic waters (Korchaskin, 1976).

Station no.	m °C m^{-1}	σ °C m^{-1}	S	K	Number of layers
Equatorial region. Probing depth 430 m:					
1	–0.04	0.08	–3.74	20.15	187
2	–0.04	0.08	–3.69	19.11	180
3	–0.03	0.05	–3.63	19.75	173
4	–0.04	0.06	–2.90	12.00	174
5	–0.04	0.07	–3.13	12.20	155
6	–0.03	0.06	–4.12	22.58	166
7	–0.04	0.08	–3.13	15.70	184
8	–0.03	0.06	–3.10	13.38	174
9	–0.03	0.07	–4.05	23.96	185
Antarctic waters. Probing depth 900 m:					
1	0.01	0.06	1.85	17.30	80
2	0.01	0.06	3.79	29.24	79
3	0.01	0.05	–0.47	8.60	87
4	0.01	0.05	1.15	19.11	82
5	0.01	0.05	0.84	11.29	76

It would be also interesting to consider the joint probability distributions of layer thickness L and temperature gradient. This may be useful, for instance, in choosing the averaging scale and t_0 for (10.1). Volochkov and Korchashkin (1977) computed the two-dimensional probability densities for L and grad $\overline{T}$ from data taken from the equatorial Pacific Ocean and used above. The isolines depicted in Figure 10.2 are curves of equal joint probability densities of layer size and temperature gradient. The most probable layer size is approximately 2.5 m, with temperature gradients of about –0.03 to –0.04 grad m^{-1}. Note that 5–8% of all layers with quasi-constant temperature gradients exhibit values of positive grad $\overline{T}$. The hydrostatic stability in such layers, where the thickness does not exceed 4–5 m, is apparently ensured by increasing salinity. A characteristic feature of the isolines of equal probability is their approach towards the mean gradient as the layer thickness increases. Besides the maximum at 2.5 m, one can also observe an increase in the joint probability density on scales of 8–9 m and 12–13 m.

Empirical integral distribution functions of $(\Delta T/\Delta z)^2$ computed from the data obtained by probing in Antarctic waters are presented on a logarithmic scale in Figure 10.3. The solid line shows the lognormal distribution, which is seen to approximate the experimental data well.

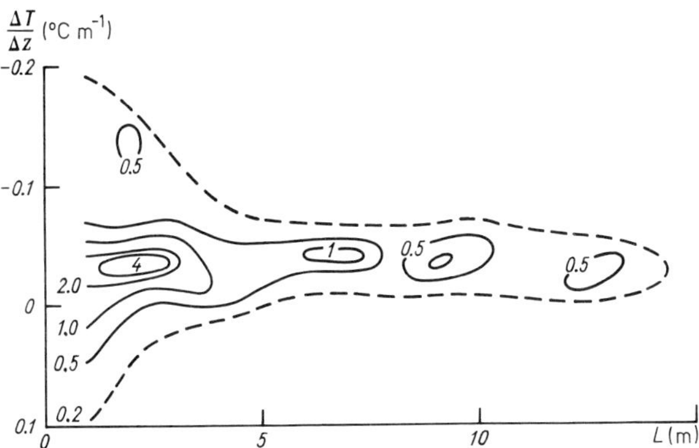

Fig. 10.2. Two-dimensional probability densities of layer thicknesses and temperature gradients derived from measurement data in the equatorial region of the Pacific Ocean (Volochkov and Korchashkin, 1977).

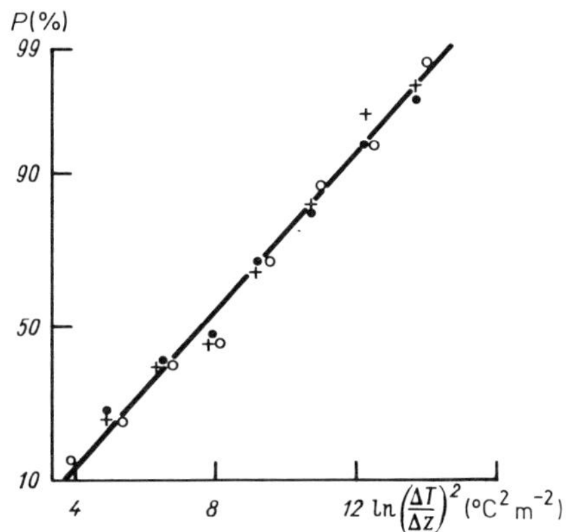

Fig. 10.3. Integral distribution function of the square of the vertical temperature gradient according to three probings (denoted by different marks) in Antarctic waters.

10.3. Variations in Fine-Structure Temperature Profiles

The space–time evolution of fine-structure temperature profiles was studied in detail during the fifteenth expedition of the 'Dmitriy Mendeleyev' in the area of the Pacific Southern subarctic front. Figure 10.4 gives an example of a temperature profile averaged over repeated probings, complemented by the standard deviation $\sigma_T(z)$ of the temperature measured in separate probings. The value of $\sigma_T(z)$ changes from 0.06 to 0.44°C. Its greatest values are concentrated in the layer from 145 to 185 m rather than in the layer with the maximum temperature gradient, which was located at 100–130 m depth. The profile of $\overline{T}(z)$ exhibits the fine structure of the temperature field, including a number of inversion layers.

As a quantitative measure of 'unevenness' in $\overline{T}(z)$ one may use the Cox number:

$$C^2(z) = \overline{\left(\frac{\partial T'}{\partial z}\right)^2} \Big/ \left(\frac{\partial \overline{T}}{\partial z}\right)^2, \tag{10.2}$$

where $T'(z) = T(z) - \overline{T}(z)$ is the temperature deviation from its mean value at the level z and the averaging is carried out over a set of $T_i(z)$ values (i is the number of the probing). Computations showed the maximum value of $C(z)$ to be 3.3, while the general value was close to unity.

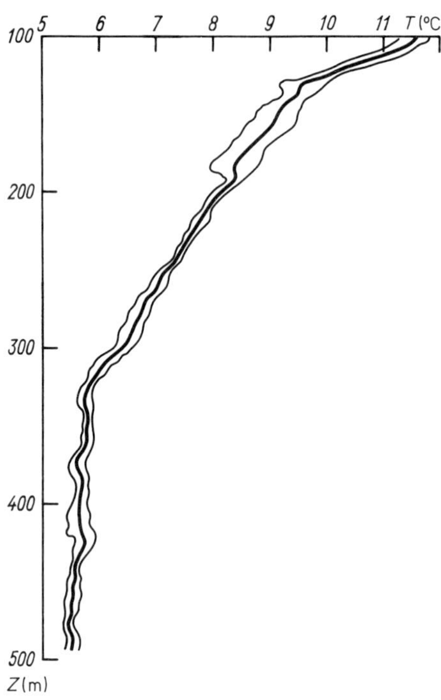

Fig. 10.4. Average vertical temperature profile and root-mean-square temperature scatter derived from repeated probings in the North-West Pacific Ocean (Beliayev and Gezentsvei, 1977).

The same probing data were used to calculate the spectra of the vertical temperature gradient for the layer at 100–500 m depth. The spectra decrease with increasing wavenumber. On a logarithmic scale, their slope is, on average, steeper when $k > 0.05$ than when $k < 0.05$ cycle m^{-1}. As an example, Figure 10.5 shows two spectra of the temperature gradient in the wavenumber range from 10^{-2} to 3×10^{-1} cycle m^{-1}, derived from data obtained in two probings. For large k-values, similar spectra were calculated from electrical conductivity fluctuations measured simultaneously with a sensor attached to the probe. The electrical conductivity fluctuations were assumed to depend linearly on the temperature fluctuations, and the effect of the salinity S was allowed for by the consideration of barometric data concerning the average profile of S. The results of the analysis for the probings mentioned above are presented in Figure 10.5 for the layers at depths of 200–210 and 480–490 m. The spectra of small-scale variations in temperature

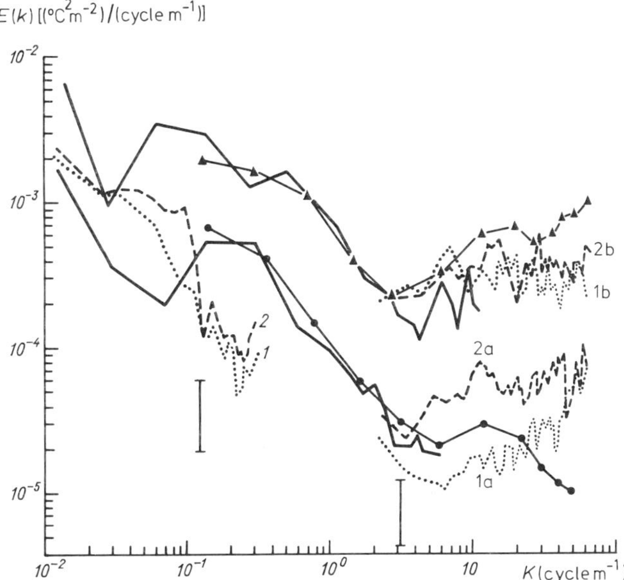

Fig. 10.5. Spectra of vertical temperature gradient variations in the ocean derived from two probings (1 and 2) in the Pacific Ocean. The letters a and b refer to the layers 200–210 and 480–490 m, respectively. Solid lines present the analog curves by Gregg *et al.* (1973). Vertical lines show the 95%-reliability interval (Beliayev and Gezentsvei, 1977).

gradients appear to be very different. This may be regarded as further hard evidence for the vertical intermittency of high-frequency processes in the ocean. The spectra obtained in this way range from 1.5 to 50 cm.

When analyzing spectra of the temperature gradient over a wide wavenumber range, attention should be paid to the existence of several wavenumber ranges with different slopes of the spectra. This is obvious, not only from the data mentioned earlier, but also from the spectra obtained by Gregg *et al.* (1973) offshore San Diego (the upper curves in Figure 10.5) and in the central part of the northern subtropic gyre of the Pacific Ocean (the lower curves). According to Gregg *et al.*, the spectra of temperature gradients can be divided into three ranges, which correspond to three different mechanisms of generation of the temperature field. For $k < 10^{-2}$ cycle m^{-1} the shape of the spectrum is determined by the exponential decrease of the mean temperature with depth. For $10^{-2} < k < 5$ cycle m^{-1} the shape of the spectrum is determined principally by regions of large temperature gradients, distributed randomly in the vertical plane. For $k >$ 5 cycle m^{-1} it is the microstructure of the temperature field, associated with medium-scale turbulence, that manifests itself in the spectrum. The relative maxima of the spectra in the microstructure range observed in Figure 10.5, however, can hardly be attributed to viscous diffusion effects. Those effects seem to be present in a shorter-wave part of the spectrum. The maxima in Figure 10.5 are perhaps caused by small-scale temperature inhomogeneities induced by internal waves.

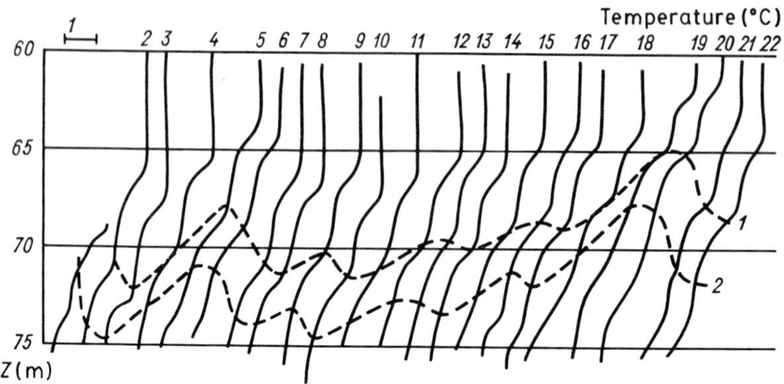

Fig. 10.6. Vertical temperature profiles obtained by repeated probings in the Japan Current (Beliayev and Gezentsvei, 1977). (1) Isotherm of 26.5°C, (2) isotherm of 27.5°C. The probing numbers are given at the top.

The substantial space–time variability of temperature gradients induced by internal waves was studied in detail during the sixtieth expedition of the research ship 'Vityaz' in the South China Sea (Beliayev and Gezentsvei, 1977). Repeated probings were carried out in a comparatively thin layer (60–75 m depth), which included the transition region between the upper mixed layer and the layer of density discontinuity. Figure 10.6 illustrates the results of the first 22 probings (out of a total of 125) carried out at intervals of 108 s. Figure 10.6 shows that the isotherms of 26.5 and 27.5°C clearly delimit the oscillating fluid motion. These oscillations (internal waves) shift the temperature profiles up and down as a whole, but change them very little. Sometimes, however, the changes can be more pronounced. This was the case in probings 38–39 and 62–63. The internal wave period was close to 6–10 min. The analysis of all 125 probings, however, revealed still stronger large-scale isotherm oscillations, with periods of several hours. The amplitudes of the short-period oscillations were 1–3 m, those of the long-period oscillations reached 12.5 m. Such oscillations must naturally result in varying temperature gradients, and hence in varying temperature fluctuations at fixed points. In the case under discussion, the distributions of the vertical shifts of the isotherms appeared to be nearly Gaussian (Figure 10.7). According to data obtained during the sixtieth expedition of the 'Vityaz', local values of the vertical temperature gradient in the discontinuity layer can reach 0.96 °C m^{-1}, the average value being 0.4–0.6 °C m^{-1}.

The Cox number defined in (10.2) can be employed to characterize the microstructural activity in the bulk ocean. Under the assumption that the generation of temperature fluctuations by vertical turbulence in a mean temperature gradient is balanced by the molecular dissipation of these fluctuations (Osborn and Cox, 1972), the relation

$$\overline{w'T'}\,\frac{\mathrm{d}\overline{T}}{\mathrm{d}z} = -\chi\,\overline{\left(\frac{\partial T'}{\partial z}\right)^2} \tag{10.3}$$

must hold, where w' is the vertical velocity component and χ is the molecular heat conductivity. If the vertical heat flux can be characterized by a vertical turbulent exchange coefficient K_z, i.e., under the assumption that

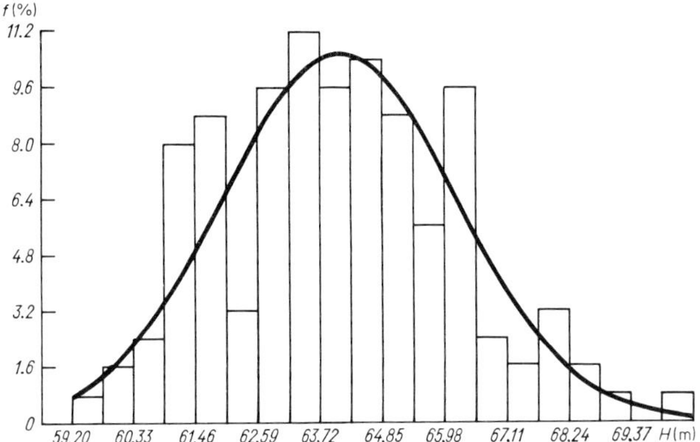

Fig. 10.7. Histogram of lower boundary depths of the upper isothermic layer from experimental data obtained during the sixtieth expedition of the 'Vitiyaz'. The histogram is approximated by a normal distribution.

$$\overline{w'T'} = -K_z \frac{\partial \overline{T}}{\partial z}, \tag{10.4}$$

then the expression (10.3) reduces to the form

$$K_z \left(\frac{\partial \overline{T}}{\partial z} \right)^2 = \chi \overline{\left(\frac{\partial T'}{\partial z} \right)^2} \tag{10.5}$$

or

$$K_z = \chi \left[\overline{\left(\frac{\partial T'}{\partial z} \right)^2} \Big/ \left(\frac{\partial \overline{T}}{\partial z} \right)^2 \right] = \chi C^2, \tag{10.6}$$

where C^2 is the Cox number. From the Cox number and its variability one can therefore judge the turbulence intensity, and even try to identify the processes that are responsible for microstructure inhomogeneities in the ocean (Gregg *et al.*, 1973; Osborn and Cox, 1972; Gargett, 1976; Gregg, 1975; Gregg and Cox, 1972; Hayes *et al.*, 1975). The parameter

$$C' = \sqrt{\overline{(T')^2}} \Big/ \frac{d\overline{T}}{dz},$$

which is closely related to the Cox number, was also calculated by Plakhin and Fedorov (1972) and by Garnich and Miropolsky (1974).

Now we shall illustrate the results of a similar analysis, made by Lozovatsky (1977) using data from a special experiment in the Baltic Sea. Repeated probings (a total of 130 profiles) were carried out every 3 min from the surface down to the seabed, in 17.5 m of water. The sampling interval used to generate the profiles was 10 cm and the time lag of the temperature sensor was 0.1 s. The vertical temperature gradient varied from 0 in the upper homogeneous layer to 4 °C m^{-1} in the temperature discontinuity layer.

Temperature gradient spectra were computed separately for the surface layer (up to 10 m in depth) and the deeper layer. In these layers the series of discrete gradient values can be considered to be homogeneous. The spectra of the vertical temperature inhomogeneities, obtained by multiplying the temperature gradient spectra by the frequency response curve of the first-order difference filter, are shown in Figure 10.8. For scales $l < 0.7$ m, the spectra can be approximated by a k^{-3}-law, while for inhomogeneities with vertical scales exceeding 1 m the slopes of the spectra increase and the spectra are better approximated by a k^{-4}–k^{-5}-law. The shapes of the spectra for the surface and deep layers are practically the same, but their levels differ significantly. The high level of temperature inhomogeneities in the lower layer can apparently be attributed to an increase in the temperature gradient.

In order to calculate the Cox number, the temperature field was divided into an average and fluctuations at the 0.7 m averaging scale, which was chosen in accordance with the point at which the curves of the spectra in Figure 10.8 exhibit a break. Figure 10.9 presents the mean gradient $d\overline{T}/dz$ in the form of vertical profiles calculated from 85 successive probings. The chart shows substantial changes in the $d\overline{T}/dz$ field that occurred during the measurement period. A stable layer with high temperature gradients, up to 3 °C m^{-1}, was present at a depth of 13–16 m by the end of the first hour of measurements. The analysis of the hydrological situation in the region led to the conclusion that this phenomenon was caused by the advection of colder, saltier water from the north along the sea bed. Between probings 65 and 85 one can see a train of short-period internal waves, with a period of approximately 12 min and a mean amplitude of 30 cm, in the layer deeper than 10 m. In the surface layer, the temperature gradient also changed noticeably with time. In the layer at a depth of 4–6 m, the gradient increased from 0.1–0.2 to 0.6–0.8 °C m^{-1}, while in the layer between 7 and 10 m, on the contrary, it decreased from 0.4–0.6 to 0.1–0.2 °C m^{-1}.

According to the measurement data from the Baltic Sea, the Cox number proved to be small and to vary negligibly, both vertically and in time. The C^2-values approached unity, which was evidence for weak microstructure activity and, consistent with (10.6), indicated an insignificant turbulence intensity during the period of measurement. Based on data obtained in a few other seas and in the open ocean, it is known that C^2-values can vary over wide ranges, i.e., from 10^{-1} to 10^3 (Gargett, 1976; Gregg and Cox, 1972; Gregg *et al.*, 1973; Hayes *et al.*, 1975; Seidler, 1974). In most cases, however, these authors report that C^2 ranges from 1 to 10. This proves that the varying intensity of vertical turbulence in the ocean and the different roles of turbulence in the generation of fine-structure temperature fields exist. In the case of the fairly stable density stratification that was observed during measurements in the Baltic Sea, the primary role in the process was apparently played by internal waves. This conclusion is confirmed by calculations of the Cox number, which can also describe the intensity of resonant interactions of internal waves in a flow (Bell, 1974). For $C^2 > 1$, the process is characterized by strong nonlinear interactions and can be termed turbulent. For $C^2 \ll 1$, there is a long resonant interaction time and temperature inhomogeneities probably result from motions caused by internal waves. Figure 10.10 presents the C^2-spectrum, calculated from data collected by probing, for the layer 10–16 m depth. The main contribution to the variance of C^2 is made by fluctuations with scales up to 1 m, while C^2 itself, calculated from the spectrum, appears to be less than one. Hence, these data also prove that turbulence

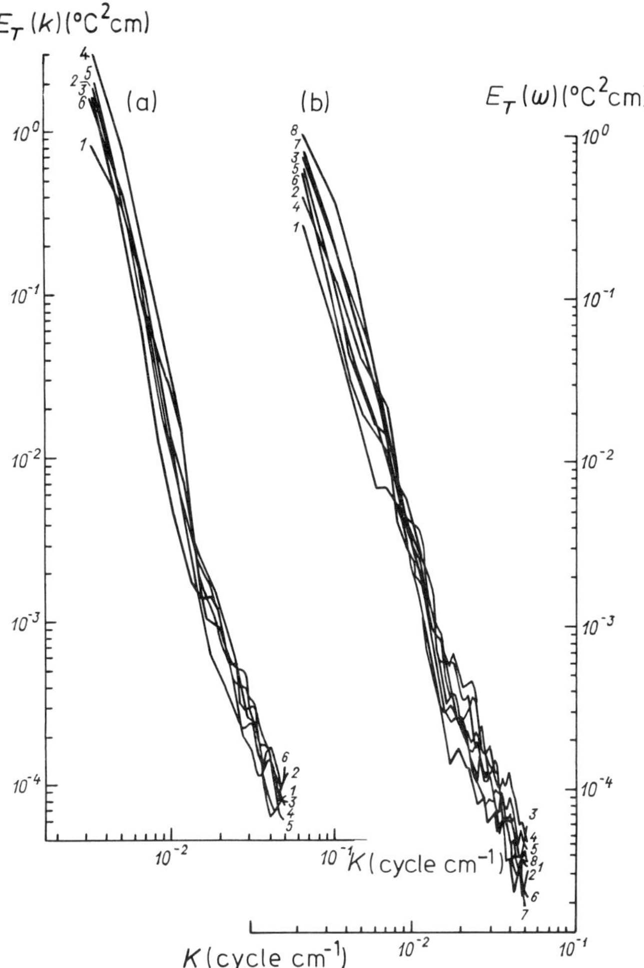

Fig. 10.8. Average spectra of temperature inhomogeneities for the upper (a) and lower (b) layers according to repeated probings in the Baltic Sea (Lozovatsky, 1977). (1) probings 1–16, (2) 17–32, (3) 33–48, (4) 49–64, (5) 65–80, (6) 86–101, (7) 102–117, (8) 118–130.

makes a negligible contribution to the generation of temperature fluctuations in the case under consideration.

For other regions of the ocean and in different hydrometeorological conditions, this conclusion is, of course, not true in general. Indeed, in the open ocean and beneath the layer of discontinuity, the mean temperature gradients, as shown above, do not usually exceed 0.01 °C m^{-1}, and in a number of cases have an even smaller order of magnitude. Nevertheless, in most cases the Cox number is large, which points to the significant contribution that turbulence makes to the creation of thermal fields. In keeping with (10.1), various combinations of mean gradients and turbulence intensities can result in different fluctuations T'. If at an averaging scale of about 1 m (and, accordingly, a value of t_0 of approximately 1 s) one chooses 1 cm s^{-1} to be a characteristic

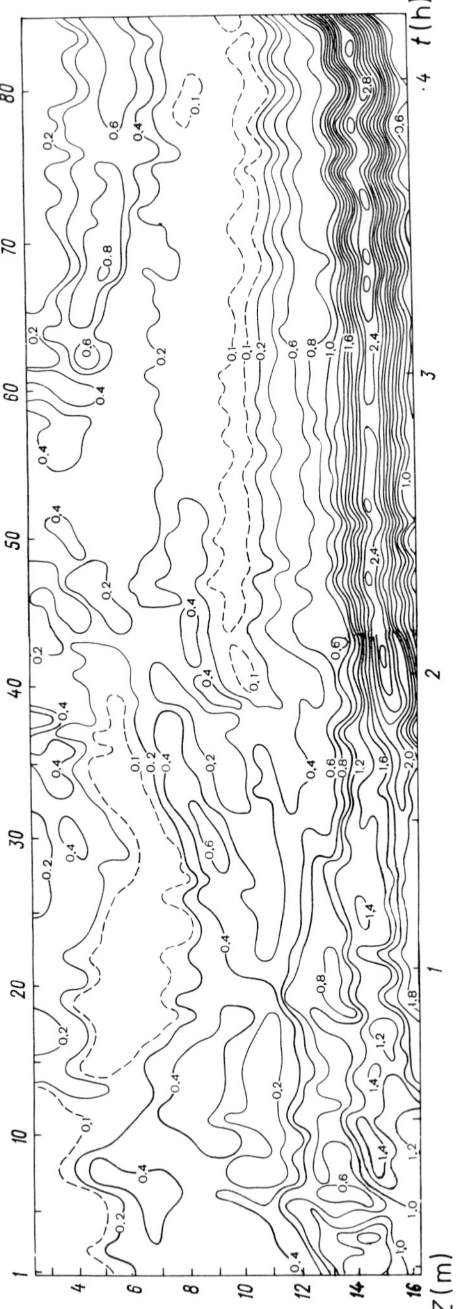

Fig. 10.9. Vertical profile of the mean temperature gradient in °C m^{-1} from data obtained by repeated probing in the Baltic Sea (Lozovatsky, 1977). The probing numbers are given at the top.

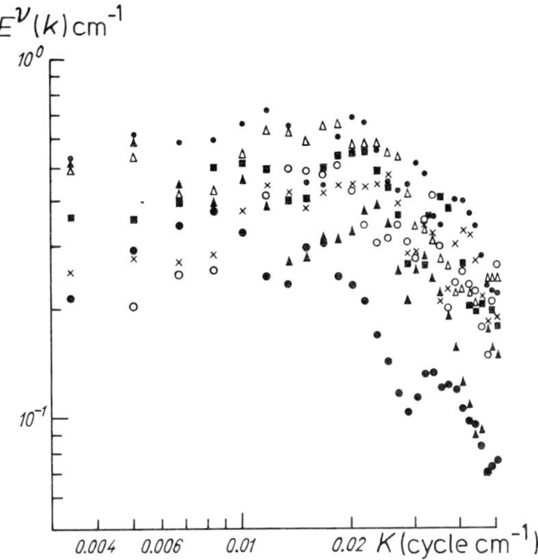

Fig. 10.10. Spectra of microstructure fluctuations in the normalized vertical temperature gradient, from measurements obtained during 16 successive probings in the Baltic Sea (Lozovatsky, 1977).

value of v' and 0.01 °C m^{-1} as a typical value of $d\bar{T}/dz$ in the bulk ocean, then T' will be about 10^{-4} °C. At the same characteristic value of the velocity fluctuations but with $d\bar{T}/dz$ = 0.1 or 1 °C m^{-1}, we obtain a value of T' equal to 10^{-3} and 10^{-2} °C, respectively. It would be interesting to compare these indirect T' estimates with those obtained by direct measurements carried out in the ocean.

10.4. Direct Measurements of High-Frequency Temperature Fluctuations

Direct measurements of high-frequency temperature fluctuations in the ocean were begun in the late 1950s and early 1960s. Unfortunately, the bandwidths and the time lag of the devices employed, as well as the averaging scales, were quite different. Moreover, this information was not always specified in publications. Using measurements obtained with 'turbulence meters' in the Caspian and Black Seas (Kontoboitseva, 1958; Kolesnikov, 1959), $\sqrt{\overline{T'^2}}$ was estimated to be 4×10^{-3}–6×10^{-2} °C. The same authors discovered an important dependence of $\sqrt{\overline{T'^2}}$ on the wind velocity over the sea (measurements were carried out in the surface layer). Similar estimates were also obtained by Speranskaya (1960). Measurements with small-inertia sensors carried out in specialized expeditions of the Institute of Oceanology in various regions of the World Ocean resulted, as a rule, in $\sqrt{\overline{T'^2}}$ estimates ranging from 10^{-3} to 10^{-1} °C. The maximum $\sqrt{\overline{T'^2}}$-values were most often discovered in the layer of temperature discontinuity, where the local temperature gradients, as shown above, can be several degrees per meter. In this case, $\sqrt{\overline{T'^2}}$ = 10^{-1} °C, which requires the vertical velocity fluctuations to be several centimeters per second. These values of w' can apparently be attributed to the turbulence-generating effect of overturning internal waves.

10.5. Turbulent Heat Fluxes

Simultaneous measurements of T' and $\mathbf{V}'$ at one point (or, more precisely, at distances much smaller than the correlation radius of the fluctuating signal) makes it possible to calculate the turbulent heat flux directly; this is determined by the expression

$$Q_i = c_p \rho \overline{T'u_i'}. \tag{10.7}$$

Here, C_p is the specific heat of water at constant pressure, ρ is the density, and $i = 1, 2, 3$. Simultaneous measurements of this kind in the high-frequency range of the spectrum are quite complex; therefore, only a few have been carried out so far. Kolesnikov (1959) was the first to try to estimate $\overline{T'w'}$ and, with $\overline{T'w'}$ and $d\overline{T}/dz$ values, to calculate the vertical exchange coefficient K_z^T from

$$K_z^T = -\frac{\overline{T'w'}}{d\overline{T}/dz}. \tag{10.8}$$

According to measurements taken under the ice of Lake Baikal, K_z^T was 0.5–0.6 cm² s⁻¹ in the layer between 5 and 15 m, where the density stratification of water was unstable, while in layers with stable stratification K_z^T sharply decreased. Having processed the data derived from measurements of fluctuations in the Counter-Current in the Pacific Ocean, Williams and Gibson (1974) estimated that Q_3 varied from 5.9×10^{-3} to 11.7×10^{-2} J cm^{-2} s^{-1} (the quantity c_p was taken as 3.9 J g °C^{-1}). The K_z^T-values estimated from these data appeared to range from 0.52 to 27 cm^2 s^{-1} (the local temperature gradients in the region were approximately 10^{-3} °C m^{-1}). Using other measurement data, Gibson *et al.* (1974) estimated K_z^T as 66–94 cm^2 s^{-1}. However, due to a high noise level of the equipment and discrepancies in its calibration, these data were revised. It was admitted that they were unreliable. The precise, direct measurement of heat fluxes (and other phenomena) induced by small-scale turbulence in the ocean still remains a problem to be solved.

10.6. Spectra of High-Frequency Temperature Fluctuations

Spectra of high-frequency temperature fluctuations in the ocean were calculated by Grant *et al.* (1968). Their estimates of the spectral density of temperature fluctuations $E^T(k)$ are presented in Figure 10.11. As can be seen, the experimental points confirm the inertial-convective range $E(k) \simeq k^{-5/3}$; the viscous-convective range where, following Batchelor's (1959) theory, $E^T(k) \simeq k^{-1}$; and, finally, the viscous-diffusive range, where the spectrum rapidly decreases with increasing k. The theoretical expressions for the spectra and the temperature correlation functions have been confirmed by a number of workers using data measured with devices of various types in different ocean regions. Thus, as far back as the 1950s, Japanese investigators plotted temperature correlation functions from measurement data for the waters offshore Japan. As found out by Inoue (1952), Nan'niti (1957, 1962), Nan'niti and Yasui (1957), and Hikosaka and Higano (1959), the correlation functions were fairly well approximated by power laws at arguments ranging from one minute to many days, with an exponent often close to the one theoretically predicted for locally isotropic turbulence. At the same time, oscillations with periods that were associated with different internal waves were also

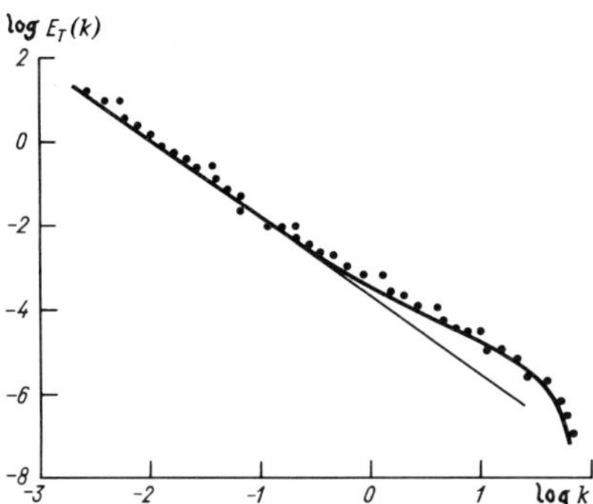

Fig. 10.11. Temperature fluctuation spectrum in the ocean, from Grant *et al.* (1968). Experimental points are approximated by universal curves.

found in the correlation functions. Similar results were obtained by Haurwitz *et al.* (1959) and by Roden (1963) from long-term temperature measurements near the Bermuda Islands and in the Pacific Ocean. The temperature spectra obtained by these workers showed peaks at the tidal and inertial frequencies, as well as peaks at internal-wave frequencies that were determined by density gradients in the region investigated. The 2/3-power law for the structure functions of large-scale temperature fluctuations was confirmed by Piskunov (1957), Nemchenko and Tishunina (1963), and some other workers. However, according to Kontoboitseva (1962) and Speranskaya (1964), the 2/3-power law does not always hold. Thus, for instance, Kontoboitseva proved that the laws of locally isotropic turbulence hold for the structure functions only at arguments smaller than 0.2–1.2 s.

Temperature spectra have been plotted for a number of polygons from measurement data obtained during expeditions of the Institute of Oceanology. Figure 10.12 presents groups of spectral curves for polygon 6, investigated during the ninth expedition of the 'Akademicik Kurchatov' (in the Atlantic Ocean, 23° S, 30° W) and for the equatorial polygon investigated during the eleventh expedition of the 'Dmitriy Mendeleyev' (in the Pacific Ocean, 165° E). In the Atlantic Ocean, measurements were carried out at levels from 30 to 77 m in depth, below the layer of density discontinuity located at a depth of 20 m. The velocity vector in the layer under study was nearly constant in amplitude and rotated through an angle of 45° counterclockwise with depth. In the Pacific Ocean, the region investigated was characterized by a thick homogeneous layer (about 170 m thick) and a marked layer of density discontinuity (2.5×10^{-1} density units per meter). The flow structure was complex, and was characterized by large velocity gradients in the layer 10–50 m and by comparatively small ones in the layer 50–350 m. Temperature fluctuations were measured at depths of 23 and 185 m.

In Figure 10.12, the temperature spectra at $k = 1$ cm^{-1} vary approximately from 10^{-6} to 10^{-4} °C^2 cm, without any monotonic dependence on depth. This can be attributed to the vertical microstructure of the temperature field and to the vertical intermittency of the velocity fluctuations. The shapes of the spectra differ from one polygon to another. The spectra obtained for the equatorial polygon exhibit two elements, with a

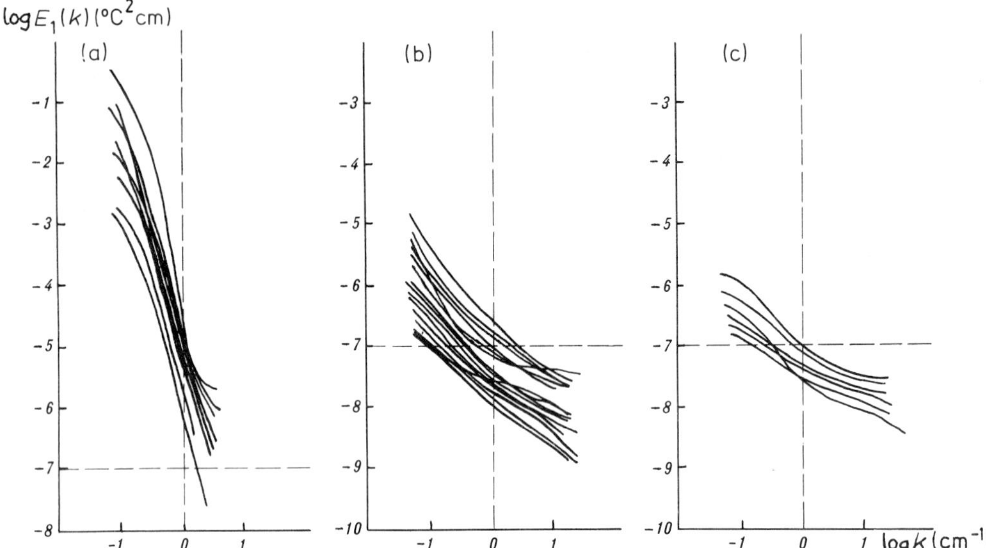

Fig. 10.12. Spectra of temperature fluctuations, from data obtained in the Atlantic Ocean (a) and in equatorial latitudes of the Pacific Ocean (b, c).

distinct difference in slope. For $k \leqslant 7 \times 10^{-1}$ cm^{-1}, the slope is close to $-5/3$, while for $k > 1$ cm^{-1} it is close to -1. The break in the curve is more pronounced in the spectra with high energy levels. The spectra exhibit parts that obey the universal laws of the inertial-convective and viscous-convective ranges. In the Atlantic polygon the spectra become steeper with increasing wavenumber, approaching $-5/3$ only at the beginning and the end of the k-range discussed, while in the middle of the range the slope is as high as -3. This behavior of the spectrum leads one to the conclusion that both the microstructure of the temperature field and internal waves are of principal importance in the generation of $E^T(k)$. When internal waves cause a comb-shaped set of fine-structure temperature jumps to pass the sensor, the signal induced will have a spectrum that is determined both by the internal wave characteristics and by the fine structure of the temperature field. Taking a few assumptions concerning these characteristics, Garrett and Munk (1971) obtained for $E^T(k)$ an expression proportional to k^{-2}. However, in more general expressions for the characteristics of the microstructure of the temperature field, the temperature spectrum can be proportional to other powers of k (Lozovatsky, 1978).

10.7. Spectral Characteristics of the Temperature Variability in the Ocean

To identify and clarify any possible contamination of the temperature fluctuation spectra from microstructure effects and internal waves, it would be expedient to consider certain features of temperature fluctuation spectra over a wider range of wavenumbers. Let us discuss this problem, as illustrated by the measurement data obtained in one of the polygons studied in the seventh expedition of the 'Dmitriy Mendeleyev' (Ozmidov *et al.*, 1974). Temperature fluctuations in the polygon were detected with a number of devices in various windows of space–time scales. The large-scale spatial structure of the temperature field in the polygon was investigated by hydrological mapping techniques. These data were used to plot the T-profile down to a depth of about 2000 m. The profile showed that the depth of isothermal layers decreased towards the south-east. This indicated an upward shift of water in the region studied. The mean temperature gradient in the surface 100 m layer was 2×10^{-2} °C m^{-1}, while in the discontinuity layer (located at a depth of approximately 150 m) it increased to a maximum of 7×10^{-2} °C m^{-1}. This is a comparatively low value; the mean temperature gradients measured in the equatorial polygon in the Indian Ocean on the same expedition of the 'Dmitriy Mendeleyev' reached 0.25 °C m^{-1}.

Large-scale variations in the temperature field in the polygon were also measured with photothermographs fitted to a buoy and stationed at depths of 100, 150, 200, and 400 m. The sampling interval of the photothermograph was 5 min and the measurement period was 63 h. The resulting series of T-values were used to calculate the temperature fluctuation spectra. At the three upper levels the functions $E^T(k)$ were practically the same, while at a depth of 400 m they were considerably lower. The functions $E^T(k)$ at depths of 100, 150, and 200 m were in good agreement with the high-frequency elements of those computed from data obtained by a radio-thermobuoy with temperature sensors located at depths of 103, 115, 126, 138, 150, 170, 212, and 232 m; the sampling interval was 12 s. Figure 10.13 presents a group of spectral curves computed from recordings taken over 4 h. As in the case of the data obtained with photothermographs, the frequency spectrum was transformed into a spatial one by the frozen-turbulence hypothesis. Data concerning the flow velocities obtained using both propeller devices fitted to a buoy station and acoustic probing techniques were used.

A characteristic feature of the fluctuations $E^T(k)$ in Figure 10.13 is their rapid decrease with increasing k. The slope of the $E^T(k)$-curves reaches -2.5 to -3. A slight maximum of the spectra in the range $k = 10^{-3}$ cm^{-1} is induced by internal waves in the region, which were discovered in layers with steep density gradients. Figure 10.14 illustrates the energy distribution as a function of wavenumber and depth. The isolines are plotted here in °C^2 s^{-1} units; in order to convert them to °C^2 cm^{-1} units, one must multiply the numbers on the isolines by $V/(2\pi)$, where V is the mean flow velocity, which approaches 10 cm s^{-1} in the case considered. Figure 10.14 shows several maxima in the spectral density of the temperature fluctuations, at which one may expect generation of temperature fluctuations and subsequent propagation of these into regions of z- and k with smaller values of $E^T(k)$. In the case considered, the principal region of energy concentration is at the 140 m-level, where the function $E^T(k)$ reaches a maximum at $k = 10^{-3}$ cm^{-1}. It is interesting to note that the steepest density gradient in the polygon was found at the same depth. Similar calculations carried out to obtain the temperature spectrum for another series of measurements, recorded 1.5 h later, resulted in a similar pattern of predominant maxima, while the remaining isolines of $E^T(k)$ underwent noticeable changes. For example, a secondary maximum with coordinates z, k at 220 m and 10^{-3} cm^{-1} was not found in the second series of measurements, while the maximum at 110 m and 8×10^{-4} cm^{-1} appeared to be somewhat shifted and lowered.

Still finer structure of the temperature field in the polygon was analyzed by repeated probing with the 'Aïst' probe and by hydrotrawl mapping. The features of the tem-

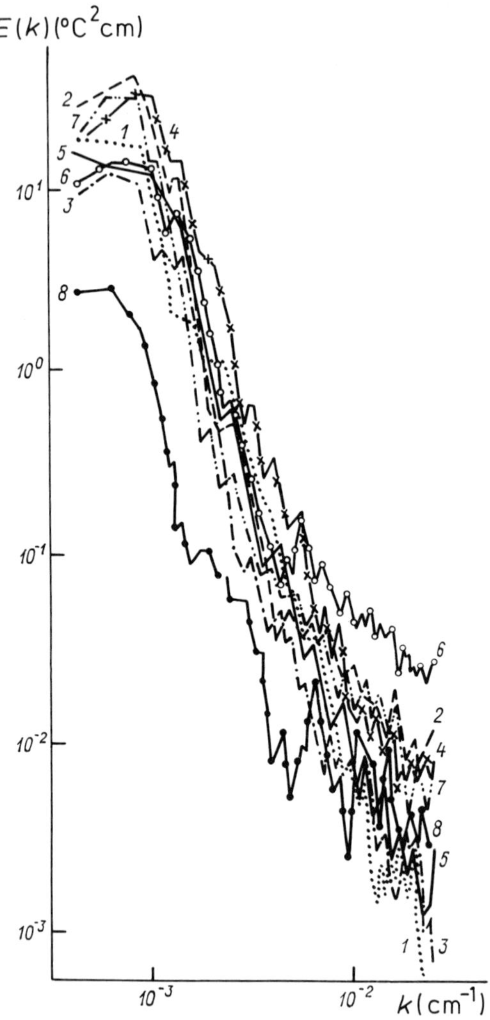

Fig. 10.13. Spectra of temperature fluctuations, from data obtained using a radio-buoy during the seventh expedition of the 'Dmitriy Mendeleyev'. (1) level 103 m, (2) 115 m, (3) 126 m, (4) 138 m, (5) 150 m, (6) 170 m, (7) 212 m, (8) 232 m.

perature field can also be easily traced on the spatial profile plotted from data obtained from ten thermotrawl sensors at 70 cm intervals (Figure 10.15). Figure 10.15 shows isotherms plotted every 0.1 °C. The sampling interval of every sensor was 3 s. The profile shows a layer with significant vertical temperature variations, which were characterized by relatively high-frequency T-fluctuations, developing against a background of slower trends in the isotherm pattern, with wavelengths of approximately 300 and 700 m.

These examples convincingly demonstrate the complexity and variability of temperature fields in the ocean. Against the background of large-scale, mean climatic structures

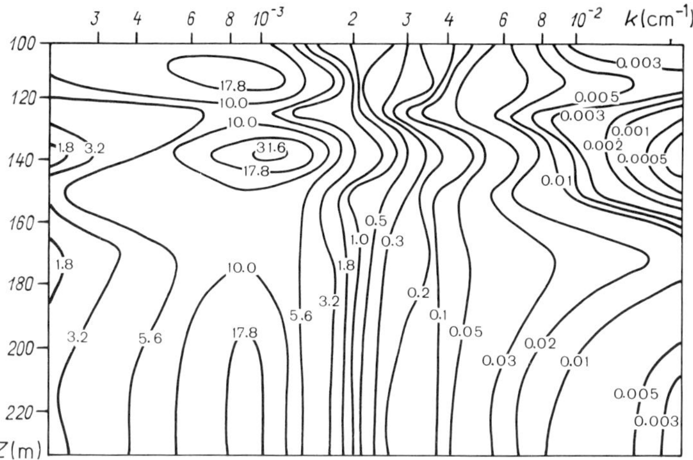

Fig. 10.14. Isolines of temperature fluctuation spectra (in K^2 s), from data obtained using a radio-buoy in a polygon during the seventh expedition of the 'Dmitriy Mendeleyev'.

there evolve medium-scale processes induced by weather and tidal forces. These are, in turn, affected by internal gravitational waves with higher frequencies, which interact with the stepwise structure of the T-profiles. And, finally, the small-scale turbulent temperature fluctuations that arise, persist, and then decay, are highly responsive to local background conditions. All these processes are interconnected directly, indirectly, or stochastically. For example, the large-scale density stratification governs the frequency limit of internal waves. Upon losing their stability, these waves give rise to spots of highly turbulent fluid, and thus to steps on the vertical temperature profiles. Small-scale turbulence, in its turn, tends to mix meighboring layers, which makes them isothermal. But this process can develop only to a certain degree, because of external forces and the flux of new amounts of energy into the bulk of the fluid. This accounts for the complex picture of non-equilibrium in the thermal field, which is constantly changing in time and space.

10.8. Dissipation Rate of Temperature Inhomogeneities

To conclude, let us consider the rate of destruction (or 'dissipation rate') of temperature inhomogeneities in the ocean, ϵ_T. Like the turbulent energy dissipation rate, this quantity can be estimated in several ways. One of these methods is based on estimating ϵ_T from spatial temperature derivatives:

$$\epsilon_T = 2\chi \sum_{i=1}^{3} \overline{\left(\frac{\partial T}{\partial x_i}\right)^2}, \tag{10.9}$$

where χ is the thermal diffusivity. If the field is assumed to be isotropic, then (10.9) reduces to

$$\epsilon_T = 6\chi \overline{\left(\frac{\partial T}{\partial x}\right)^2}. \tag{10.10}$$

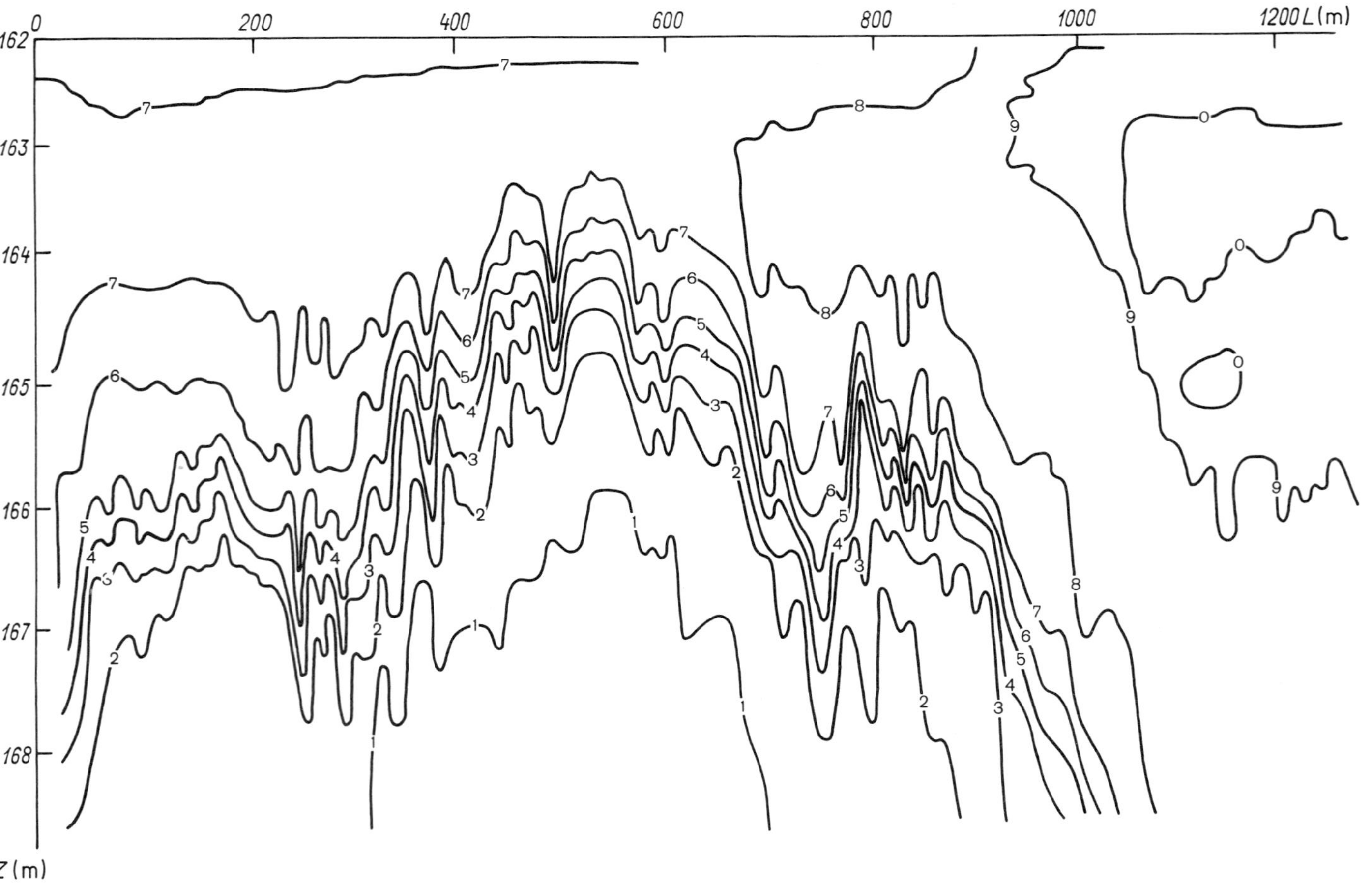

Fig. 10.15. Spatial temperature profile from data obtained with a hydrotrawl in a polygon studied during the seventh expedition of the 'Dmitriy Mendeleyev'. Figures indicate isotherms plotted per 0.1° C.

When time derivatives are substituted for spatial ones, (10.10) becomes

$$\epsilon_T = 6\chi V^{-2}\left(\frac{\partial T}{\partial t}\right)^2, \tag{10.11}$$

where V is the mean velocity at the point of measurement.

The following two methods of determining ϵ_T employ expressions for the fluctuation spectra in the inertial-convective range,

$$E^T(k) = c_1\epsilon^{-1/3}\epsilon_T k^{-5/3}, \tag{10.12}$$

and in the convective-diffusive range,

$$E^T(k) = c_2\epsilon_T \nu^{1/2}\epsilon^{-1/2}k^{-1}. \tag{10.13}$$

Here, ϵ is the turbulent energy dissipation rate and c_1, c_2 are universal constants. In order to use these expressions, one must know the value of ϵ in the same space–time region.

One further way of estimating ϵ_T is based on the integration of the dissipation spectrum of temperature fluctuations:

$$\epsilon_T = 6\chi\int_0^\infty k^2 E^T(k)\,\mathrm{d}k. \tag{10.14}$$

Estimating ϵ_T from fluctuation measurements in a tidal flow, Grant *et al.* (1968) found values that varied from 9.7×10^{-5} to 2.9×10^{-6} °C^2 s^{-1}. At the International Symposium on Ocean Turbulence (Vancouver, 1968) Grant (see Monin, 1969a) reported ϵ_T to be 5×10^{-4}–7×10^{-8} °C^2 s^{-1}. According to Gregg *et al.* (1973), who investigated the northern Pacific Ocean using a free-sliding probe, ϵ_T varied from 10^{-5} to 10^{-7} °C^2 s^{-1} at different depths. Having reviewed all the available measurements of ϵ_T, Gibson *et al.* (1974) concluded that $\epsilon = 10^{-5}$–10^{-9} °C^2 s^{-1}. From their own measurements taken using low-inertia sensors, Williams and Gibson (1974) obtained values between 7×10^{-5} and 8×10^{-6} °C^2 s^{-1} in the equatorial Counter–Current in the Pacific Ocean and 4×10^{-8} °C^2 s^{-1} in the California Current. From measurement data obtained from the anchored 'Flip' platform, Gibson *et al.* (1974) estimated ϵ_T, computed using (10.12), to be 1.62×10^{-5} and 3×10^{-5} °C^2 s^{-1}. Later, having made the calibration of the device more precise, they reduced these estimates to 5 - 6×10^{-8} °C^2 s^{-1}. They pointed out, however, that the latter value was not quite reliable either.

The quantity ϵ_T was estimated from (10.11) using data obtained during expeditions of the Institute of Oceanology, with a sampling interval for temperature fluctuations of 1/300 s (the derivative $\mathrm{d}T/\mathrm{d}t$ was replaced by the ratio of finite differences). Table 10.2 presents computed values of ϵ_T, averaged over a number of record lengths at different measurement levels, and the standard deviations of ϵ_T. These numbers are based on data obtained in one of the polygons in the Atlantic Ocean. The significant scatter in the ϵ_T-values from level to level and the large standard deviation σ_{ϵ_T} are evidence of the considerable intermittency of temperature fluctuations in the ocean. Due to the considerable intermittency of ϵ_T, its mean value also appears to be variable over comparatively small averaging periods. That is why no unambiguous relation between $\bar{\epsilon}_T$ and the

TABLE 10.2

Computed rate of temperature inhomogeneity dissipation in a polygon in the Atlantic Ocean (Beliayev *et al.*, 1973)

Depth (m)	$\bar{\epsilon}_T$ °C^2 s^{-1}	σ_{ϵ_T} °C^2 s^{-1}
30	6.8×10^{-8}	3.4×10^{-8}
40	4.2×10^{-6}	1.6×10^{-6}
50	8.4×10^{-7}	3.0×10^{-7}
60	2.6×10^{-6}	8.0×10^{-7}
77	8.6×10^{-6}	2.2×10^{-6}

large-scale hydrological conditions in the polygon has been found. However, simultaneous measurements of high-frequency temperature fluctuations and microstructure T-profiles have revealed a dependence of ϵ_T on local background conditions. Measurements of this kind were carried out under the leadership of Beliayev during the fifteenth expedition of the 'Dmitriy Mendeleyev' (1975) in the north-west Pacific Ocean. Figure 10.16 illustrates the results of three temperature probings carried out from a drifting ship. The time intervals between the probings were 20 and 15 min, respectively. The temperature scale is plotted only for probing III; the vertical marks on the other profiles correspond to 10°C. As seen from Figure 10.16, the temperature field in the polygon had a pronounced microstructure, which varied slightly from probing to probing (note that between probings I and III, the ship drifted as far as 1 mile). The layers of temperature inversion are also of interest; the depths of these changed from measurement to measurement.

The quantity ϵ_T was computed from (10.11) using data derived from measurements with fluctuation sensors during the same probings. The results are presented in Figure 10.17 as continuous ϵ_T profiles. On average, ϵ_T varied from 3×10^{-7} to 7×10^{-7} °C^2 s^{-1}

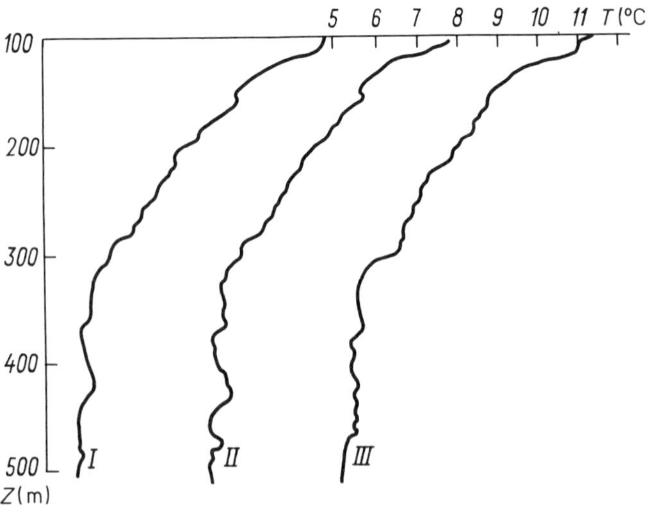

Fig. 10.16. Vertical temperature profiles, derived from data of three successive probings (I, II, III) carried out during the fifteenth expedition of the 'Dmitriy Mendeleyev'.

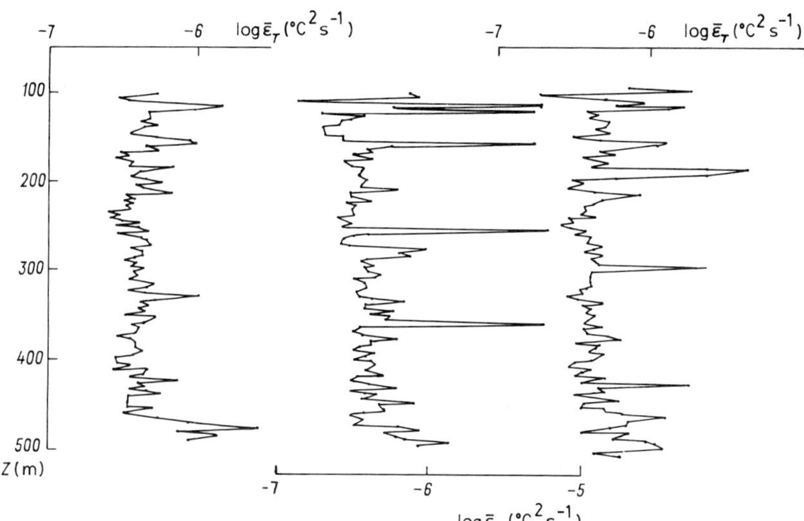

Fig. 10.17. Vertical profiles of the dissipation rate of temperature inhomogeneities, according to three successive probings in the fifteenth expedition of the 'Dmitriy Mendeleyev'.

in the layer between 100 and 500 m depth. At certain levels, however, it deviated considerably from these estimates. It is interesting to note that these deviations, just like those in the turbulent energy dissipation rate, are primarily found in temperature inversion layers. This phenomenon can apparently be associated with hydrostatic instability and with turbulence generation in temperature inversion layers, provided that for some reason the resulting decrease of density with depth is not compensated for by an increase in water salinity.

11. FLUCTUATIONS OF ELECTRICAL CONDUCTIVITY AND SALINITY

11.1. Fundamentals

Fluctuations of electrical conductivity and salinity, just like those of temperature, are secondary compared with velocity fluctuations. In the presence of gradients of electrical conductivity C and salinity S, the fluctuations can be approximately described by the expressions

$$C' = (\text{grad}\,\bar{C} \cdot \mathbf{u}')t_0; \qquad S' = (\text{grad}\,\bar{S} \cdot \mathbf{u}')t_0. \tag{11.1}$$

Here, as in (10.1), $\mathbf{u}$ is the velocity fluctuation vector and t_0 is the longest period of velocity fluctuations.

The electrical conductivity of ocean water is a complex function of temperature, salinity, and pressure. Nevertheless, comparatively small fluctuations of C can be approximated by a linear combination of the T- and S-fluctuations, with an accuracy sufficient for practical purposes:

$$C' = \alpha_1 T' + \beta_1 S'. \tag{11.2}$$

Here, the coefficients α_1 and β_1 are, in general, functions of pressure (and hence depth). When measurements are carried out at a fixed (or negligibly changing) depth, α_1 and β_1 may be assumed to be constant. They can be found in tables, by referring to the known mean T- and S-values for the corresponding measurement level. Moreover, if we assume that the temperature and salinity fluctuations at the point of measurement are proportional to their mean gradients, then the contributions of T and S to the electrical conductivity fluctuations will be determined by the corresponding gradients of these fields. Under average thermochalinic conditions in the ocean, a 1°C change in temperature results in an electrical conductivity fluctuation that is approximately equivalent to a 1‰ salinity variation. This explains why, in most ocean regions, the C-fluctuations are mainly governed by temperature fluctuations, while the contribution from S-fluctuations is usually comparatively small. However, there may, of course, be conditions in which it is the salinity fluctuations that cause most of the C-variations, as, for instance, in isothermal, but not isochaline, layers.

Measurements of fluctuations of the electrical conductivity of water are comparatively simple, since electrical conductivity sensors of the contact and induction types can be readily manufactured (see §7). As regards direct measurements of salinity fluctuations in the ocean, these have remained impracticable to date. It is only C' that is usually measured in the ocean. The salinity fluctuations can then be calculated from the simultaneous measurement of temperature and electrical conductivity. In this case, the temperature and electrical conductivity sensors must be mounted on the same probe, as close to one another as possible, to ensure equal pressure conditions for both measurements. This is important in the subsequent calculations. The accuracy of the T- and C-measurements, necessary for calculating the S'-values, must be sufficiently high. As stated above, the main term in the function $C(T, S)$ can become the temperature term. In that case, the system of equations for S'-calculations appears to be close to degenerate.

11.2. Local Gradients of *C* and *S*

The electrical conductivity and salinity gradients in the ocean can vary over a fairly wide range. The probability distributions of the thickness of layers with constant gradients, and that of the gradients themselves, were calculated by Korchashkin (1976) from data collected by means of probing techniques during the eleventh expedition of the 'Dmitriy Mendeleyev'. His calculations were based on criteria similar to those used in §10 for the analysis of the fine structure of temperature fields in the ocean. Figure 11.1 presents an example of a salinity profile in which layers with constant gradient $d\bar{S}/dz$ have been selected by a computer. Empirical probability distributions of layer thicknesses and salinity gradients are given in Figure 11.2 for Antarctic waters and equatorial latitudes of the Pacific Ocean. In contrast with the corresponding distributions for the fine structure of the temperature field, the symmetrical shape of the function $P(d\bar{S}/dz)$ is typical, and is related to the salinity profiles. The statistical characteristics of vertical salinity profiles suffer less from space–time variations than the corresponding characteristics of

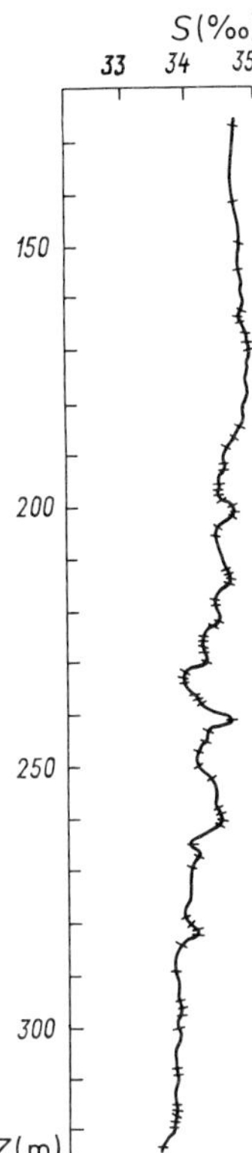

Fig. 11.1. Salinity profiles subdivided into sections with constant gradients of S.

the temperature field. Table 11.1 lists computer-calculated moments of salinity gradient distributions in Antarctic waters and in equatorial latitudes of the Pacific Ocean. As can be seen from Table 11.1, the mean salinity gradients in both polygons proved to be zero, accurate to 0.01 ‰ m^{-1}. The root-mean-square deviation of the distributions, however, was 0.02–0.03 ‰ m^{-1}.

Since the value of $d\bar{S}/dz$ has a variable sign, the probability of detecting salinity gradients equal in magnitude to several one-hundredths of a pro mille per meter in a thin layer is sufficiently great.

The two-dimensional probability densities of salinity gradients and layer-thicknesses, $P(dS/dz, L_s)$, qualitatively resemble the corresponding probabilities for the fine structure of the temperature field (Volochkov and Korchashkin, 1977). Figure 11.3 presents the function $P(dS/dz, L_s)$ computed from data measured at one of the stations in equatorial latitudes of the Pacific Ocean. The maxima of $P(\partial S/\partial z, L_S)$ are located along the mean salinity gradient (which was close to zero), with L_S equal to 4, 6 and 12 m.

Local gradients of the electrical conductivity profile can be determined, using (11.2), from data concerning the corresponding salinity and temperature gradients:

$$\frac{\partial \bar{C}}{\partial z} = \alpha_1 \frac{\partial \bar{T}}{\partial z} + \beta_1 \frac{\partial \bar{S}}{\partial z}\,. \tag{11.3}$$

TABLE 11.1
Statistical characteristics of salinity gradients

Station, no.	$\frac{d\bar{S}}{dz}$ ‰ m^{-1}	δ‰ m^{-1}	*S*	*K*	Number of layers on profiles
Antarctic waters, probing depth 480 m:					
1	0	0.02	0.38	3.65	80
2	0	0.02	0.36	4.81	118
3	0	0.02	0.37	4.13	109
4	0	0.02	0.20	7.40	113
5	0	0.02	0.19	4.65	117
6	0	0.02	0.48	5.00	101
Equatorial region, probing depth 900 m:					
1	0	0.03	0.09	7.87	183
2	0	0.03	0.98	10.05	170
3	0	0.02	−0.21	7.59	138
4	0	0.02	−0.66	8.66	131
5	0	0.03	−1.13	9.88	164
6	0	0.03	−0.37	7.30	202
7	0	0.03	−1.44	10.06	183
8	0	0.03	−0.35	5.05	165
9	0	0.03	−0.39	6.13	164

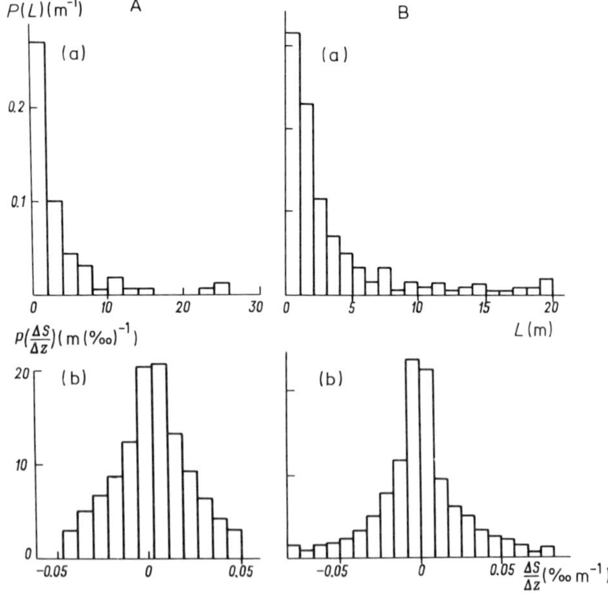

Fig. 11.2. One-dimensional probability densities of layer thickness (a) and vertical salinity gradient (b) at the stations 826 (A) and 848 (B) made during the eleventh expedition of the 'Dmitriy Mendeleyev'.

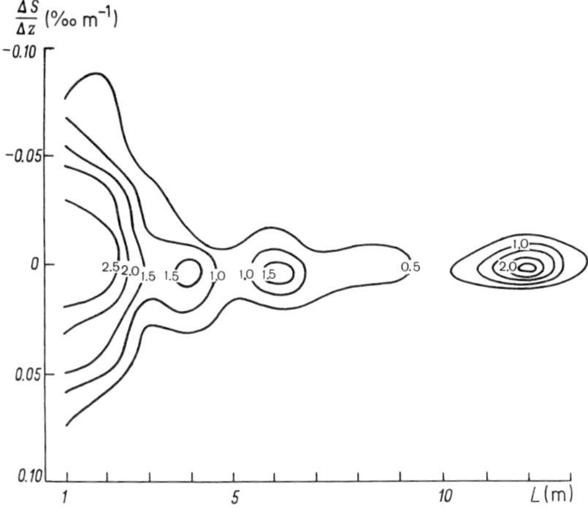

Fig. 11.3. Two-dimensional probability distribution of vertical scales and fine-structure gradients of the salinity field at station 848 made during the eleventh expedition of the 'Dmitriy Mendeleyev'.

Equation (11.3) shows that the expression that relates electrical conductivity fluctuations to temperature fluctuations and to the gradients of S and T has the form

$$C' = T'\left[\alpha_1 + \beta_1\left(\frac{\partial\bar{S}}{\partial z}\Big/\frac{\partial\bar{T}}{\partial z}\right)\right]. \tag{11.4}$$

The contribution of temperature and salinity gradients to the gradient of electrical conductivity (and to the density gradient) can have different signs. As the temperature drops and the salinity increases with increasing depth, their contributions to $\partial\bar{C}/\partial z$ appear to be different. Under certain conditions, they can even compensate for each other. This can also happen when the depth dependence of T and S is the other way around. For the other two combinations of the signs of $\partial\bar{T}/\partial z$ and $\partial\bar{S}/\partial z$, their contributions to $\partial\bar{C}/\partial z$ will add to each other. The joint occurrence of different signs and values of $\partial\bar{T}/\partial z$ and $\partial\bar{S}/\partial z$ in fine thermochaline structures in the ocean has been investigated by a number of authors. Of course, the most common sign combinations of $\partial\bar{T}/\partial z$ and $\partial\bar{S}/\partial z$ are those at which the water layer remains hydrostatically stable, i.e., when an increase of temperature with depth is compensated for by an increase in salinity or, vice versa, when decreasing salinity is compensated for by decreasing temperature. In this case, the correlation between $\partial\bar{T}/\partial z$ and $\partial\bar{S}/\partial z$ clearly will be positive. However, in conditions of stable density stratification, one can also observe a negative correlation between the gradients of T and S. This takes place when a decrease in temperature with depth is accompanied by an increase in salinity. The opposite situation, in which the salinity decreases and the temperature increases (i.e., again a negative correlation between their gradients) corresponds to conditions of unstable stratification. This is observed in the ocean only occasionally. The other two cases of instability occur for positive correlations between the gradients of T and S, but only at such values that the density decrease due to changes in one parameter is not compensated for by the changes induced by corresponding variations in the other.

The gradients $\partial\bar{T}/\partial z$ and $\partial\bar{S}/\partial z$, measured simultaneously with a scanning system in the north Pacific Ocean, were analyzed in detail by Gargett (1976). He used an averaging scale of 0.5 m. Figure 11.4 gives the joint probability distribution of the gradients. The second quadrant in Figure 11.4 corresponds to mean conditions in the region of measurements, i.e., to decreasing T and increasing S with an increase in depth (absolutely stable state of the layers). This quadrant incorporates 34.2% of all computed gradient pairs. The first quadrant, with the inverse temperature gradient but a 'stable' $\partial\overline{S}/\partial z$, incorporates about 56.5% of these results. The third quadrant, which is 'stable' with respect to temperature but 'unstable' with respect to salinity, includes only 7.2% of the results. Finally, the absolutely unstable fourth quadrant includes 1.5% of the results. A straight line, corresponding to $R = \beta_1\ \Delta S/(\alpha_1\ \Delta T) = 1$, runs through the first and third quadrants. Hence, hydrostatic instability must occur for pairs of gradients that are located below this line. As seen from Figure 11.4, there are only a few of these points. Moreover, as noted by the author, in most cases they are located at distances from the straight line, $R = 1$, that do not exceed the calculation accuracy, ± 0.004 °C m^{-1} and ± 0.010 ‰ m^{-1} for temperature and salinity, respectively. The distribution of $\partial\overline{T}/\partial z$ and $\partial\overline{S}/\partial z$ over the quadrants allow one, to a certain extent, to judge possible mechanisms of turbulence generation that are not associated with shear instability. The small number of reliable points below the line $R = 1$ shows that convective instability occurs relatively seldom in some parts of the bulk ocean. The instability that is associated with the phenomenon of double diffusion (instability in either temperature or salinity, with a total instability in density) is more probable. However, the occurrence of double diffusion does not necessarily mean that it will induce turbulence. According to Gargett, turbulence then arises only in about 10% of the cases (this estimate needs refinement). Moreover, the turbulence that is generated could also be induced by other factors (for instance, it might be generated by shear instability).

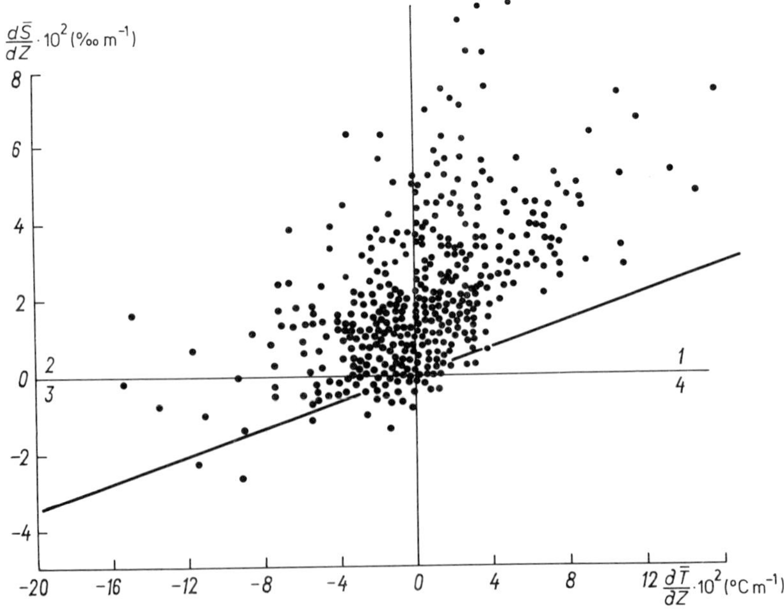

Fig. 11.4. Relative distribution of local temperature and salinity gradients in the North Pacific Ocean (Gargett, 1976).

11.3. Spectral Characteristics

High-frequency electrical conductivity fluctuations were measured repeatedly during expeditions of the Institute of Oceanology of the U.S.S.R. Academy of Sciences. Figure 11.5 presents groups of spectra of electrical conductivity fluctuations plotted from data collected in a number of polygons in various regions of the World Ocean. The reference lines on the plots correspond to a spectral density $E_C(k) = 10^{-12}\ \Omega^{-2}\ \text{cm}^{-1}$ and a wave-number $k = 1\ \text{cm}^{-1}$. At this value of k the spectral density of C varies over five orders of magnitude, i.e., from several $10^{-11}\ \Omega^{-2}\ \text{cm}^{-1}$ in the eighth polygon studied during the seventh expedition of the 'Dmitriy Mendeleyev' to about $10^{-16}\ \Omega^{-2}\ \text{cm}^{-1}$ in the fifth polygon studied during the same expedition. The scatter in the $E_C(k)$ values obtained in some polygons was also rather different. At $k = 1\ \text{cm}^{-1}$ the difference reached two, or even three, orders of magnitude. The spectral level changed with depth non-monotonically; this was also true for the fluctuations of the other hydrophysical field variables. The shapes of the spectral curves were different in each case. Perhaps this can be attributed to the variety of local background conditions against which the electrical conductivity fluctuations occurred.

The spectrum of the electrical conductivity fluctuations and that of the dissipation rate of electrical conductivity inhomogeneities, i.e., the functions $kE_C(k)$ and $k^2E_C(k)$, were computed for a number of polygons studied during the seventh expedition of the 'Dmitriy Mendeleyev' (Beliayev *et al.*, 1974a). In most cases, $kE_C(k)$ proved to increase nearly monotonically with decreasing k throughout the investigated range of k-values, from 6.3×10^{-2} to $6.7\ \text{cm}^{-1}$. At certain levels, however, the $kE_C(k)$-curves displayed maxima. This suggests that there are possible mechanisms for generating electrical conductivity inhomogeneities with scales of only several tens of centimeters. The dissipation spectra of electrical conductivity inhomogeneities displayed no maxima, even at the largest k, the only exception being the $k^2E_C(k)$ spectra obtained from polygon 6.

The most pronounced maxima of the functions $kE_C(k)$ and $k^2E_C(k)$ were detected at deep levels when taking measurements using a lowered probe. Figure 11.6 presents, as an example, the $kE_C(k)$- and $k^2E_C(k)$-curves obtained with the 'Sigma' probe in polygon 7 on the same expedition of the 'Dmitriy Mendeleyev'. Such a diversity in the values and shapes of these functions can, of course, be associated with the great variety of local background conditions that is responsible for the generation and dissipation of electrical conductivity inhomogeneities. In these mechanisms, a major part is played by buoyancy forces and local velocity gradients, which are caused, for example, by high-frequency internal waves. The latter are, presumably, also responsible for the steep slope of $E_C(k)$ commonly observed in the ocean.

11.4. Dependence on Local Background Conditions

The most convenient method for the studying the dependence of the statistical characteristics of electrical conductivity fluctuations against the local background conditions is to use measurement data obtained using probes equipped with fluctuation and noise sensors. A considerable amount of this kind of complex information was obtained on the eleventh expedition of the 'Dmitriy Mendeleyev' and the eighteenth expedition of the 'Akademicik Kurchatov'. A probe equipped with velocity and electrical conductivity

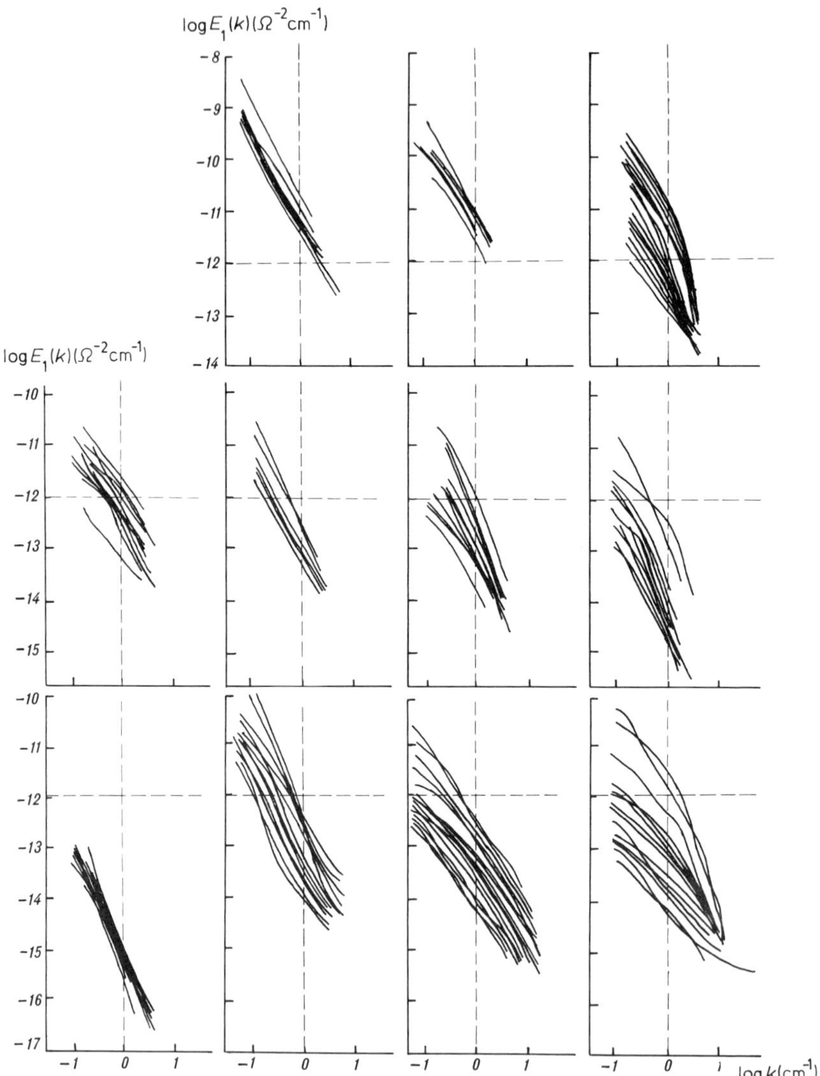

Fig. 11.5. Spectra of electrical conductivity fluctuations, derived from measurements obtained at a number of polygons on specialized expeditions of the Institute of Oceanology of the U.S.S.R. Academy of Sciences.

fluctuation sensors (bandwidths 1–250 Hz), a mean electrical conductivity sensor (0–1 Hz), a mean temperature sensor (0–10 Hz), and a depth sensor was used. As a rule, the measurements were carried out at probing velocities of 1.4 and 1.7 m s^{-1}, whilst the probing depth reached 2000 m. The information was recorded on analog tape recorders and a high-frequency five-channel automatic recorder. The automatic recorder also registered the mean-square values of velocity and electrical conductivity fluctuations

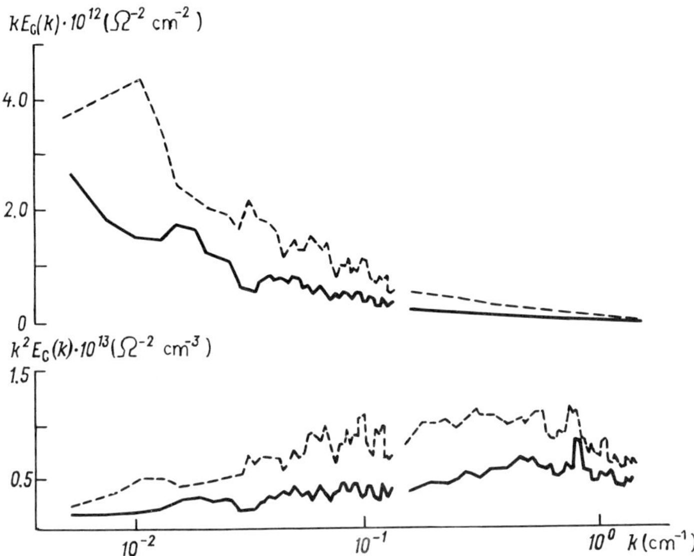

Fig. 11.6. Variance and dissipation spectra of electrical conductivity fluctuations, from data obtained with the 'Sigma' probe in polygon 7 during the seventh expedition of the 'Dmitriy Mendeleyev' (Ozmidov *et al.*, 1974).

over an averaging period of about 1 s. The statistical characteristics were computed from record segments that were selected in accordance with the properties of the profile at the point of probing. The depths of layers with characteristic profile features (called gradients, inversions, and homogeneous interlayers) were marked on the records, thus correlating the fluctuation measurements and the specific local background conditions.

A large discrepancy in the shapes of the temperature profile was observed in the meridional traverse from Tasmania to Antarctica. This section crossed the frontal zone of the Antarctic convergence, which separates the sub-Antarctic and Antarctic water structures. Vertical temperature profiles obtained by probing at a number of stations in the sub-Antarctic water structure are presented in Figure 11.8(a)–(f). The most characteristic profile of this structure is that shown in Figure 11.8(b), the features include a distinct upper mixed layer, a sharp seasonal temperature jump with a 2.3°C temperature drop in the layer 70–100 m, an intermediate quasi-homogeneous layer of 450 m and a main thermocline between 560 and 1200 m. Fine-structure details in the temperature profile are observed only in the main thermocline. Measurements taken at a more northern station (Figure 11.8(a)) showed that the upper homogeneous layer is practically absent, while the seasonal layer of discontinuity is significantly thinned and exhibits a highly developed stepwise structure. Repeated probings at this station revealed great variability in the fine structure. This suggested appreciable dynamic activity over the measurement period. Measurements taken at station 813 showed that the lower boundary of the intermediate quasi-homogeneous layer had several temperature inversions, apparently associated with horizontal water movements. The quasi-homogeneous layer

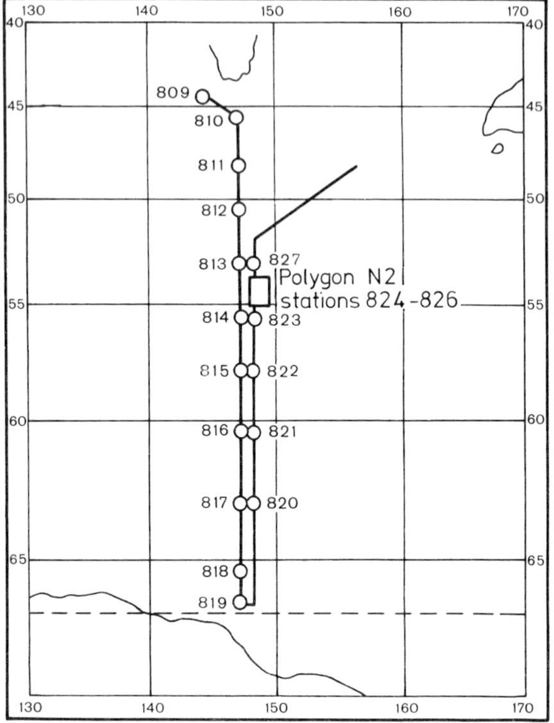

Fig. 11.7. Location of stations along the meridional section studied during the eleventh expedition of the 'Dmitriy Mendeleyev' (147° E., 27 January–12 February 1974).

itself was formed by means of wind mixing and convective mixing in the preceding seasons. This layer is, so to speak, a 'fossil' mixed surface layer.

The spectral densities of electrical conductivity fluctuations were computed for a number of characteristic segments of the record. The function $E_c(k)$ proved to have its highest values in layers with significant temperature gradients. Measurements taken at station 811 show that the $E_c(k)$-curves form two groups whose mean levels differ by nearly an order of magnitude. The curves in the first group are computed for layers with a pronounced stepwise structure in the temperature profile, while the curves in the second group correspond to quasi-homogeneous parts of the profile. A similar situation is observed for stations 812 and 823. Turbulence in a quasi-isothermal intermediate layer can be attributed to convection processes. The $E_c(k)$-curves then can be approximated by the expression $E_c(k) \simeq k^{-5/3}$ over a comparatively wide range of wavenumbers. This can be accounted for by a large outer scale of turbulence (the total thickness of the layer is several hundred meters), and thus by large Reynolds numbers, even with small characteristic changes in the mean flow velocities in the layer. The comparatively low level of the spectra here can be attributed to the small gradient of the mean electrical conductivity, which cannot generate large values of C', even in conditions of fully developed turbulence.

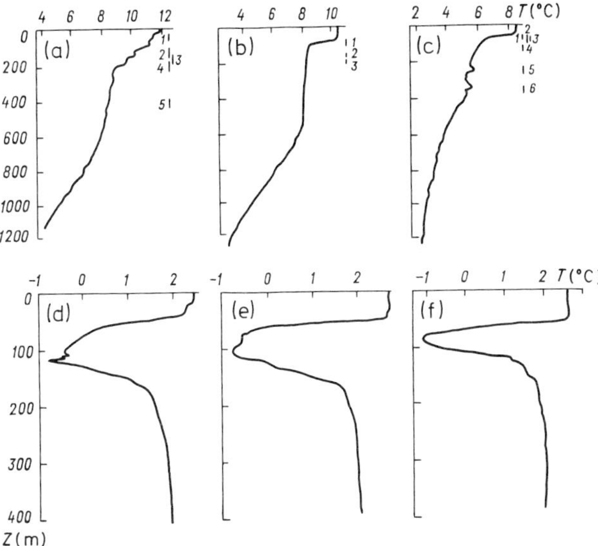

Fig. 11.8. Vertical temperature profiles at stations 811(a), 812(b), 813(c), 821(d), 822(e), and 823(f) located along the meridional section studied during the eleventh expedition of the 'Dmitriy Mendeleyev'.

Specific features of the electrical conducitivity fluctuations in layers with pronounced temperature gradients can be deduced from changes in the shape of $E_c(k)$ at different wavenumbers. In some of the curves obtained from measurements at stations 811 and 813, one can distinguish two ranges of wavenumbers, which correspond to different spectral slopes. For instance, at $k = 1\ \mathrm{cm}^{-1}$ the slope of the two curves obtained at station 813 is close to -1; it increases noticeably at high values of k. At the same time, the other curve obtained from measurements at the same station has a slope close to -2.5 throughout the k-range considered. This behavior of the spectra can be accounted for by the fact that in layers with pronounced temperature gradients the velocity fluctuations are suppressed so that the inner scale of turbulence increases considerably and the dissipation range shifts to smaller wavenumbers. In this case, the electrical conductivity spectra can exhibit viscous-convective and viscous-diffusive ranges, with corresponding slopes of $E_c(k)$. In layers that exhibit a pronounced fine structure in the temperature field, which changes noticeably from probing to probing, active dynamic processes generate inertial-convective ranges in the spectra, with an upper limit at $k \simeq 1\ \mathrm{cm}^{-1}$. The elements of the spectra with a slope of approximately -2 can reflect the properties of electrical conductivity fluctuations in the presence of fine structure.

Typical temperature profiles obtained for stations in Antarctic waters are presented in Figure 11.9(d)–(f). Here, below the upper homogeneous layer, there existed distinct layers of temperature and salinity discontinuity. Within the halocline there was a cold intermediate layer, with a temperature that sometimes decreased to $-1.7°$C. Below this minimum, the temperature was observed to rise to a maximum in a warm intermediate layer at some 400 m depth. The fine structure of the temperature field was best developed in the cold intermediate layer. It changed most in the range of minimum temperatures, in the vicinity of the point where the profile bends. The spectra of the function $E_C(k)$ for this water structure were computed for layers located at the tip of the cold tongue, for layers with steep temperature gradients, and for layers below the density discontinuity. The spectra in the thermocline had the highest levels, while the layers below the density discontinuity exhibited pronounced spectral ranges obeying a 5/3-power law. Fully developed turbulence below the thermocline is induced by convection that arises in layers with unstable temperature distributions. In this layers, the limits of the inertial-convective and buoyancy ranges can shift due to extremely low water temperatures. Indeed, in the case of water at 20°C and with 35‰ salinity, the dynamic viscosity of seawater μ is 0.01 g cm^{-1} s^{-1} and the thermal expansion coefficient α is 2.6×10^{-4} °C^{-1}; however, when $T = -1.7$°C, and the salinity is the same, $\mu \simeq$ 0.02 g cm^{-1} s^{-1} and $\alpha \simeq 0.3 \times 10^{-4}$ °C^{-1}. For these values of the molecular coefficients, the inner scale of turbulence becomes approximately twice as large and the buoyancy scale increases by more than an order of magnitude. This may result in an expansion of the inertial-convective range and in a shift toward larger scales.

The shapes of the spectra proved to be most variable in the layers with the lowest temperatures that lie in the cold intermediate water in the vicinity of the bend in the density profile. These layers are characterized by very favorable conditions for the development of fine structure in the density field (see, e.g., Turner and Stommel, 1964).

Interesting data concerning the variability in the spectral characteristics of electrical conductivity fluctuations as a function of the rearrangements taking place in the temperature profiles in stormy conditions were obtained by taking measurements at the same point of the Antarctic section before and after stormy weather (stations 813 and 827). Figure 11.9 gives the temperature profile obtained at that station during a heavy storm. The profile differs significantly from the one that was obtained before the storm (Figure 11.8(c)). The storm caused the erosion of the thermocline at its lower boundary. The temperature difference in the thermocline decreased from 2 to 1°C. In an intermediate quasi-homogeneous layer, with a mean background gradient of 0.01 °C m^{-1}, a distinct fine structure developed, whose elements had scales ranging from 1 to 30 m and gradients $\partial\bar{C}/\partial z$ that reached $(3-4) \times 10^{-4}$ Ω^{-1} cm^{-1} in some interlayers; (this corresponds to temperature gradients of several tenths of a degree per meter). In this case, the root-mean-square electrical conductivity fluctuation was 2×10^{-5} Ω^{-1} cm^{-1} (which is equivalent to some 2×10^{-2} °C).

The spectra computed from the data obtained prior to, and during, the storm at stations 813 and 827 are given in Figure 11.10. Before the storm, the value of $E_C(k)$ at $k = 1$ cm^{-1} varied over one order of magnitude, but in stormy conditions the variability of $E_C(k)$ exceeded two orders of magnitude at the same range of depths. When the ocean is calm, the turbulent fluctuations are generated by local processes, which are associated with the fine structure. This manifests itself in the absence of any dependence of $E_C(k)$ on the depth of the measurement. The level and the shape of the spectra then depend only on the local gradients; in particular, on $\partial\bar{T}/\partial z$. In stormy conditions, when the primary energy flux comes from the ocean surface, the level of $E_C(k)$ monotonically decreases with depth and the spectra have distinct parts that satisfy a 5/3-power law. In other words, in a storm the type of turbulence changes from local to fully developed, associated with the large-scale 'stormy' mixing of the upper ocean. This example is in

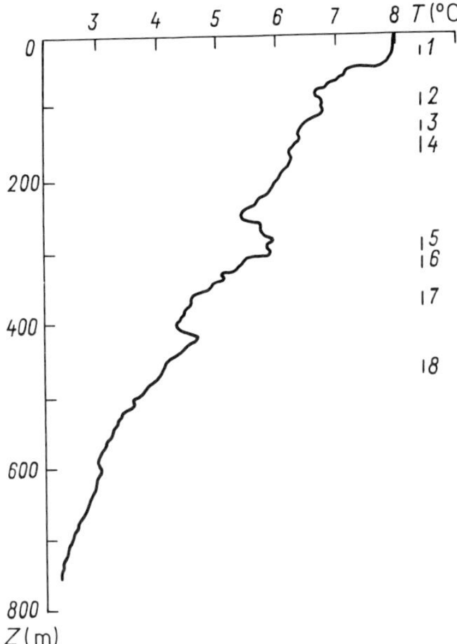

Fig. 11.9. Vertical temperature profile obtained under stormy conditions at station 827 during the eleventh expedition of the 'Dmitriy Mendeleyev'. Vertical bars denote profile segments with computed electrical conductivity spectra.

good agreement with the earlier statement concerning the relation between small-scale fluctuations and the background processes that generate them. These background processes often have a local fine-structure character. In a number of cases, however, due to external factors, they turn into large-scale phenomena (stormy mixing, winter convection, etc.).

The dependence of the characteristics of electrical conductivity on background conditions was qualitatively estimated from observational data obtained at station 821, located in Antarctic waters. Figure 11.11 exemplifies simultaneous profiles of the average value and the fluctuations of the electrical conductivity at this station. The fluctuating signal C' is seen to depend directly on the value and the sign of $\partial\overline{C}/\partial z$. An increase in C' was observed in layers with a high value of $\partial\overline{C}/\partial z$. The spectral densities of the electrical conductivity fluctuations were determined from results that corresponded to different, typical parts of the $\partial\overline{C}/\partial z$-profile. The moving variance $\sigma_C^2 = \overline{C'^2}$ of the electrical conductivity fluctuations and the moving structure functions $D_C(\tau)$, where τ is the time shift, were also computed. These quantities were averaged over 32 readings, with steps of 1/64 s. This corresponded to a spatial averaging scale of 0.7 m. A quantitative relationship was sought between the gradient $\partial T/\partial z$, on the one hand, and the quantities σ_C^2 and $D_C(r_n)$ on the other. Here, $r_n = V_3 r^{n-1}\,\Delta t$, where V_3 is the probing velocity and Δt is the time step for the recordings. Also, $n = 1, 2, 3$; hence, $r_1 = 2.2$ cm, $r_2 = 4.4$ cm, and $r_3 = 8.8$ cm. For the pairs of series $\log D_C^i(r_n) \sim \Delta T_k^i/\Delta z$ and $\sigma_C^2 \sim \Delta T_k^i/\Delta z$, with $i = 1, 2, 3, \ldots$, the normalized joint correlation functions $R_D^k(z)$ and $R_\sigma^k(z)$ were computed, where z is the shift along the vertical axis and the index k denotes the size of a

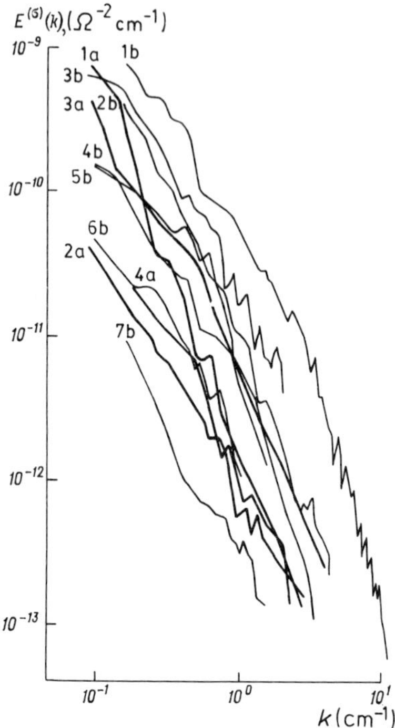

Fig. 11.10. Spectra of electrical conductivity fluctuations, computed for the marked segments of the profile in Figure 11.9.

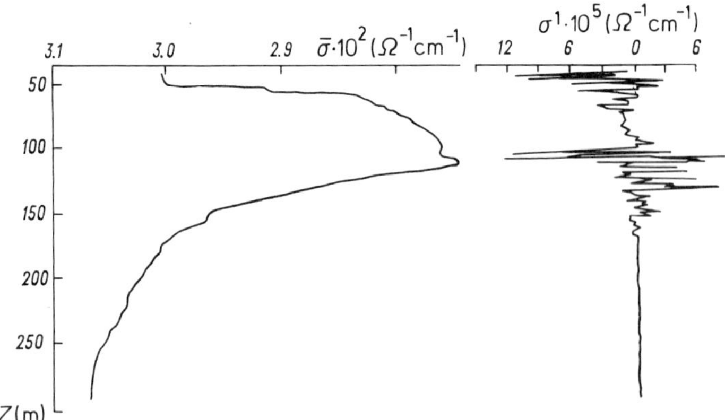

Fig. 11.11. Records of high-frequency and mean profiles of the electrical conductivity, obtained using a probing device at station 821 of the Antarctic section (Beliayev *et al.*, 1975).

finite-difference step in the temperature-gradient calculations. As an example, Figure 11.12 presents a combined plot of a group of functions $R_D^k(z)$. As can be seen, the functions $R_D^k(z)$ take on the highest values at a small shift z. The maximum value, equal to 0.47, belongs to the function $R_D^8(z)$ when the temperature gradient was calculated with a step $r = 5.6$ m. The joint correlation functions $R_\sigma^k(z)$ are of a similar type. Their maximum is 0.57; it belongs to the function $R_\sigma^8(z)$, too. Computing the temperature gradients with a smaller (2.8, 1.4, 0.7 m) or greater (11.2, 22.4 m) step, Δz, resulted in a decrease in the maxima of the joint correlation function. Vertical inhomogeneities of the temperature field with scales L equal to two times 5.6 m therefore proved to exert the greatest influence on the fluctuations. The mean correlation radius between the fluctuating signal characteristics and the temperature gradient was about 15 m. Of course, these estimates are valid only for the experimental conditions mentioned. However, this kind of connection between electrical conductivity fluctuations and local background conditions seems to be typical of the ocean as a whole. This conclusion was confirmed by similar calculations carried out using measurement data obtained at the equatorial polygon during the eleventh expedition of the 'Dmitriy Mendeleyev' and measurements taken in the Mediterranean during the eighteenth expedition of the 'Akademicik Kurchatov'. In the equatorial polygon, the joint correlation function between the variance of the high-frequency electrical conductivity fluctuations and the fine-structure gradients of the profiles of mean temperature and electrical conductivity proved to be greatest (0.57) at a 0.5–1 m downward shift of the σ_C^2 series relative to the gradient series. The generation of turbulence above the gradient interlayers is due to interaction with

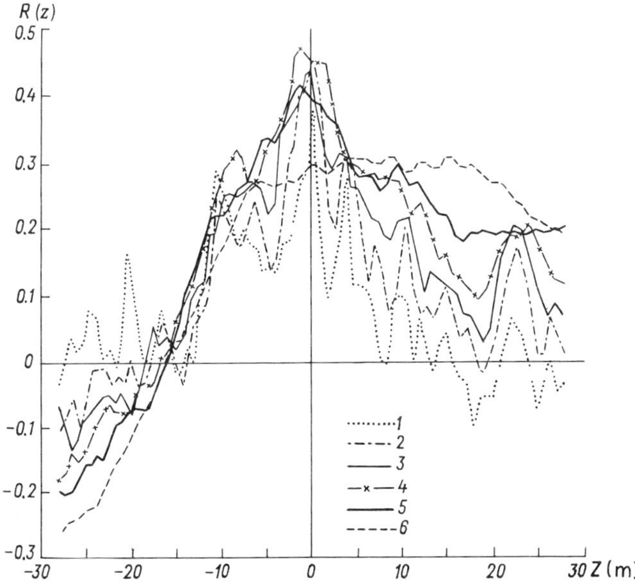

Fig. 11.12. Reciprocal normalized correlation functions of the logarithm of the structure functions for electrical conductivity fluctuations and temperature gradients at $\Delta z = 0.7$ m (1), 1.4 m (2), 2.8 m (3), 5.6 m (4), 11.2 m (5) and 22.4 m (6) (Beliayev *et al.*, 1975, pp. 1078–1083).

descending internal waves that are reflected by the interlayers (Delisi and Orlanski, 1975). A noticeable correlation between the different series considered was also observed at a 1 m shift of the σ_C^2- series above, and a 3 m shift below, the gradient series. In the Mediterranean, similar studies were carried out using data obtained in the Gulf of Tunis from a large number of fluctuation and noise measurements. The normalized joint correlation functions with the largest values were observed here for inhomogeneity scales of up to 8–12 m.

11.5. Intermittency of Electrical Conductivity Fluctuations

Moving structure functions of the electrical conductivity field, $D_C(\tau_n)$, are useful as a means to judge the intermittency of the fluctuating signal. Figure 11.13 presents plots of log $D_C(\tau_n)$ that correspond to the probing data obtained at station 819 in the Antarctic. The argument varied from 0.2 cm (points numbered as 1) to 12.5 cm (points numbered as 7). On the X-axis the ordinal numbers of the parts of the initial series are given; the figures on the Y-axis (instead of points) denote log $D_C(\tau_n)$ at shifts $\tau_n = 2^{n-1}\ \Delta t$ ($n = 1, 2, \ldots, 7$, $\Delta t = 1/512$ s). As seen, $D_C(\tau_n)$ changes by more than an order of magnitude at fixed τ_n (for the sake of visualization, the values of log $D_C(\tau_n)$ for $n = 1$ and $n = 7$ are connected with heavy lines). Variations in the lengths of the line segments that connect points with the same numbers are, in all probability, indicative of changes in the shape of the structure function. In a number of cases, the structure function becomes rapidly saturated (small distances between points 1–7); sometimes, however, no saturation is observed at the available τ_n shifts. Remember that, for a stationary random process with a zero mean value, the structure function is related to the correlation function by $D(\tau) = 2\,[R(0) - R(\tau)]$. In most cases, $R(\tau) \to 0$ as $\tau \to \infty$, and hence $\lim_{\tau \to \infty} D(\tau) = 2R(0)$. For $\tau = 0$, the correlation function equals the variance of the process; hence variations in the values of $D(\tau)$ when close to saturation can be considered to be variations in the variance (energy) of the fluctuating signal. The spacing of the points in logarithmic plots of the moving structure function is proportional to the tangent of the slope of the corresponding parts of the structure function. Since the shape of the structure function determines that of the spectrum, the plots of lg $D(\tau)$ can be used to judge variations not only in the level of fluctuations but also in their spectrum.

The moving structure functions $D(\tau_n)$ were computed at a 0.002 s shift (corresponding to a spatial shift of 0.34 cm) and at a 0.128 s shift (corresponding to a spatial shift of 21.8 cm) from data obtained during the fifteenth expedition of the 'Dmitriy Mendeleyev' in the layer from 100 to 500 m depth. The structure functions computed for layers of thickness 1.7 m proved to vary over a considerable range. Here, the distance between the $D(\tau_n)$-points at two chosen τ_n-values also varied with depth, which is evidence for the variability in the spectral composition of the small-scale electrical conductivity fluctuations along the direction of probing. Figure 11.14 illustrates the variance σ_C^2 as a function of depth, according to the same data. The variance of the signal changes over relatively smaller distances than the corresponding structure functions. This suggests large relative variations of the high-frequency part of the spectrum as compared with the low-frequency components. As can be seen, in the ocean there exist comparatively thin layers (as thick as several meters) with a raised fluctuation level and a quasi-periodic

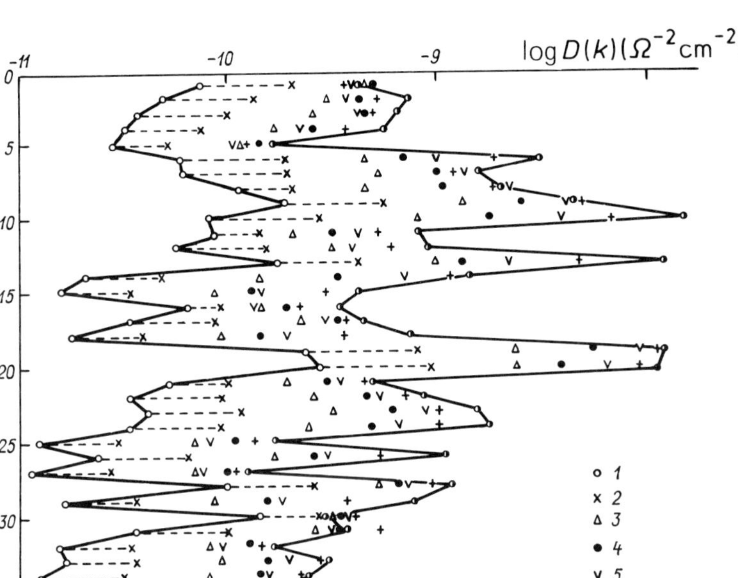

Fig. 11.13. Specimen of a structure function for electrical conductivity fluctuations, derived from measurements taken at station 819 of the Antarctic section.

distribution. As stated above, this reflects the dependence of turbulence processes on local background conditions.

The variations in the variances of the velocity fluctuations and of those in the electrical conductivity are not always related. Thus, at one of the stations in the Tunis Strait, the maximum value of σ_u^2 was observed in a layer with high mean velocity gradients, while no maximum of σ_C^2 was detected. An increase in the σ_u^2-values in the layers 135–145 and 190–200 m at the other station was not accompanied by any changes in the intensity of electrical conductivity fluctuations. These phenomena are, first of all, associated with the variable electrical conductivity gradients in these layers. For example, in the thermocline, when σ_u^2 decreased to 4×10^{-2}–8×10^{-2} cm s^{-1}, σ_C^2 increased to 3×10^{-5} Ω^{-1} cm^{-1} in a number of cases. This was due to a high local gradient of C, corresponding to 0.2 °C m^{-1} in the temperature field. Similar data were obtained during the sixtieth expedition of the 'Vityaz' in the Pacific Ocean. From measurement data taken from the station located at 28°40′ N and 155°10′ E, the variance σ_u^2 varied between 0.03 and 0.34 cm^2 s^{-2} and σ_C^2 between 0.17×10^{-6} to 3.2×10^{-6} Ω^{-2} cm^{-2}. The measurements were carried out during towing at a depth of 30 m. The variations of σ_u^2 and σ_C^2 along the direction of measurements can be weakly correlated. The C'-signal showed sharp peaks more often than the u'-signal. An increase in the electrical conductivity fluctuations was, as a rule, observed in the gradient interlayers. The absence of pronounced peaks in the u'-signal, while the turbulent C'-fluctuations in layers with a

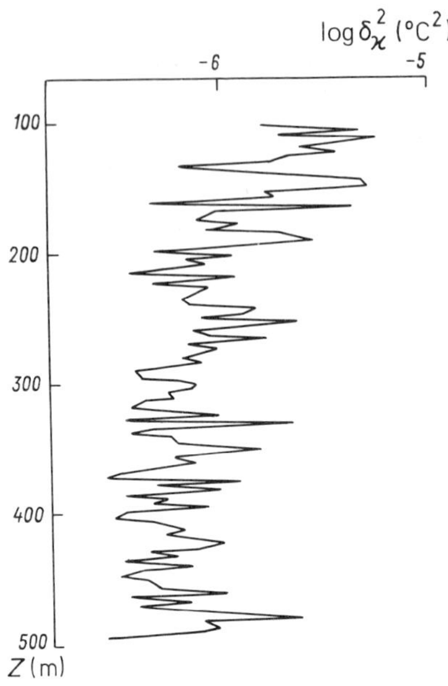

Fig. 11.14. Vertical profile of the variance of electrical conductivity fluctuations, derived from measurements taken during the fifteenth expedition of the 'Dmitriy Mendeleyev'.

constant gradient increase can be attributed to the interesting phenomenon of 'fossil turbulence'. This phenomenon arises because turbulent velocity fluctuations decay faster than the fluctuations of temperature and salinity (and hence of electrical conductivity). This is due to the difference in the molecular exchange coefficients. The thermal diffusivity and the diffusivity of salt are roughly one and three orders of magnitude smaller than the kinematic viscosity, respectively. Before they decay, the velocity fluctuations then induce inhomogeneities of T, S and C. Fossil turbulence of these quantities can thus exist in a laminar medium, until molecular forces destroy it. The converse situation, i.e., the absence of peaks in C' while u' increases, can be observed in the case when sensors are located in an interlayer with a homogeneous electrical conductivity or in a thermocline at the first moment of turbulence spot formation.

11.6. Deep-Sea Measurement Data

Electrical conductivity fluctuations at great depths were studied in detail during the seventh expedition of the 'Dmitriy Mendeleyev' employing the 'Sigma' probe (Beliayev *et al.*, 1974c). The probing data down to a depth of 1250 m obtained at polygon 7 (10°57′ N, 55°40′ E) were thoroughly processed. According to data describing the average profile C, there exists an upper quasi-homogeneous layer as thick as about 100 m which, after a slight inversion, transforms into a layer of discontinuity (100–210 m). This is followed by multiple steps and thin inversion layers (especially at depths of 500–800 m) with respect to the $\bar{C}$-distribution. Whilst probing, the device was held steady at 7 levels, in order to carry out 10-minute recordings under 'hovering' conditions. The fluctuation data recorded in both probing and hovering conditions revealed large variations of C' in space and time. The vertical profile of the mean absolute value of C' exhibits independent fluctuations in the signal level, reaching several tenths of an equivalent degree. At the hovering level of 835 m depth, the electrical conductivity fluctuations proved to be small, while at the 720 m level a number of sharp peaks in the C'-fluctuations were discovered. The greatest C'-fluctuations were registered at the 520 m level, where a local jump in the value of C' was observed. The level at 478 m depth also showed a few pronounced jumps in C'-values, while in the upper quasi-homogeneous layer the C'-values proved to be small. The spectral densities of C', plotted for all the hovering levels, also exhibited different intensities of the C'-fluctuations. The value of $E_C(k)$ at fixed wavenumbers differed from measurement to measurement by one or two orders of magnitude, without any monotonic dependence on the depth of measurement. The shapes of the spectra also varied from measurement to measurement. This can be attributed to variations in local background conditions in the temperature and salinity field. Indeed, let us assume that the T'- and S'-fluctuations obey independent diffusion equations and have high-frequency spectra which are universal, but, generally speaking, still correlated. With a standard stratification of the ocean, i.e., when temperature decreases and salinity increases with depth, we have $\overline{T'S'} < 0$. The fluctuations of T' and S' then make contributions of the same sign to the density fluctuations ρ'. This promotes ρ' and hence also the fluctuations of the buoyancy force, which are proportional to it. The spectral density of the electrical conductivity fluctuations then can be reliably described by (5.18). This shows that the shape of the high-frequency spectrum of the electrical conductivity depends on the relationships between the parameters $\alpha_1^2 \epsilon_T$, $\beta_1^2 \epsilon_S$, and $\alpha_1 \beta_1 \epsilon_{TS}$. These parameters exhibit a large variability; they are determined by simultaneous recordings of T and S (or T and C) fluctuations at the same point, as has already been stated above.

11.7. Determination of Salinity Fluctuations

A detailed joint analysis of the fluctuation spectra for flow velocity, temperature, and salinity in a stratified fluid was carrie'd out by Nozdrin (1974). He employed a system of equations relating motion, heat conduction and salt diffusion with terms that accounted for the effects of buoyancy forces and thermal diffusion. The parameters included in the system make it possible, in much the same way as in (4.17)–(4.18), to present the four length scales

$$L_\rho = \left(\frac{\rho_0}{g}\right)^{3/2} \frac{\epsilon^{5/4}}{\epsilon_\rho^{3/4}}; \qquad L_* = \frac{\epsilon^{5/4}}{(\alpha g)^{3/2} \epsilon_T^{3/4}},$$

$$L_{**} = \frac{\epsilon^{5/4}}{(\beta g)^{3/2} \epsilon_S^{3/4}}; \qquad \eta = \left(\frac{\nu^3}{\epsilon}\right)^{1/4}. \tag{11.5}$$

Here ϵ_ρ is the dissipation rate of density fluctuations, α is the coefficient of thermal expansion, β is the coefficient of saline expansion, and the other notations are the same as above. When the outer scale of the turbulence (L) exceeds all scales in (11.5), then dimensional arguments permit us to write explicit expressions for the spectra of the field variables in different wavenumber ranges. In particular, the following expressions hold for the salinity: for scales $L > l > L_{**}$, i.e., in the buoyancy range we have $E_S(k) \simeq k^{-7/5}$, in the inertial-convective range (for $L_{**} > l > \eta$) $E_S(k) \simeq k^{-5/3}$, and for smaller scales $E_S(k) \simeq k^{-1}$ (diffusive-convective range). After that the spectrum decays exponentially due to molecular forces in the viscous-diffusive interval. Since in the ocean the scales L_ρ, L_*, L_{**}, and η vary greatly and the boundaries between the buoyancy, inertial-convective, viscous-convective, diffusive-convective and viscous-diffusive ranges are mobile, they can shift and may not coincide in different circumstances. The scales L_ρ, L_*, and L_{**} were estimated by Nozdrin (1975) from data concerning velocity, temperature, and electrical conductivity fluctuations obtained during the eleventh expedition of the 'Dmitriy Mendeleyev' and appeared to lie between 8 and 20 cm. The position of the boundaries between the buoyancy and inertial-convective ranges in the temperature and electrical conductivity spectra made it possible to determine the roles of salinity and temperature fluctuations in the electrical conductivity fluctuations observed in certain of the measurements. The salinity spectra were calculated from data on T- and C-fluctuations that were obtained in the equatorial polygon during the eleventh expedition of the 'Dmitriy Mendeleyev'. Examples of the first two $E_S(k)$ spectra obtained are shown in Figure 11.15. The levels of the spectra are somewhat different and their slopes at same values of k approach -1. For large k, the lower curve shows a sharp decrease of spectral density, which is characteristic of a viscous-diffusive range. The left-hand parts of the curves apparently can be attributed to a diffusive-convective range in the universal spectrum $E_S(k)$. The root-mean-square of the salinity fluctuations over the wavenumber range considered proves to be close to 10^{-3}‰, and the dissipation rate of salinity inhomogeneities appears to be about 10^{-7}–10^{-8} (‰)^2 s^{-1}.

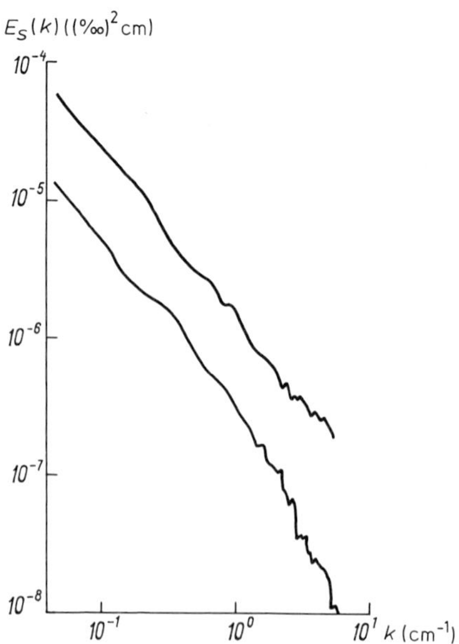

Fig. 11.15. Examples of the spectra of salinity fluctuations in the ocean (Nozdrin, 1975).

11.8. Density Fluctuations and Turbulent Mass Flux

Simultaneous information describing T- and S-fluctuations enables one to judge fluctuations of the water density, ρ'. If these data are added to those on velocity component fluctuations, then one can directly estimate the mass flux in the ocean. This is an important parameter in the understanding of processes involved in oceanologic field formation. In conditions of stable density stratification, the vertical turbulent mass flux, equal to $m_z = \overline{\rho' w'}$, results in a loss of turbulent energy to the buoyancy forces. This can be expressed by the relation (Monin, 1970b):

$$\frac{\partial K}{\partial t} = -\frac{g}{\rho} m_z. \tag{11.6}$$

Here K is the kinetic energy per unit mass, g is the acceleration due to gravity, and t is time. In conditions of unstable stratification, when the water density decreases with depth, the mass flux is directed downward and the buoyancy forces, now generating convective turbulence, will contribute to the kinetic energy.

The vertical mass flux can be estimated indirectly from the boundary condition for m_z at the ocean surface:

$$(m_{z'})_0 = (P - E)S - \frac{\alpha}{c_p}(LE + Q) + \alpha P(T_p - T_w). \tag{11.7}$$

The first term of (11.7) describes changes in the salinity of the upper layer due to precipitation P and evaporation E. For average values of P and E this term is of the order of 10^{-7} g cm^{-2} s^{-1}, but during intensive precipitation or evaporation it can increase by one, or even two, orders of magnitude. The second term describes density changes of the upper water layer due to heat exchange with the atmosphere (c_p is the heat capacity of water, L is the latent heat of evaporation, and Q is the sum of the turbulent and radiant heat fluxes in the air at the water surface). Since $\alpha L/c_p \simeq 0.12$, the contribution of evaporation is four times as large as that in the first term. Finally, the third term describes density changes due to the heat capacity of the precipitation (T_p and T_w are the temperatures of the precipitation and of the surface water, respectively).

The value of m_z in internal layers can be estimated in the following way. Monin *et al.* (1970) proved that at depths exceeding 1.5–2 km the Brunt–Väisälä frequency $N = (g/\rho)(\partial\rho/\partial z)$ varies with depth according to $N = A(H - z)$, where H is the total depth of the ocean, and A is a constant that varies from 10^{-7} to 10^{-6} m^{-1} s^{-1}. This constant can be also represented as $A = \Gamma/L$, where L is a typical scale of pronounced turbulence inhomogeneities and Γ is the mean velocity gradient at larger distances from the bottom (compared with L). According to the similarity theory for turbulence in stratified flows, the quantities L and Γ can be also represented as

$$L = u_*^3 \left(\kappa \frac{g}{\rho} m_z\right)^{-1}; \qquad \Gamma = \frac{u_*}{\kappa R L}, \tag{11.8}$$

where u_* is the friction velocity, κ is the Karman constant, and $R \simeq 0.1$ is the critical value of the flux Richardson number. Using these expressions, for m_z (Monin, 1970b):

$$m_z = (\rho/g) u_*^{5/2} \left(\frac{RA}{\kappa}\right)^{1/2}. \tag{11.9}$$

If we assume that u_* in the ocean is 1 cm s^{-1}, then m_z will be 10^{-8} g cm^{-2} s^{-1}. This estimate is approximately two orders of magnitude smaller than that in the stably stratified atmosphere. In spots of intermittent turbulence in the ocean this value, of course, increases in accordance with the 'spottiness' of the turbulence.

Empirical data prove that A decreases with increasing total ocean depth H. In this case, according to (11.9), the mass flux in internal layers will also be H-dependent, i.e., it will decrease with increasing H. Let us now use one more empirical dependence, the 'depth law' $Nz = w_*$, which relates N to z (w_* is a constant close to 2.2 m s^{-1}). This law is valid down to a depth of 500–5000 m. Equating the expressions for N at mid-ocean depths ($z = H/2$) given by both laws yields $w_* = A(H/2)^2$. The expression for A then results in the following equation

$$\frac{L}{H} = \left(\frac{1}{4\kappa R}\,\frac{u_*}{w_*}\right)^{1/2} = \delta, \tag{11.10}$$

where δ can be assumed to be 0.1. Hence,

$$m_z = \frac{\rho}{g}\,\frac{u_*^3}{\kappa\delta H}. \tag{11.11}$$

If we now employ the following obvious relations

$$u_*^2 \simeq r_{uw} w'^2, \qquad M \simeq r_{\rho w} w'\rho', \tag{11.12}$$

where r_{uw} and $r_{\rho w}$ are the corresponding correlation coefficients, (11.11) yields the following expression for density fluctuations in turbulent ocean layers:

$$\frac{\rho'}{\rho} \simeq \frac{r_{uw} u_*^2}{\kappa\delta r_{\rho w} g H}. \tag{11.13}$$

Using reasonable estimates for the quantities involved in (11.13), one obtains values of ρ' around 10^{-6} g cm^{-3}. Computations of ρ' based on the linearized expression for its dependence on T and S yields the following estimates. The contribution from temperature fluctuations $T' = 10^{-2}$–10^{-4} °C cm results in $\rho' = 10^{-6}$–10^{-8} g cm^{-3}, and the contribution from salinity fluctuations $S' = 10^{-3}$–10^{-4} ‰ cm^{-1} results, correspondingly, in $\rho' = 10^{-6}$–10^{-7} g cm^{-3} (in preliminary calculations the constants in the linearized equation of state were taken from Mamayev, 1964). Besides, to a certain extent ρ' can be affected by pressure fluctuations, data concerning which are scarce at present. Thus, density fluctuation estimates derived from data concerning the fluctuations of other variables result in ρ'-values that are in agreement with the indirect estimates mentioned above.

Density fluctuations of ocean water induce fluctuations in other hydrophysical parameters, in particular, in the speed of sound c_0 and the refraction coefficient n. Indeed, these quantities are functions of ρ; hence, any variation of ρ results in corresponding variations of c_0 and n. Studies of c_0- and n-fluctuations are very important for computations of the distribution and diffusion of sound and light in the ocean. Therefore, both direct measurements and calculations of c_0' and n are of great practical importance. The

c_0- and n-fluctuations can be estimated from the expressions that relate c_0' and n' to the ρ-fluctuations or to the T-, S- and p-fluctuations. For example, to estimate fluctuations of the speed of sound, one can employ the well-known Wilson formula (Wilson, 1960), which relates c_0 to T, S and p, or the simpler (but fairly accurate) expression suggested by Leroy (1969):

$$c_0 = 1492.9 + 3(T - 10) - 6 \times 10^{-3}(T - 10)^2 - 4 \times 10^2(T - 18)^2 + \\ + 1.2(S - 35) - 10^{-2}(T - 18)(S - 35) + z/61,$$

where the speed of sound is given in meters per second, the temperature in degrees centigrade, the salinity in pro mille and the depth z in meters. For the temperature and salinity fluctuations given above, the c_0-fluctuations can be several tenths of a meter per second. The major contribution to them, then, is made not by salinity, but by temperature fluctuations. In a similar way data concerning T and S can be employed to estimate fluctuations of the refraction coefficient of ocean water; these parameters receive their major contribution from temperature fluctuations.

11.9. Climatology of Electrical Conductivity Fluctuations

Generally speaking, the dependence of the statistical characteristics of particular values of small-scale electrical conductivity fluctuations against the large-scale background conditions is not deterministic. We have encountered this problem earlier, for velocity fluctuations. The dependence on background conditions can reveal itself only for the parameters of a set of statistical characteristics of particular values that is obtained in fixed conditions in one polygon. This can be illustrated by the variability of the mean spectral level obtained at fixed wavenumbers in polygons with different mean hydrometeorological conditions. Thus, in polygon 8 studied during the seventh expedition of the 'Dmitriy Mendeleyev', measurements were carried out in stratified layers with steep velocity gradients. These layers, in particular, a layer that was characterized by a sharp change in the slope of the density profile, had sufficiently high $E_c(k)$-levels. In the layer from 50 to 200 m the flow direction changed by $\pi/2$. In polygon 6, the mean density gradient varied significantly with depth, and the vertical velocity distribution also presented a complex picture. This resulted in a fan-shaped group of spectra, with levels that varied over more than three orders of magnitude. In polygon 2, the levels of the spectra at $k = 1$ were somewhat lower than those obtained for previous polygons. This may be attributed to a lower turbulence intensity in the polygon because of steeper density gradients. The comparatively low mean level of the spectra obtained in polygon 4 is caused by a shallow gradient of mean electrical conductivity and by a moderate generation of turbulence by the mean-flow gradient. The small $E_c(k)$-values obtained for polygons 3 and 5 can be accounted for by steep density (and electrical conductivity) gradients, by a vertically nearly uniform velocity in polygon 3, and by comparatively shallow velocity gradients in polygon 5.

The probability distribution of the spectrum $E_c(k)$ at a fixed wavenumber was plotted for a set of $E_c(k)$-curves obtained for seven polygons studied during the seventh expedition of the 'Dmitriy Mendeleyev' (Figure 11.16). The spectra were plotted from data obtained by towing a probe at depths of 23 to 217 m. At $k_0 = 1$ cm^{-1}, the $E_c(k)$-values in

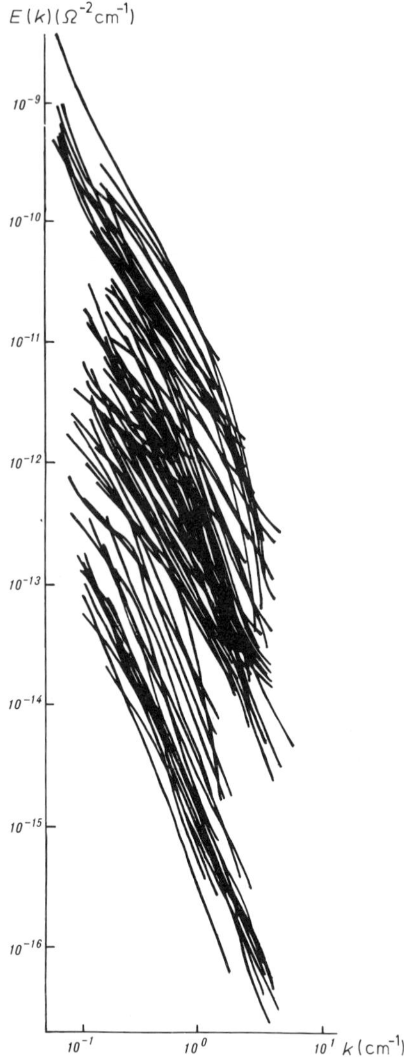

Fig. 11.16. Curves of one-dimensional probability densities of electrical conductivity fluctuations in the ocean (Beliayev *et al.*, 1974).

the set of curves varied by nearly five orders of magnitude (from 3.29×10^{-16} to 2.30×10^{-11} Ω^{-2} cm^{-1}). Figure 11.17 presents the histogram of $\log E_c(k_0)$ and its empirical integral probability distribution; the histogram is plotted with an interval of 0.5. Table 11.2 lists the mean value m, variance D, the standard deviation σ, the skewness S, the kurtosis K, and the variances of S and K. When estimated with Kolmogorov's criterion, the distribution proved to differ appreciably from the normal one. This can also be seen from the moments of the distribution. Thus, unlike the distribution of $\log E_u(k_0)$, which is fairly accurately described by a normal law, the distribution of electrical

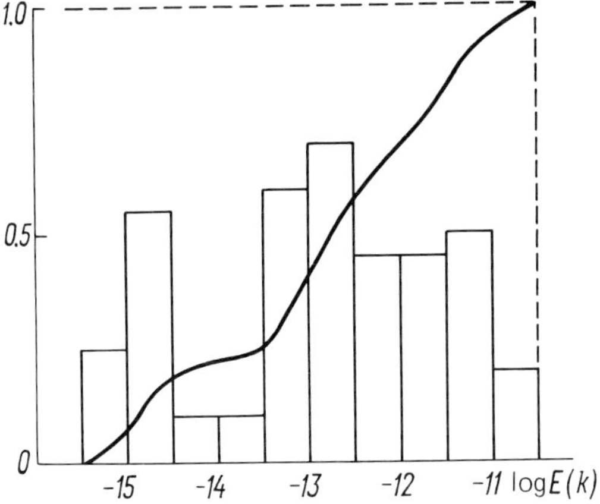

Fig. 11.17. Histogram and empirical integral function of $\log E_c(k_0)$ for $k_0 = 1\ \text{cm}^{-1}$ (Beliayev *et al.*, 1974).

conductivity fluctuations is more complicated. This can be attributed to the fact that the C'-fluctuations depend not only on u', but also on the gradients $d\bar{C}/dz$. For the probability distribution of C', this results in the superposition of the probability distributions of u' and dC/dz.

TABLE 11.2

Parameters of the probability distribution of $\log E_c(k_0)$ for $k_0 = 1\ \text{cm}^{-1}$

m	D	σ	S	K	$D(S)$	$D(K)$
−12.87	1.72	1.31	−0.35	−0.92	0.07	0.25

The fact that some segments of the $E_c(k)$-curves can be approximated by various power laws is illustrated by the two-dimensional probability densities of the exponents α at different wavenumbers k. To plot these probabilities, the range of $\log k$ (from −1 to 0.5) was divided into six equal subranges; in each, α was determined for all spectra in the set considered. The values of α proved to range from 0.5 to 4.5. Histograms of the absolute values of the spectral slopes were plotted for each subrange of wavenumbers. The entire range of α was subdivided into classes with a width of 0.5 units. Table 11.3 lists the mean value m, the standard deviation σ, the skewness S, and the kurtosis K of the distribution of α. In each of the subranges, these distributions were used to estimate the two-dimensional probability densities of the slope of $E_c(k)$ (Figure 11.18). The numbers on the isolines denote the probability density of the spectral slope at the corresponding scales of turbulence. The isolines were constructed with the aid of a rectangular grid with a 0.25 step along the $\log k$-axis and a 0.5 step along the α-axis. The probability density estimates in each of the subranges were related to the centers of the corresponding grid meshes. The isolines were plotted using a linear interpolation between neighboring points.

TABLE 11.3

Parameters of the probability distribution of α for the function $E_C(k)$

Distribution parameters	Number of subrange and $\log k$ boundary values					
	1	2	3	4	5	6
	−1 −0.75	−0.5	−0.25	0	0.25	0.5
m	1.90	1.82	1.75	2.00	2.15	2.07
σ	0.63	0.66	0.61	0.63	0.67	0.82
S	0.12	0.31	0.09	0.44	0.37	0.71
K	−1.06	−0.63	0.57	−0.38	−0.30	−0.33

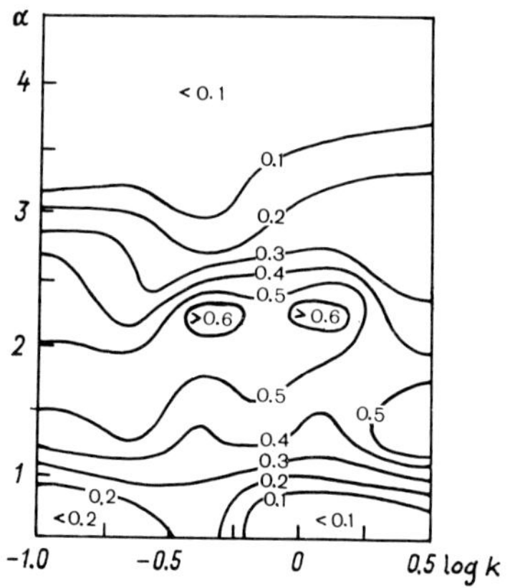

Fig. 11.18. Isolines of the probability density of the spectral slopes of the electrical conductivity fluctuations (Beliayev *et al.*, 1976).

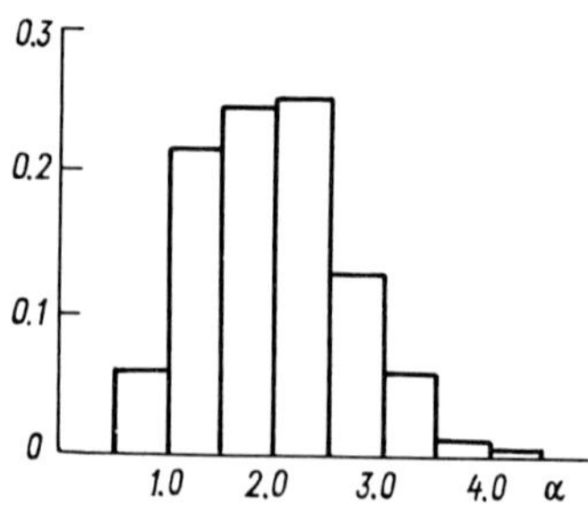

Fig. 11.19. Histogram of the distribution of spectral slopes of the electrical conductivity fluctuations in the ocean (Beliayev *et al.*, 1976).

The two-dimensional distribution of the slope of $E_c(k)$ has a much more complex shape than that of $E^u(k)$ (Figure 11.18). In subranges 1 and 2, the value of α for electrical conductivity fluctuations most often lies between 1.5 and 2.0. This can be attributed to the presence, in the electrical conductivity fluctuation spectra, of either an inertial range or a buoyancy range, with slopes (on a logarithmic scale) equal to $-5/3$ and $-7/5$, respectively. In subranges 3–5, α occurs most frequently in the range from 2 to 2.5, while in subrange 6 it varies from 1.0 to 1.5. Such a decrease in α for electrical conductivity fluctuations at small scales proves the predominance of viscous-convective turbulence.

A summary histogram of the distributions of α for $E_c(k)$-spectra in all subranges considered is shown in Figure 11.19. A characteristic feature of the histogram is its skewness. The most reliable slope of $E_c(k)$ in the ocean appears to occur in the range between 1.5 and 2.5.

CHAPTER III

Large-Scale Horizontal Turbulence

12. LARGE-SCALE TURBULENCE AND NEGATIVE EDDY VISCOSITY

Large-scale turbulence (macroturbulence) is defined here as turbulence with characteristic linear sizes that significantly exceed the thickness of the upper homogeneous ocean and buoyancy scale. The orientation of the vorticity axis of this kind of turbulence is close to the vertical. Large-scale turbulence in the ocean can result from circulatory movements that are directly induced in the water by a turbulent wind field, barotropic and baroclinic instabilities of large-scale flows, vortices originating over topographical irregularities of the sea bed, etc.

One of the manifestations of large-scale turbulence in the ocean is the generation of synoptic vortices with characteristic sizes of an order of magnitude similar to the Rossby deformation radius $hNf^{-1} \simeq 100$ km, where h is the thickness of the baroclinic layer in the ocean, N is the mean Brunt–Väisälä frequency in this layer, and f is the Coriolis parameter. The available potential energy associated with the slope of isopycnic surfaces in the ocean may serve as an energy source for these vortices (Gill *et al.*, 1974). Large-scale turbulence has aroused considerable interest, mainly because of its high energy (Bernstein and White, 1974; Thompson, 1971; Wunsh, 1972) and its important role in the general dynamics of the ocean.

Large-scale turbulence in the ocean can be characterized by additional scales that pertain to various ranges of turbulent motion and depend on the Earth's rotation (Woods, 1974). One of these scales can be written as

$$L_f = \epsilon^{1/2} f^{-3/2}, \tag{12.1}$$

while another scale, allowing for the latitude-dependent Coriolis parameter, can be written as

$$L_\beta = \epsilon^{1/2} \beta^{-3/5}. \tag{12.2}$$

Here $\beta = (2\Omega/R) \cos\varphi$, Ω is the Earth's angular velocity, R is the Earth's radius, and φ is latitude.

Several generation mechanisms of large-scale turbulence and the possibility of supplying it with energy from external sources (e.g., the wind) over a wide range of scales were mentioned by Stommel (1946) and Ichiye (1951). This variety greatly hampers experimental and theoretical investigations. Indeed, even to choose a turbulence scale or an averaging scale for the equations of motion is quite a problem. In this respect, an urgent problem arises: to what extent is the ocean supplied with energy at various scale ranges, or, in other words, are there continuous scale ranges of energy supply or can it be assumed to be localized in the vicinity of discrete scales? This problem can be tackled by processing available data on the energy spectra of large-scale turbulence. The analysis of those spectra and of possible mechanisms of energy transfer from external sources has provided a means for the hypothesis that there are discrete zones of energy supply for turbulence in the ocean (Ozmidov, 1965b). According to this hypothesis, it is possible to single out some scales L_i of intensive energy supply. In the ranges between these scales, the energy supply is considerably lower; in a number of cases, it can be assumed to be negligible. The energy and the enstrophy transferred from external sources to ocean waters during processes with a characteristic scale L_i and a characteristic velocity v_i, can then be transferred to the motions at other scales that arise because the energy-supplying motion is unsteady. In the case of three-dimensional turbulence, energy is transferred to smaller-scale motions, while in the two-dimensional case the energy flow can be in the opposite direction.

Since there are several specific scales of energy supply in the ocean, the velocity components and other hydrodynamic quantities should be divided into mean values and fluctuations, on the basis of an averaging scale that is determined by the scale of the energy-supplying process. The average equations of motion will then describe the smooth hydrodynamic variables (of scale L_i) that result from the external forces under consideration, while random field fluctuations with scales smaller than L_i will serve as fluctuations. As a result, the fluctuations in the Reynolds stresses $\tau_y = -\rho\overline{u_i'u_j'}$ can be associated with rather large-scale motions. Therefore, the stresses can reach high values. To close the Reynolds equations, one can introduce turbulent exchange coefficients that serve as coefficients of proportionality between the Reynolds stress tensor and the strain-rate tensor, $\Phi_{ij} = \partial\bar{u}_i/\partial x_j + \partial\bar{u}_j/\partial x_i$.

The general form of the dependence of τ_{ij} on Φ_{ij} can be written (Monin and Yaglom, 1965) as

$$\tau_{ij} = -\tfrac{2}{3}\bar{\rho}b\delta_{ij} + \bar{\rho}K_{ij\alpha\beta}\Phi_{\alpha\beta}, \qquad (12.3)$$

where b is the turbulent kinetic energy (per unit mass) and δ_{ij} is the Kronecker delta. The components of the fourth-rank tensor $K_{ij\alpha\beta}$ in (12.3) are turbulent exchange coefficients. Since (12.3) contains only six equations, it is, generally speaking, impossible to determine all the components of $K_{ij\alpha\beta}$ from τ_{ij} and $\Phi_{\alpha\beta}$: additional hypotheses on the structure of $K_{ij\alpha\beta}$ are needed. The symmetry of the tensors τ_{ij} and $\Phi_{\alpha\beta}$ and the incompressibility of the fluid make it possible to reduce the number of independent components of $K_{ij\alpha\beta}$ to 29. To decrease this number still further, Kamenkovich (1967) assumed that the tensor $K_{ij\alpha\beta}$ is symmetric with respect to the vertical axis in the ocean. This reduces the number of independent components to three. However, even in this case the mean equations of motion are rather cumbersome, because the experimental determination of the three coefficients is quite a problem. Much simpler, but less justifiable, assumptions concerning the form of the turbulent viscosity tensor were made by Saint-Guily (1956) and analyzed by Fofonoff (1962). They introduced only two turbulent viscosities: a vertical one, K_z, and a horizontal one, K_l. This simplified assumption concerning the structure of $K_{ij\alpha\beta}$ has become widespread in oceanology.

The turbulent exchange coefficients of momentum, heat and contaminants (the latter two are used in the average heat and diffusion equations) can be determined in two ways. First, the so-called direct method is based on measuring the fluctuations of velocity, temperature, and contaminant concentration, followed by the calculation of the Reynolds stresses and turbulent fluxes of heat and contaminant. In the so-called indirect methods, the turbulent exchange coefficients are estimated by comparing the average observed fields with calculations of simplified average equations of motion, thermal conduction, and contaminant diffusion. In this case, the exchange coefficients are chosen to provide the best agreement between theory and experiment.

Even the first estimates of the eddy viscosity for large-scale motion in the atmosphere and the ocean led to very high K_l-values. For example, Defant (1921b) obtained a value of about 10^{10} and even 10^{11} cm^2 s^{-1} for the overall circulation of the atmosphere. He used the zonal mean flow as the average velocity field and cyclones and anticyclones as the large-scale eddy fluctuations. The idea of a statistical description of these large-scale disturbances put forward by Defant proved to be quite fruitful. Cyclones and anticyclones, which are the eddies of the large-scale turbulence, are important not only as heat carriers but also as vital dynamic factors that promote the general circulation of the atmosphere. When converted into energy of large-scale turbulence, the high energy of the mean flow than creates large-scale eddy friction with large exchange coefficients. It is, however, conceivable that cyclones and anticyclones receive energy locally from the potential energy of an inhomogeneous temperature field and from local heat sources. In this case, the statistical average of the energy conversion over the ensemble of large-scale eddy disturbances may prove to be other than zero, i.e., there may arise smooth air motions on a global scale, at the expense of the energy of the irregular motion. The eddy viscosity K_l determined from empirical data then must be negative. The feasibility of such atmospheric phenomena is proved by data concerning energy budgets, momentum, and angular momentum of the general circulation of the atmosphere. These budgets cannot be balanced without assuming that the energy of irregular atmospheric disturbances is converted into that of the mean flow, at least in certain regions of the Earth.

Large-scale turbulence in the atmosphere and its role in the energy balance were analyzed in detail by Monin (1956a). The expressions for the divergence of the components of the Reynolds stress tensor and for the energy conversion between the average flow and the fluctuations, expressed by Monin with reference to a spherical coordinate system, were estimated from aerological charts. In the case of two-dimensional atmospheric motion with zonal and meridian velocity components v_θ and v_λ, the rate of energy conversion is

$$A = -\frac{1}{R}\left[\frac{\partial \bar{v}_\theta}{\partial \theta}\rho\overline{v_\theta'^2} + \left(\frac{\partial \bar{v}_\lambda}{\partial \theta} - \bar{v}_\lambda \operatorname{ctg}\theta + \frac{1}{\sin\theta}\frac{\partial \bar{v}_\theta}{\partial \lambda}\right)\rho\overline{v_\theta' v_\lambda'} + \right.$$
$$\left. + \left(\bar{v}_\theta \operatorname{ctg}\theta + \frac{1}{\sin\theta}\frac{\partial \bar{v}_\lambda}{\partial \lambda}\right)\rho\overline{v_\lambda'^2}\right]. \tag{12.4}$$

Here, θ and λ are geographical coordinates (complements to latitude and longitude) and R is the Earth's radius. If the average atmospheric motion is close to zonal (i.e., if $\bar{v}_\theta$ = 0 and $\bar{v}_\lambda$ is λ-independent), then (12.4) reduces to

$$A \simeq -\left(\frac{\sin\theta}{R}\frac{\partial}{\partial\theta}\frac{\bar{v}_\lambda}{\sin\theta}\right)\rho\overline{v_\theta' v_\lambda'}. \tag{12.5}$$

Assuming that the turbulent exchange coefficient tensor is isotropic, we obtain $\tau_{ij} = K\Phi_{\alpha\beta}$ from (12.3). Hence, (12.5) takes the form

$$A = \frac{K \sin^2 \theta}{R^2} \left[\left(\frac{\partial}{\partial \theta} \frac{\bar{v}_\theta}{\sin \theta} - \frac{1}{\sin^2 \theta} \frac{\partial \bar{v}_\lambda}{\partial \lambda} \right)^2 + \right.$$

$$\left. + \left(\frac{\partial}{\partial \theta} \frac{\bar{v}_\lambda}{\sin \theta} + \frac{1}{\sin^2 \theta} \frac{\partial \bar{v}_\theta}{\partial \lambda} \right)^2 \right]. \qquad (12.6)$$

Equation (12.6) demonstrates that $A < 0$ corresponds to a negative sign of the turbulent exchange coefficient K.

The quantity A and other large-scale eddy characteristics for the Northern Hemisphere were estimated by Monin with the help of the aerological charts of 500, 700, and 1000 mbar for the period 20 September–19 October, 1951. He determined $\overline{v_\lambda'^2}$, $\overline{v_\theta'^2}$, $\overline{v_\theta' v_\lambda'}$, the total turbulent energy, the horizontal exchange coefficients of momentum and heat, the components of the divergence of the Reynolds stress tensor, and A. It turned out that $\overline{v_\theta'^2}$ and $\overline{v_\lambda'^2}$ in the atmosphere are of the same order of magnitude, and that the velocity correlation tensor is nearly diagonal. In the lower troposphere the large-scale eddy heat flux from the equator to the pole is close to 0.7 J cm^{-2} s^{-1}, the root-mean-square fluctuations amount to 9 m s^{-1}, and the friction velocity v_* is about 2 m s^{-1}. The horizontal turbulent heat exchange coefficient proves to be 5×10^9–10^{10} cm^2 s^{-1}, i.e., somewhat lower than that reported by Defant (1921). This can be attributed to a larger averaging scale employed by Defant in his calculations with climatological data. The rate of energy exchange between the mean motion and the large-scale eddies is about 10^{-1}–10^{-2} cm^2 s^{-3} per unit mass. This value is negative at latitudes θ = 40–50°, i.e., the energy of the turbulence is converted here to that of the mean motion. As estimated by Starr (1953), A is, on average, negative throughout this hemisphere and amounts to some 10^{20}–10^{21} erg s^{-1}.

An important effect of the negative eddy viscosity on the generation of the zonal circulation in the outer layers of the Sun was pointed out by Rossby (1947), who estimated the large-scale eddy transfer of absolute vorticity along meridians. In the force balance of the Sun's photosphere, magnetic forces are of vital importance. This must be taken into account when calculating the large-scale turbulent energy of the photosphere. Starr (1971) published a diagram of the mean energy balance of five types of photospheric energy (nuclear sources, available potential energy of the turbulence, turbulent kinetic energy, kinetic energy of the zonal flow, and turbulent energy of the magnetic field). According to this diagram, the turbulent kinetic energy is converted to kinetic energy of the zonal flow and to turbulent energy of the magnetic field. The plot of the conversion rate of turbulent kinetic energy to zonal-flow energy versus latitude (Starr, 1971), shows that the negative viscosity effect is greatest at latitudes close to 45°.

Gruza (1961) analyzed a large body of empirical information on the large-scale eddy exchange processes in the Earth's atmosphere. The characteristics of the large-scale turbulence were calculated at a pressure level of 700 mbar from daily synoptic charts of 1953 and 1956, and pressure levels of 1000, 850, 700, 500, 300, 200, and 100 mbar with data from January and July of 1956. The calculations demonstrated that the energy of the large-scale turbulence varies negligibly with latitude. On average, the values of A

at temperate latitudes prove to be positive, while in the zone where the mean zonal flow is strongest, the energy of the fluctuations is transferred mainly to the mean flow, i.e., A is negative. Large-scale eddies are responsible for the convergence of warm and cold air masses and for the generation of local regions with large temperature gradients. In these regions the potential energy is converted rapidly into kinetic energy of the large-scale velocity fluctuations. The statistical average of the newly-induced irregular velocities gives a non-zero meridional heat flux.

Starr (1971) considered various physical aspects of phenomena with a negative viscosity. He substantiated the concept of a negative viscosity, which at the time seemed unusual, and even paradoxical. However, still earlier (Gruza, 1961; Monin, 1956a) it was stated that the turbulent exchange coefficients for momentum, heat, and matter characterize the statistical properties of the turbulent fluid motion rather than the physical properties of the fluids themselves. As a result, turbulent exchange coefficients are not necessarily positive; this situation contrasts with that of the corresponding molecular coefficients (which are positive because of the laws of irreversible thermodynamic processes). In his book, Starr gives a systematic presentation of data concerning negative viscosity effects in the Earth's atmosphere and those obtained in model experiments. He also considers applications of the concepts developed to the motion in the Sun's photosphere and in space, and also to the rotation of spiral galaxies and gaseous nebulae. Figure 12.1, taken from Starr's book, depicts some characteristics of the motion in the Earth's atmosphere, including the flux of angular momentum. The negative sign of this flux corresponds with a negative eddy viscosity.

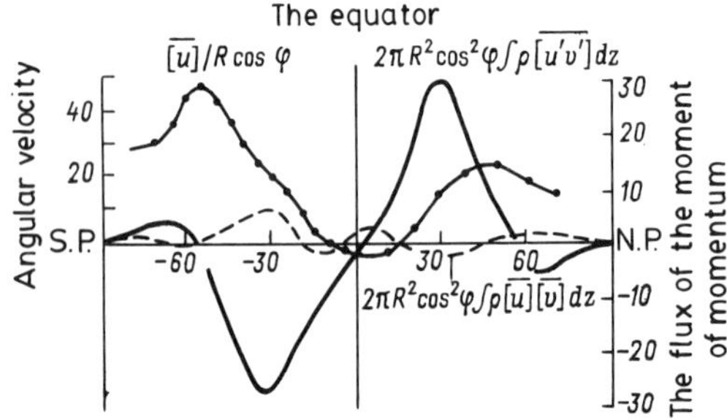

Fig. 12.1. Latitudinal profiles of the mean relative angular velocity of the Earth's atmosphere around the polar axis, $u/R \cos \varphi$ (dotted line), the mean flux of angular momentum $2\pi R^2 \cos^2 \varphi \int \rho \bar{u}\bar{v}\, dz$ that is induced by the mean meridional circulation (dashed line), and of the mean turbulent flux of angular momentum $2\pi R^2 \cos^2 \varphi \int \rho \overline{u'v'}\, dz$ in the northward direction (solid line) (Starr, 1971).

One of the first theoretical calculations of the conditions in which a negative eddy viscosity is possible was carried out by Lorenz (1953). He considered a two-dimensional turbulent flow with a stream function Ψ consisting of two terms, $\bar{\Psi}$ for the mean flow and ψ for the turbulent fluctuations. The mean kinetic energy of the flow per unit area was assumed to be constant in time. The overall energy of the flow E also has two components: $\bar{E}$ is the energy of the mean flow and e is the energy of the turbulent disturbances. The mean motion is considered to be unstable if $\partial\bar{E}/\partial t < 0$ or $\partial\bar{E}/\partial t = 0$; however, $\partial^2\bar{E}/\partial t^2 < 0$. In general, the value and the sign of $\partial\bar{E}/\partial t$ depend not only on

$\bar{\Psi}$, but also on ψ. In a statistical approach to this problem one should consider $\bar{\Psi}$ to be the average over an ensemble M of possible instantaneous values of Ψ, with corresponding instantaneous values of ψ. The statistical properties of the ensemble can be described by multidimensional probability densities P_n, where the increment $P_n(x_1, y_1; x_2, y_2; \ldots x_n, y_n; \psi_1, \psi_2, \ldots, \psi_n)\, d\psi_1, d\psi_2 \ldots$ is the probability of finding ψ between ψ_1 and $\psi_1 + d\psi_1$ at the point (x_1, y_1), between ψ_2 and $\psi_2 + d\psi_2$ at the point (x_2, y_2), etc. In many cases it is sufficient to determine only the first two moments, namely, the average over the ensemble of ψ at the point (x, y), $[\psi(x, y)]$ and the correlation function $F(x, y, x', y') = [\psi(x, y)\psi(x', y')]$. More detailed information on the ensemble M can be gained from ensemble – averaged products at three, four, and more points of the plane (x, y).

Lorenz assumed that the ensemble M is reversible, homogeneous, and isotropic. Such an ensemble can be characterized by a spectrum that is the Fourier transform of the spatial correlation function F taken over the ensemble. The integral of the spectrum over the entire area of wavenumbers is the overall energy of the system; hence it must be limited. As a result, the spectrum cannot be constant at all wavenumbers. In the flow there must be some prevalent frequencies characterized by substantial energies. To investigate the stability characteristics of the flow, Lorenz analyzed the vorticity equation. He assumed homogeneity for ψ, but not for Ψ. In this case, the main flow proves to be stationary if the wavelengths of all turbulent disturbances are shorter than those of all mean-flow harmonics (i.e., if their spectra do not overlap). If there are short-wave components in the mean flow (e.g., in the case of narrow jets) and the spectra of the mean and fluctuating motions overlap, then the flow is unstable. The energy of the fluctuations can then be transferred to the mean flow. This is equivalent, as mentioned above, to a negative eddy viscosity.

Rossby (1936) seems to have been the first worker who pointed to the vital importance of large-scale turbulence in the ocean. Later, some investigators estimated the horizontal turbulent exchange coefficients induced by large-scale turbulence in the ocean. Rossby (1939) investigated the dynamics of the north Atlantic Ocean, which is characterized by the well-known anticyclonic circulation. Rossby discovered that benthic friction did not balance the tangential friction on the ocean surface that results from the anticyclonic wind system. As a result, the wind-induced torque must be compensated by the frictional torque at the continential slope, which is transferred to the water by horizontal turbulent friction. The condition that the two torques must balance was used by Rossby to obtain the horizontal turbulent exchange coefficient for the north Atlantic, $K_l = 10^7$–10^8 cm^2 s^{-1}.

During the following years, the large-scale eddy exchange coefficients in the ocean were estimated several times, both by indirect and by direct methods. In the former case, the distribution of any variable in the ocean – momentum, salt, temperature – was compared with a semi-empirical transfer equation for this variable. The best possible agreement between the calculated and observed fields was used to select the turbulent exchange coefficients involved. In the latter case, the exchange coefficients were calculated from measurements of turbulent fluctuations of velocity, temperature, and salinity in the ocean. For example, using indirect methods based on salinity distributions, Montgomery (1939) obtained $K_l = 4 \times 10^7$ cm^2 s^{-1} for the Counter-Current in the Atlantic Ocean, while Sverdrup (1939) obtained $K_l = 10^8$ cm^2 s^{-1} for the south Atlantic

Ocean. Observations on the diffusion of a radioactive contaminant in Bikini Atoll gave $K_l = 1.5 \times 10^5$ cm^2 s^{-1} (Munk *et al.*, 1949).

Pioneering investigations devoted to determining the large-scale eddy exchange in the ocean were made by Shtokman. He carried out long-term observations of flows in the Caspian Sea (Shtokman and Ivanovsky, 1937). Flow velocities were measured at levels of 5, 10, 20, and 30 m at two-hour intervals for 23 days from an anchored ship. The horizontal exchange coefficients estimated from the data obtained proved to be 10^6–10^7 cm^2 s^{-1}. The large-scale eddy exchange was more intensive along the shoreline than in the perpendicular direction (Shtokman, 1940, 1941).

A large number of horizontal turbulent exchange coefficients were estimated from data concerning the diffusion of discrete particles (indicators) or continuously distributed contaminants (radio-isotopes or dyestuffs) introduced for that purpose into the ocean. Richardson and Stommel (1948, 1949) were the first to experiment with buoyant indicators. With the particles spaced at a few centimeters to some kilometers, K_l proved to range from a few to several hundred square centimeters per second. In experiments with continuously distributed detectors, in which the diffusive currents and the dye spots had sizes ranging from tens to hundreds of meters, K_l also was several square centimeters per second (Ozmidov *et al.*, 1973).

These estimates of K_l, and a great deal more that are available in the literature, demonstrate that in the ocean (and in the atmosphere) there may be a wide range of horizontal turbulent exchange coefficients, which show a strong dependence on the scale of the phenomenon. This dependence seems to be quite natural, since an increase in the scale of the phenomenon must result, generally speaking, in an increase of the fluctuations of the hydrological variables, and in a decrease of the spatial derivatives of the mean variables.

In the ocean, negative viscosity effects in conditions of large-scale turbulence were first discovered in the Gulf Stream. To estimate the rate of transfer of energy between the mean flow and the eddies, Webster (1961, 1965) used data obtained with an electromagnetic meter that was repeatedly towed across the Gulf Stream off Cape Hatteras and in the Florida Strait. The average profile of the longitudinal flow velocity $\bar{v}$ was calculated by averaging all measurements, and the velocity fluctuations were obtained by subtracting the average velocity from the individual v values. In this way, Webster managed to estimate only one term in the expression for A, namely $-\rho \overline{u'v'}\, \partial\bar{v}/\partial x$ (the X-axis was oriented in the direction of the flow, the Y-axis perpendicular to it). The profile of A had positive and negative segments; the latter prevailed (Figure 12.2). As may be seen from Figure 12.2, the energy transfer rate from the eddies to the mean flow is about 79×10^{-4} g cm^{-1} s^{-3} off Cape Hatteras and 3×10^{-4} g cm^{-1} s^{-3} in the Florida Strait.

The values of A estimated by one term, however, cannot be sufficiently reliable. In the general case of three-dimensional flow, the expression for A in the Cartesian coordinate system must consist of 9 terms:

$$A = -\rho \left\{ \overline{u'^2}\frac{\partial\bar{u}}{\partial x} + \overline{u'v'}\frac{\partial\bar{v}}{\partial x} + \overline{u'w'}\frac{\partial\bar{w}}{\partial x} + \overline{v'u'}\frac{\partial\bar{u}}{\partial y} + \right.$$

$$\left. + \overline{v'^2}\frac{\partial\bar{v}}{\partial y} + \overline{v'w'}\frac{\partial\bar{w}}{\partial y} + \overline{w'u'}\frac{\partial\bar{u}}{\partial z} + \overline{w'v'}\frac{\partial\bar{v}}{\partial z} + \overline{w'^2}\frac{\partial\bar{w}}{\partial z} \right\}, \qquad (12.7)$$

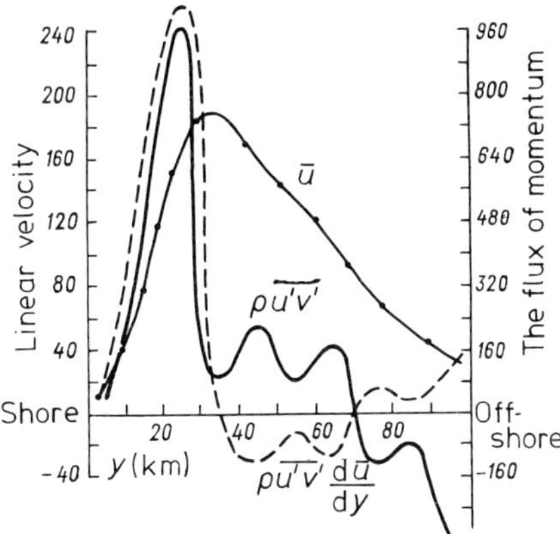

Fig. 12.2. Profiles of mean velocity, $\bar{u}$ (dotted line, cm s^{-1}), turbulent momentum flux from the shore per unit area, $\rho\overline{u'v'}$ (solid line, g cm^{-1} s^{-2}), and of the rate at which the energy of the large-scale turbulence is converted into kinetic energy of the mean flow, $\rho\overline{u'v'}(\partial\bar{u}/\partial y)$ (dashed line); after Webster (1965) in the Gulf Stream north of Cape Hatteras.

where u, v, w are the respective velocity components along the X-, Y- and Z-axes. To measure all the fluctuation and average velocities in the ocean is quite a problem. It is especially difficult to measure the vertical velocity component, which is approximately five orders of magnitude smaller than typical values of horizontal velocity components. As a result, it proves to be extremely difficult to estimate the five terms in (12.7) that involve the component w. However, for large-scale processes in the ocean, at least three of the terms (the third, the sixth, and the ninth) seem to be negligible compared with those containing no w. With less reliability one can also neglect the terms with vertical derivatives of the horizontal mean velocity components. However, since vertical scales in the ocean are only three orders of magnitude smaller than the horizontal ones, and since the ratio of the corresponding velocity components is close to 10^{-5}, the seventh and eighth terms of (12.7) seem also to be of no importance. To estimate the remaining four terms, one needs information concerning $\overline{u'^2}$, $\overline{v'^2}$, $\overline{u'v'}$ and the horizontal derivatives of the mean horizontal velocity components. This information can be obtained by long-term observations of flow velocities at a number of closely spaced points. The idea of such observations, later referred to as polygon observations, was put forward by Shtokman as far back as the pre-war years and carried out in 1956 in the Black Sea (Ozmidov, 1962), in 1958 in the Atlantic Ocean (Ozmidov and Yampolsky, 1965), and in 1967 in the Arabian Sea (Shtokman *et al.*, 1969). In the first case, the measurements were carried out at two points during a period of 18 h, in the second case at three points over one month, in the third case at seven points located at the apex and along the sides of a right-angled triangle over a period of two months at 20-minute intervals. The flow meters were located at several levels, from 25 to 1200 m depth. The buoy stations along the sides of the triangle were 10, 15, 35 miles apart.

The most appropriate data for calculations of A, based on (12.7), proved to be those obtained in the Arabian Sea (Ozmidov *et al.*, 1970). The mean velocity components at the points of measurement were obtained by averaging over the entire measurement period; the velocity fluctuations were calculated by subtracting the mean values from

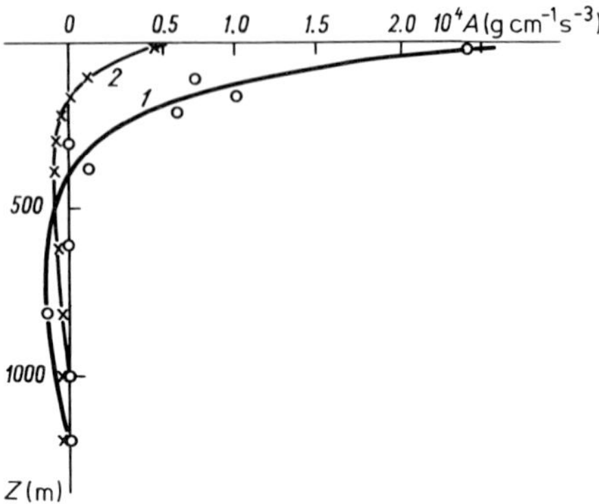

Fig. 12.3. Vertical distribution of the energy conversion rate between the large-scale turbulence and the mean flow, for two averaging scales (curves 1 and 2), based on data from the Arabian polygon (Ozmidov *et al.*, 1970).

the measured ones. The moments $\overline{u'^2}$, $\overline{v'^2}$ and $\overline{u'v'}$ were averaged over the entire observation period. The spatial derivatives of the mean velocity components were calculated with finite-difference expressions. The calculations demonstrated that, on average, the moments decrease with depth and that the diagonal terms substantially exceed the non-diagonal ones. This proved that approximate horizontal isotropy of the velocity field in the polygon. Figure 12.3 shows estimates of A for some levels. The difference in the points used to plot curves 1 and 2 lies in different spatial steps employed to calculate the derivatives (10 and 60 miles, respectively) and in the fact that, in the former case, the moments are calculated from data measured at the central buoy station, while curve 2 is obtained using data measured at all seven stations. Thus, the averaging scale used in the case of curve 2 can be considered to be 6 to 7 times higher than that of curve 1. As seen from Figure 12.3, A is positive in the surface layer; in the case of the first (smaller) averaging scale, the positive sign is maintained down to substantial depths; the quantity A then reaches high values. However, at abyssal levels A is negative for both averaging scales and reaches several thousandths of g cm^{-2} s^{-3} in magnitude. In other words, it is below the value obtained by Webster in the Gulf Stream.

This result can be accounted for, since negative viscosity effects, as shown above, must be much stronger in jet flows. In the Arabian Sea, which is characterized by a comparatively calm flow field, a 'normal' flux of energy from the mean flow to the fluctuations was observed in the upper layers at both averaging scales. However, this energy flux decreased with increasing averaging scale. An energy transfer from the large-scale turbulence to the mean flow was observed for the second (larger) averaging scale in the layer from 150 to 400 m depth, while for the first averaging scale the energy is transferred in the opposite direction. Hence, in this layer the energy of motions with scales intermediate between the two averaging scales is transferred towards both more and less energetic motions, i.e., within the range of scales considered there must be an external source of energy. At depths greater than 400 m, all flows with characteristic sizes that exceed the first averaging scale are 'fed' by small-scale water motions. These features are, however, characteristic of the region of observation; they may be different in other ocean areas with differing typical hydrometeorological conditions.

Information concerning the energy exchange between the mean flow and the large-scale eddies was also obtained in the unique Polygon-70 investigations in the Atlantic Ocean. These experiments lasted for 6 months (Brekhovskih *et al.*, 1971). A 200 × 200 km polygon was fitted with 17 buoy stations with flow-meters at levels from 25 to 1500 m depth (4500 m depth at one of the stations). The data obtained demonstrated a large variability of the velocity field, both in time and in space. The large-scale turbulent energy E was calculated by Vasilenko and Krivelevich (1974) for the velocity field averaged over 25 days. It was discovered that E decreased with depth, the maximum drop in E corresponding to 200–300 m depth. Figure 12.4 shows a typical depth distribution of the large-scale eddy energy obtained from the polygon. In the upper level E is twice that obtained in the Arabian polygon (Beliayev and Ozmidov, 1971). The shapes of the curves $E(z)$ and E, at 300–400 m depth, are almost the same for both polygons.

Lozovatsky (1974) estimated the terms of the large-scale turbulent energy equation from data acquired in Polygon-70. The calculations were performed on the basis of measurements taken at 10 points at 300 m depth. The averaging was realized over successive 5-day periods and over the entire measurement period. The distribution of E proved to be inhomogeneous throughout the polygon (Figure 12.5). It is interesting to note that, in regions with crowded isolines of E, the large-scale eddy energy increased at

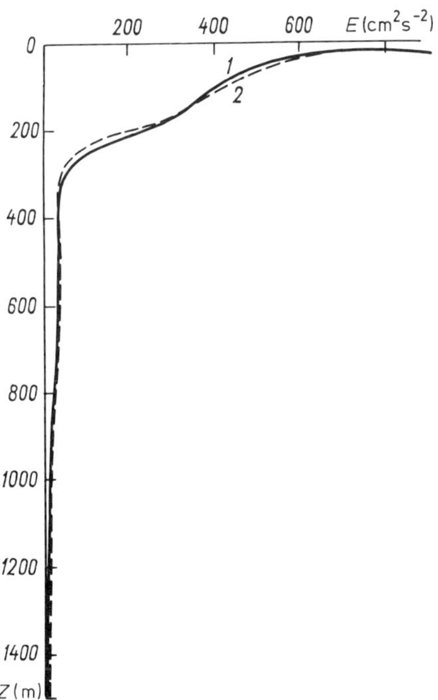

Fig. 12.4. The distribution of large-scale turbulent energy with depth, derived from data measured in 'Polygon-70' (Vasilenko and Krivelevich, 1974). (1) Zonal velocity component, (2) meridional velocity component.

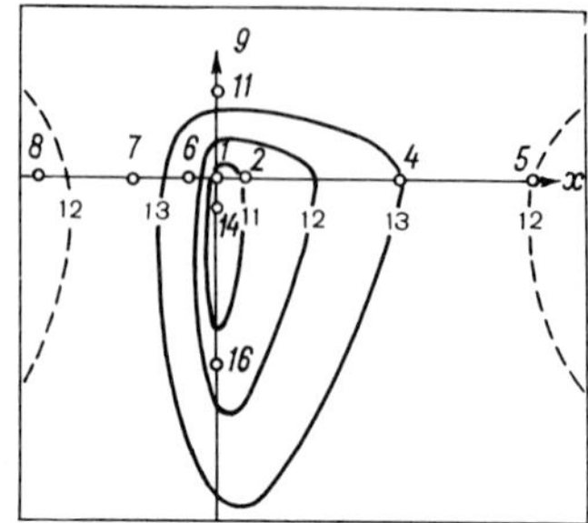

Fig. 12.5. Isolines of twice the large-scale turbulent energy (cm^2 s^{-2}, 'Polygon-70'). Figures at the points denote buoy stations (Lozovatsky, 1974).

the expense of that of the mean flow; therefore, A was positive. Also, in regions with negative values of A, the field E was more homogeneous. The energy transfer by the mean flow proved to be much lower than that by velocity fluctuations. Finally, for processes operating at the scales considered, the spatial turbulent energy transfer exceeded the energy transfer between flows at different scales.

The whole complex of statistical characteristics was first calculated with data obtained in the Black Sea (Ozmidov, 1962). The root-mean-square velocity components of the large-scale turbulence were about 10 cm s^{-1}, the turbulence intensity (the ratio of the root-mean-square fluctuations to the mean velocity components) was about unity. The non-diagonal elements of the stress tensor were less than the diagonal ones. However, they were not zero, which proves the incomplete nature of the isotropy of large-scale turbulence in the sea. The empirical probability densities of the large-scale fluctuations of the horizontal velocity components were plotted by Beliayev and Ozmidov from data obtained in the Arabian Sea (1971). The calculations were carried out for levels at 25 and 1200 m depth. The probability densities obtained were not Gaussian ones; the moments of the distribution varied with depth in a systematic way. The skewness (S) and kurtosis (K) of the probability distribution changed sign in the layer of density discontinuity: S and K were, as a rule, negative above this discontinuity and positive below it. This variation in the probability distributions with depth can, most likely, be attributed to different velocity generation mechanisms at different depths. In particular, the situation is different below the layer of density discontinuity where the water is protected from direct wind effects by a layer with a high density gradient. Possible factors that may be responsible for the variations in the probability density can be judged from the probability-density equations (Monin, 1967b), which include advection, diffusion, and non-linear terms.

The histograms of velocity components measured in Polygon-70 also differ significantly from the standard ones (Vasilenko and Krivelevich, 1974). Various shapes of the histograms of velocity components and flow directions have been reported by the Woods Hole Oceanographic Institute (1965, 1966, 1967, 1970, 1971, 1975). The histograms were plotted from data obtained at a long-term station in the north Atlantic Ocean at 39°29′ N, 70° W (station D).

Subsequent to the Polygon-70 experiments, large-scale turbulence was investigated in the Atlantic Ocean by American and English scientists in the MODE program (Koshliakov and Monin, 1978). The eddy characteristics obtained differed noticeably in the main thermocline layer and in the interior (Robinson, 1975). The existence of large-scale eddies in deep layers was discovered, with the help of neutrally buoyant buoys, by Swallow and Worthington (1961), Pochapsky (1963, 1966), Swallow (1969), and others. In contrast with stationary measurements taken in polygons, one can determine the Lagrangian characteristics of large-scale turbulence by tracing the trajectories of such buoys. This is of particular importance for diffusion problems. The observation of horizontal large-scale turbulence from space seems to be extremely promising. Turbulent eddies, several kilometers to tens of kilometers in diameter, were observed in various ocean regions from the orbiting satellite Skylab (Stevenson, 1974, 1975). Ocean eddies were also frequently observed by Soviet astronauts in orbit. Eddies can be identified from space by surface temperature gradients, by differences in water color and surface roughness between the eddies and the surrounding water, by plankton and foam gathered at the boundaries of water areas with different temperature, etc.

A large amount of information concerning the structure of large-scale turbulence in the ocean was obtained by the Soviet–American experiment POLYMODE, completed in the autumn of 1978. The year-long measurements of flows and other hydrophysical variables yielded new data concerning the parameters of large-scale turbulence and their variability in a very interesting region of the Atlantic Ocean: the vicinity of the Bermuda triangle. These data (at present being processed) will undoubtedly contribute to a better understanding of the kinematics and dynamics of large eddies in the ocean.

Before discussing the available data describing the spectrum of large-scale turbulence in the ocean, let us now review the various theoretical interpretations of two-dimensional turbulence. This is the subject of the following section.

13. THEORY OF TWO-DIMENSIONAL TURBULENCE

In the inertial range of three-dimensional turbulence, there occurs a spectral energy cascade from small to large wavenumbers. This cascade exhibits a constant (independent of wavenumber) spectral energy flux, ϵ. According to the second similarity hypothesis by Kolmogorov (see §5), ϵ appears to be the only constant parameter that determines the statistical characteristics of turbulence components in the inertial range.

This statement refers in particular to such statistical characteristics as the induced eddy diffusivity for passive contaminants, K. Turbulent diffusion of a contaminant cloud of size l is mainly contributed to by turbulent vortices with dimensions of the order of l, so that K is a function of l. Providing that l belongs to the inertial range, this function can depend only on the dimensional parameter ϵ. For dimensional reasons, it must have the form

$$K(l) = C_K \epsilon^{1/3} l^{4/3}, \tag{13.1}$$

where C_K is a numerical constant of order one. This 4/3-power law was first stated by Richardson (1926); for applications to the theory of turbulent diffusion, see Monin (1956b) and Ozmidov, (1958). This law has been confirmed in a number of experiments (see §14). In the atmosphere and the ocean it may be expected to be valued at scales either smaller than, or comparable with, the effective heights over which the turbulence may be said to be three-dimensional. At scales which exceed the effective height of the atmsophere and the ocean, the turbulence is quasi-two-dimensional. Its statistical properties then will reveal the specific features of two-dimensional turbulence that distinguish it from its three-dimensional counterpart.

The most important of these features is that, in two-dimensional flows of incompressible and inviscid fluids, the particles conserve not only their kinetic energy, but also their vorticity ω. The *enstrophy*, which is one-half the squared vorticity, then is also conserved. In a viscous fluid, the kinetic energy $E = \frac{1}{2}\overline{|\mathbf{u}|^2}$ and the enstrophy $\Omega = \frac{1}{2}\overline{\omega^2}$ decay according to

$$\frac{\partial E}{\partial t} = -2\nu\Omega \quad (= -\epsilon); \qquad \frac{\partial \Omega}{\partial t} = -\nu\overline{|\nabla\omega|^2} \quad (= -\epsilon_\omega). \tag{13.2}$$

Hence, for all finite initial values of $\overline{|u|^2}$ and $\overline{\omega^2}$, the value of $\overline{\omega^2}$ can only decrease in time. Also, at $\nu \to 0$, we have $\epsilon \to 0$, so that at large Reynolds numbers the kinetic energy will be nearly constant and the spectral transfer of considerable energy from small to large wavenumbers will be impossible. According to Batchelor (1969), the energy spectrum in he energy cascade range then must have the form

$$E(k, t) = E^{3/2}\, t f(E^{1/2} k t).$$

Hence, $\Omega = \frac{1}{2}At^{-2}$ and $\epsilon_\omega = At^{-3}$, where $A = \int_0^\infty x^2 f(x)\, dx$ is a numerical constant. However, for $\nu \to 0$, an enstrophy cascade from small to large wavenumbers and a non-zero limit of ϵ_ω are quite possible. In the inertial range of wavenumbers k, or, more precisely, for

$$L^{-1} \ll k \ll \min(\epsilon^{1/4}\nu^{-3/4}, \epsilon_\omega^{1/6}\nu^{-1/2}),$$

the energy spectrum $E(k)$ then will be determined, generally speaking, by both the parameters ϵ and ϵ_ω. In this case, the constant length scale $L_\omega = (\epsilon/\epsilon_\omega)^{1/2}$ arises. For dimensional reasons, the energy spectrum then must have the form

$$E(k) = C_1(kL_\omega)\epsilon^{2/3}k^{-5/3},$$

where $C_1(kL_\omega)$ is a function of the dimensionless wavenumber kL_ω. Assuming that the parameter ϵ is essential only at one end of the inertial range, and that the parameter ϵ_ω is essential only at the opposite end, we then have $C_1(kL_\omega) \simeq$ const. The energy spectrum then obeys a conventional 5/3-power law in the first case, while in the latter case $C_1(kL_\omega) \simeq (kL_\omega)^{-4/3}$, so that the energy spectrum must be expressed as a minus-three law:

$$E(k) = C_\omega \epsilon^{2/3} k^{-3}. \tag{13.3}$$

Here C_ω is a numeric constant. Over the range of applicability of this law, the turbulent diffusivity $K(l)$ does not obey Richardson's 4/3-power law (13.1), but the square law

$$K(l) = C_K' \epsilon_\omega^{1/3} l^2. \tag{13.4}$$

Last, but not least, in the inertial-convective range, or, more precisely, for

$$L^{-1} \ll k \leqslant \min(\epsilon^{1/4}\nu^{-3/4}, \epsilon^{1/4}\chi^{-3/4}, \epsilon_\omega^{1/6}\nu^{-1/2}),$$

the spectrum $E_T(k)$ of the passive contaminant will be determined by the three parameters ϵ, ϵ_ω, and ϵ_T. For dimensional reasons, this spectrum must have the form

$$E_T(k) = B_1(kL_\omega)\epsilon_T\epsilon^{-1/3}k^{-5/3},$$

where $B_1(kL_\omega)$ is a function of kL_ω. At the end of the inertial-convective range, where the parameter ϵ_ω is not essential, we have $B_1(kL_\omega) \simeq$ const. The conventional 5/3-power law (5.12) then holds for $E_T(k)$. At the opposite end of this range, where the parameter ϵ is not essential, we have $B_1(kL_\omega) \simeq (kL_\omega)^{2/3}$. The spectrum $E_T(k)$ then is expressed as a minus-one power law,

$$E_T(k) = B_\omega\epsilon_T\epsilon_\omega^{-1/3}k^{-1}, \tag{13.5}$$

where B_ω is a numerical constant (Gavrilin *et al.*, 1972; Kraichnan, 1974; Saunders, 1972).

It was Ogyra (1952) who first discovered the minus-three law (13.3) during the examination of large-scale meteorological spectra (see also Ogyra, 1958, 1962). Kraichnan (1967) interpreted this law as being a consequence of spectral enstrophy transfer (see also Kraichnan, 1971, 1975). Batchelor (1969) supplemented this interpretation with the results of preliminary theoretical experiments on the evolution of two-dimensional turbulence that were carried out by his graduate student Bray in the period 1961–64 and presented as a thesis in 1966.

To discover a relation between the spectral cascades of kinetic energy and the enstrophy, let us consider, following Kraichnan (1967), two-dimensional isotropic turbulence which obeys the following spectral energy equation

$$\left(\frac{\partial}{\partial t} + 2\nu k^2\right) E(k) = \frac{1}{2}\int_0^\infty \int_0^\infty T(k, p, q)\, \mathrm{d}p\, \mathrm{d}q. \tag{13.6}$$

This is derived from (5.23) in a similar way as (5.24). In (13.6) the function $T(k, p, q)$ is expressed as the Fourier transform of the third-order two-point correlation function of the velocity field. It has the following properties: it differs from zero only if k, p, q can form a triangle, it is symmetric with respect to the second and the third arguments, and it obeys the equations $T(k, p, q) + T(p, q, k) + T(q, k, p) = 0$ and $k^2 T(k, p, q) + p^2 T(p, q, k) + q^2 T(q, k, p) = 0$. These equations express energy and enstrophy conservation due to interactions between any three turbulence components with the wavevectors **k**, **p**, and **q**. According to these laws, we have

$$\frac{T(k, p, q)}{T(q, k, p)} = \frac{p^2 - q^2}{k^2 - p^2} \quad \text{and} \quad \frac{T(p, q, k)}{T(q, k, p)} = \frac{q^2 - k^2}{k^2 - p^2}.$$

Now consider a spectral region where the functions $E(k)$ and $T(k, p, q)$ demonstrate the following similarity properties:

$$\frac{E(ak)}{E(k)} = a^{-n}; \qquad \frac{T(ak, ap, aq)}{T(k, q, p)} = a^{-(3n+1)/2},$$

with arbitrary a. The exponent in the latter condition is the same as that in the function $[E(k)]^{3/2}k^{-1/2}$, which has the same dimensions as T. These properties of the function

$T(k, p, q)$ are used to derive the following expressions for the energy and enstrophy fluxes $\Pi(k)$ and $Q(k)$ from wavenumbers smaller than k to those larger than k:

$$\Pi(k) = \frac{1}{2} \int_k^\infty dk' \int_0^k \int_0^k T(k', p, q)\, dp\, dq -$$

$$- \frac{1}{2} \int_0^k dk' \int_k^\infty \int_k^\infty T(k', p, q)\, dp\, dq$$

$$= k^{(5-3n)/2} \int_0^1 dv \int_1^\infty T(1, v, w) W_1(v, w, n)\, dw, \tag{13.7}$$

where

$$W_1(v, w, n) = -(w^2 - v^2)^{-1} \left[(1 - v^2) \int_1^w u^{(3n-7)/2}\, du - \right.$$

$$\left. - (w^2 - 1) \int_v^1 u^{(3n-7)/2}\, du \right],$$

and

$$Q(k) = \frac{1}{2} \int_k^\infty k'^2\, dk' \int_0^k \int_0^k T(k', p, q)\, dp\, dq -$$

$$- \frac{1}{2} \int_0^k k'^2\, dk' \int_k^\infty \int_k^\infty T(k', p, q)\, dp\, dq$$

$$= k^{(9-3n)/2} \int_0^1 dv \int_1^\infty T(1, v, w) W_2(v, w, n)\, dw, \tag{13.8}$$

where

$$W_2(v, w, n) = -(w^2 - v^2)^{-1} \left[(1 - v^2) w^2 \int_1^w u^{(3n-11)/2}\, du - \right.$$

$$\left. - (w^2 - 1) v^2 \int_0^1 u^{(3n-11)/2}\, du \right].$$

Hence, we see that in the energy cascading range, which obeys the 5/3-power law, i.e., for $n = 5/3$, $\Pi(k)$ is independent of k (it equals ϵ), and $W_2(v, w, 5/3)$ is zero, Therefore, $Q(k) = 0$, too. On the other hand, in the enstrophy cascading range, which corresponds to the minus-three law (13.3), i.e., for $n = 3$, the quantity $Q(k)$ is independent of k (it equals ϵ_ω), and $W_1(v, w, 3)$ equals zero so that $\Pi(k) = 0$. Moreover, one can prove that $W_1(v, w, 5/3) > 0$ and $W_2(v, w, 3) < 0$, so that the sign of ϵ in the energy cascading range and the sign of ϵ_ω in the enstrophy cascading range are opposite.

In conditions of three-dimensional isotropic turbulence, the absolute statistical equilibrium at zero viscosity corresponds to a homogeneous energy distribution over all degrees of freedom (all wavenumbers k). This leads to a constant three-dimensional spectrum, $F(k)$ = const, and a spectrum $E(k) = 4\pi k^2 F(k) \simeq k^2$. For a spectrum $E(k) \simeq k^{-5/3}$, the energy levels at large wavenumbers then prove to be much lower than those for absolute equilibrium conditions at small wavenumbers. It would be natural to expect that the interactions of wavenumbers then must tend to restore the absolute equilibrium, i.e., to create a spectral energy flux from small to large wavenumbers, $\epsilon > 0$.

In conditions of two-dimensional isotropic turbulence at zero viscosity, the quantity

$$\int_0^\infty (\alpha + \beta k^2) F(k)\, \mathrm{d}k$$

is conservative (α and β are constants). Under conditions of absolute statistical equilibrium, or conditions of homogeneous energy distribution over all the degrees of freedom, the two-dimensional spectrum is given by $F(k) \simeq 1/(\alpha + \beta k^2)$ and the spectral enstrophy density (Obukhov, 1969; Kraichnan, 1967) by

$$k^2 E(k) = \pi k^3 F(k) \simeq \frac{\pi k^3}{\alpha + \beta k^2}\,.$$

For $k^2 E(k) \simeq k^{-1}$, the enstrophy levels at large wavenumbers prove to be much lower than those of the absolute equilibrium at small wavenumbers. The interaction between wavenumbers must tend to restore the absolute equilibrium, i.e., it creates a spectral enstrophy flux from small to large wavenumbers, $\epsilon_\omega > 0$. In (13.8) the condition $T(l, v, w) < 0$ pertains to this situation; hence, (23.7) yields $\epsilon < 0$. Thus, if, in the vicinity of a wavenumber k_1, there is an external energy flux at a rate ϵ and an enstrophy flux at a rate ϵ_ω, and if k_1 is within the inertial range, one can expect most of the incoming enstrophy to be transferred toward large wavenumbers by the minus-three law. At $k \gg k_1$ the enstrophy dissipates due to the viscosity of the fluid. At the same time, most of the incoming energy is transferred toward small wavenumbers by the 5/3-power law, and it reaches wavenumbers $k \ll k_1$ in time $t \simeq (\epsilon k^2)^{-1/3}$. When the volume occupied by the fluid is limited, and corresponds to a minimum wavenumber $k_0 \ll k_1$, the energy accumulates in turbulence components with the wavenumber k_0.

Thus, the absolute statistical equilibrium mentioned above is attainable only in inviscid fluids. Let us consider inviscid fluid motion inside a square with sides D. If we expand the velocity field in Fourier series, and distinguish between the real and imaginary parts of the Fourier amplitude $u_j(\mathbf{k}) = a_j(\mathbf{k}) + ib_j(\mathbf{k})$, we conclude that all values of k that belong to the semi-plane form the phase space of the moving fluid (Lee, 1952). Also, the flux of phase points is non-divergent inside this space. In other words, the detailed Liouville theorem holds:

$$\frac{\partial \dot{a}_j(\mathbf{k})}{\partial a_j(\mathbf{k})} + \frac{\partial \dot{b}_j(\mathbf{k})}{\partial b_j(\mathbf{k})} = 0,$$

where the dots denote time derivatives. This is also valid when all Fourier amplitudes with wavenumbers k outside the range $k_0 \leqslant k \leqslant k_{\max}$ are set equal to zero. The total energy $\hat{E}$ and the total enstrophy $\hat{\Omega}$ are integrals of the equation of motion, so that

$$\sum_{\mathbf{k}} \dot{u}_j(\mathbf{k}) \frac{\partial \hat{E}}{\partial u_j(\mathbf{k})} = 0 \quad \text{and} \quad \sum_{\mathbf{k}} \dot{u}_j(\mathbf{k}) \frac{\partial \hat{\Omega}}{\partial u_j(\mathbf{k})} = 0.$$

These equalities, the Liouville theorem, and the continuity equation

$$\frac{\partial P}{\partial t} + \sum_{\mathbf{k}} \frac{\partial}{\partial u_j(\mathbf{k})} \left[\dot{u}_j(k) \cdot \frac{\partial P}{\partial u_j(\mathbf{k})} \right] = 0,$$

show that all probability densities of the form $P(\hat{E}, \hat{\Omega})$ do not vary with time in phase space.

The termal equilibrium is consistent with the canonical probability distribution $P \simeq \exp(-\alpha\hat{E} - \beta\hat{\Omega})$, with 'temperatures' α^{-1} and β^{-1}. It is stable with respect to all interactions inside the system that obey the integrals of motion $\hat{E}$ and $\hat{\Omega}$, and it has the two-dimensional spectrum $F(k) = 1/2(\alpha + \beta k^2)^{-1}$. Kraichnan (1967, 1971, 1975) states that, depending on the signs of α and β and on $k_1^2 = \langle\hat{\Omega}\rangle\langle\hat{E}\rangle^{-1}$ (where the angular brackets denote averaging over the canonical distribution), three types of absolute equilibrium are possible. First, a standard equilibrium with positive temperatures, when $\alpha > 0$, $\beta > 0$, and $k_a^2 < k_1^2 < k_b^2$. Second, an equilibrium with a negative temperature, when $\beta > 0$, $0 > \alpha > -\beta k_a^2$, and $k_0^2 < k_1^2 < k_a^2$. Third, an equilibrium with a negative secondary temperature, when $\alpha > 0$, $0 > \beta k_{\max}^2 > -\alpha$, and $k_b^2 < k_1^2 < k_{\max}^2$, where

$$k_a^2 = \frac{1}{2} \frac{k_{\max}^2 - k_0^2}{\ln(k_{\max}/k_0)} \quad \text{and} \quad k_b^2 = \tfrac{1}{2}(k_{\max}^2 + k_0^2).$$

Spatially unlimited flows exhibit only the standard type of equilibrium with positive temperatures.

These equilibria are, in many respects, similar to the equilibrium states in the classical limit of a perfect quantum boson gas, with a number of particles $\hat{E}$, a kinetic energy $\hat{\Omega}$, and $k_0 \leqslant k \leqslant k_{\max}$. For this gas, β^{-1} is temperature and $\alpha\beta^{-1}$ is a chemical potential. In particular, for $\alpha < 0$, the condensation of kinetic energy of two-dimensional turbulence at the minimum wavenumber k_0 is similar to the condensation of two-dimensional free boson gas particles to the ground state over the range of finite sizes. In the case of a three-dimensional boson gas, this condensation is also plausible in an infinite space. The difference between turbulence and a boson gas lies in the fact that a fluid can approach equilibrium because of internal interactions, while in the case of a boson gas the interactions must be imposed from the outside. Moreover, the cut-off at $k_{\max}$ results from the dissipative effect of viscosity, which violates the equilibrium in real fluids, and from the non-dissipative quantum effect which modulates, rather than violates, the equilibrium in a boson gas.

The predictions concerning the relationship between the spectral fluxes of energy and enstrophy given above were confirmed by calculating the functions (13.7)–(13.8) for two-dimensional Gaussian turbulence (Kraichnan, 1967) and the functions $\Pi(k)$ and $Q(k)$ in the so-called diffusion approximation (Leith, 1968)::

$$\Pi(k) = -bk^{-1} \frac{\partial}{\partial k} k^{9/2} E^{3/2}(k); \qquad Q(k) = -bk^3 \frac{\partial}{\partial k} k^{5/2} E^{3/2}(k). \tag{13.9}$$

Here, b is a positive numerical constant, so that $\Pi = \text{const} < 0$ and $Q = 0$ in the case of the 5/3-power law, while $Q = \text{const} > 0$ and $\Pi = 0$ in the case of the minus-three law.

Note that, in the case of the minus-three law (13.3), the total enstrophy $\int_{k_1}^{k_{\max}} k^2 E(k)\,\mathrm{d}k$ logarithmically diverges both at $k_1 \to 0$ and $k_{\max} \to \infty$ for the energy spectrum in the inertial range. To avoid this divergence, the minus-three law must be supplied with a logarithmic correction. Kraichnan (1971) proposed this correction, based on the following considerations. The spectral energy flux $\Pi(k)$ can be presented, in order of magnitude, as the product of the total energy $kE(k)$ of the motion at wavenumber k and the root-mean-square velocity gradient of the large-scale motion (with wavenumbers below k). This is determined, as in (5.26) and (5.27), by the expression

$$\omega(k) \simeq \left[\int_0^k k'^2 E(k')\,\mathrm{d}k'\right]^{1/2}. \tag{13.10}$$

In this case, the condition $\Pi(k) = \text{const}$ ensures only a 5/3-power law, because the integral in (13.10) then is mainly contributed to by wavenumbers $k' \simeq k$. This demonstrates the locality of the spectral energy transfer. In a similar way, the spectral enstrophy flux can be presented as $Q(k) \simeq k^3 E(k) \cdot \omega(k)$. With a minus-three law, we would then have $\omega(k) \simeq [\ln(k/k_1)]^{1/2}$; also, $Q(k)$ would increase with k. To make sure that $Q(k) = \text{const}$, we must put, instead of (13.3),

$$E(k) = C_\omega \epsilon_\omega^{2/3} k^{-3} \left[\ln \frac{k}{k_1}\right]^{-1/3}. \tag{13.3'}$$

For $k \gg k_1$, the integral in (13.10) then is primarily contributed to by wavenumbers k' that belong to the range $k_1 \ll k' \ll k$. Spectral enstrophy transfer thus appears to be a much less localized process than spectral energy transfer. Using (13.3′) and the so-called almost-Markovian Galilean-invariant turbulence model, Kraichnan estimated $C_\omega = 1.74$, while similar calculations for the coefficient C_1 in the 5/3-power law gave, for the three-dimensional and two-dimensional cases, the estimates $C_1 = 1.40$ and $C_1 = 6.69$, respectively. Later, Kraichnan (1975) proposed that in the enstrophy cascading range the intermittency of spatial enstrophy derivatives increases with k. This is not so for enstrophy itself, since enstrophy is preserved in every fluid particle, provided that the effect of viscosity is negligible. As a result, this intermittency does not affect the spectral shape (13.3′). Saffman's theory (1971), which in the enstrophy cascading range results in the law $E(k) \simeq k^{-4}$ rather than in the law (13.3′), cannot be valid because enstrophy concentration in thin boundary layers between large vortices is not possible. That is why the numerical corroboration of Saffman's theory (Deem and Zabusky, 1971) appears to be accidental. On the other hand, in the energy cascading range the intermittency of non-negative hydrodynamic characteristics increases with k. This results in the substitution of the 5/3-power law by a law having a lower exponent, and in the tendency of large-scale vortices with the same sign to group together on the plane.

So far we have considered two-dimensional random vorticity fields, $\omega(\mathbf{x}, t) = \mathrm{rot}_z\, \mathbf{u}(\mathbf{x}, t)$. For a number of calculations a model with a discrete distribution of N point vortices is of interest:

$$\omega(\mathbf{x}, t) = \sum_{j=1}^{N} q_j \delta[\mathbf{x} - \mathbf{r}_j(t)]. \tag{13.11}$$

Here, q_j is the vortex strength and $\mathbf{r}_j = (x_j, y_j)$ are the vortex coordinates. Statistical mechanical investigations of this model were carried out by Lin (1943), Onsager (1949) and Fjörtöft (1953), who discovered the possibility of energy transfer from small to large scales. In inviscid fluids, the vorticity field (13.11) forms a Hamiltonian system with the Hamiltonian

$$H = -\frac{1}{2\pi} \sum_{j>l} q_j q_l \ln|\mathbf{r}_j - \mathbf{r}_l|. \tag{13.12}$$

The equations of motion then assume the form

$$\frac{dx_j}{dt} = -\frac{1}{2\pi} \sum_{l \neq j} \frac{q_l(y_j - y_l)}{|r_j - r_l|^2}; \qquad \frac{dy_j}{dt} = \frac{1}{2\pi} \sum_{l \neq j} \frac{q_l(x_j - x_l)}{|r_j - r_l|^2}. \tag{13.13}$$

Apart from the Hamiltonian itself, the functions

$$\sum_j q_j x_j = \text{const} \quad \left(= X \sum_j q_j\right); \qquad \sum_j q_j x_j = \text{const} \quad \left(= Y \sum_j q_j\right);$$
$$\sum_j q_j [(x_j - X)^2 + (y_j - Y)^2] = \text{const} \quad \left(= R^2 \sum_j q_j\right) \tag{13.14}$$

are also integrals of the equations of motion.

A stationary point (X, Y) is referred to as a *vorticity centre*. Analytical solutions of (13.13) are known for $N = 1, 2, 3$. A solitary vortex is stationary. Two vortices rotate with a constant velocity around the vorticity centre, the distance between them being constant. Except for one particular case, three vortices always perform a periodic relative motion (Novikov, 1975). In the wave approximation, the vorticity field (13.11) is described by the function

$$\tilde{\omega}(\mathbf{k}, t) = \frac{1}{2\pi} \int \omega(\mathbf{x}, t)\, e^{-i\mathbf{k}\mathbf{x}}\, d\mathbf{x} = \frac{1}{2\pi} \sum_j q_j\, e^{-i\mathbf{k}\mathbf{x}_j(t)}. \tag{13.15}$$

This field corresponds to the following spectral energy density (Novikov, 1975)

$$E(k) = \frac{\pi}{k} \overline{|\tilde{\omega}(\mathbf{k})|^2} = \frac{1}{4\pi k} \left[2 \sum_{j>l} q_j q_l J_0(k|\mathbf{r}_j - \mathbf{r}_l|) + \sum_j q_j^2 \right], \tag{13.16}$$

where the bar designates averaging over all directions of the wave-vector. The first term in (13.16) corresponds to the vortex interaction energy, the second to the vortex energy. A continuous vorticity distribution has no energy of its own. A continuous distribution can approximate the discrete system of vortices (13.11), provided the continuum is cut off at the wavenumber $k_{\max} \simeq n^{1/2}$, where $n = ND^{-2}$ is the particle number density in the discrete system, and provided it is of no importance when vortices approach each other at distances $a \ll n^{-1/2}$ (the so-called supercondensation at negative temperatures), see Kraichnan (1975).

Equation (13.16) yields

$$\int_{L^{-1}}^{l^{-1}} E(k)\,\mathrm{d}k = \frac{1}{4\pi}\left(\ln\frac{L}{l}\right)\sum_j q_j^2 + \frac{1}{2\pi}\sum_{j>l} q_j q_l \ln\frac{LC}{|\mathbf{r}_j - \mathbf{r}_l|}; \qquad (13.17)$$

$$\int_{L^{-1}}^{l^{-1}} kE(k)\,\mathrm{d}k = \frac{l^{-1} - L^{-1}}{4\pi}\sum_j q_j^2 + \frac{1}{2\pi}\sum_{j>l}\frac{q_j q_l}{|\mathbf{r}_j - \mathbf{r}_l|}, \qquad (13.18)$$

where $C = 2e^{-a}$ and $a \simeq 0.577$ is Euler's constant. The ratio of the second expression to the first is the mean wavenumber $\bar{k}(t)$ with respect to energy. The ratio is mainly affected by the second term in (13.18); in the case of similar vortices, this term is proportional to the relative velocities (13.13) induced by them. The evolution of the vortex system then slows down with decreasing $\bar{k}(t)$. In other words, for most of the time the system is characterized by small values of $\bar{k}$. This implies a statistically irreversible tendency to transfer energy from large to small k. Moreover, one can readily prove the inequality

$$\bar{k}(t)\bar{\bar{k}}(t) \geqslant k_1^2 = \langle\hat{\Omega}\rangle\langle\hat{E}\rangle^{-1}, \quad \text{with } \bar{k}(t) \leqslant k_1, \quad \bar{\bar{k}}(t) \geqslant k_1. \qquad (13.19)$$

Here, $\bar{\bar{k}}(t)$ is the mean wavenumber with respect to enstrophy (Novikov, 1974). This inequality demonstrates that the energy transfer towards small k (decreasing $\bar{k}$) must be accompanied by enstrophy transfer towards large k (increasing $\bar{\bar{k}}$).

The dynamics of the discrete system of point vortices (13.11) is quite similar to that of a plasma with two main centers that consist of charged filaments arranged parallel to a constant magnetic field and that moves perpendicular to the direction of the field due to Coulomb interactions (Edwards and Taylor, 1974; Joyce and Montgomery, 1972, 1973; Montgomery, 1972; Montgomery and Joyce, 1974; Montgomery *et al.*, 1972; Montgomery and Tappert, 1972; Seyler, 1974; Taylor, 1972; Taylor and McNamara, 1971; Vahala, 1972; Vahala and Montgomery, 1971). In this case, one must define

$$q_j = -\frac{4\pi c e_j \mathbf{B}}{l B^2},$$

where $\mathbf{B}$ is the strength of the magnetic field, e_j is the charge of the jth filament, l is its length, c is the speed of light, and the Hamiltonian (12.13) is proportional to the Coulomb interaction energy

$$E = -2\sum_{j>l}\frac{e_j e_l}{l^2}\ln|\mathbf{r}_j - \mathbf{r}_l|.$$

Two-dimensional turbulence dynamics can be investigated by computer simulation. The very first numerical experiments of this kind, carried out by Bray (see Batchelor, 1969), proved the validity of the inverse-cube law (13.13). More detailed data were obtained by Lilly (1969), who numerically integrated the two-dimensional vorticity equation

$$\frac{\partial\,\Delta\psi}{\partial t} + \frac{\partial(\psi, \Delta\psi)}{\partial(x, y)} - \nu\,\Delta\Delta\psi = F. \qquad (13.20)$$

He found the value of the exciting function F at the nth time step, F_n, by solving the first-order difference equation of the Markovian process

$$F_{n+1} = R_n F_n + (1 + R_n^2)^{1/2} \hat{F}_{n+1}, \quad \text{where } R_n = \frac{1 - \Delta t/2\tau}{1 + \Delta t/2\tau}.$$

Here, $\hat{F}_{n+1}$ is given by a random set of Gaussian amplitudes for all the Fourier components with wave-vectors $\mathbf{k} = (k_x, k_y)$ on the sides of a square of size $2k_1$ (a statistically stationary random excitation at wavenumber k_1). In the main experiment, the following dimensionless values were used: $k_1 = 2\pi$, $\nu = 2.5 \times 10^{-4}$, $\tau = 1/2$, and $\Delta x = 1/8$. A 64×64-point spatial grid was chosen. After 2200 time steps, the spectrum $E(k)$ obeyed the 5/3-power law with a constant $C_1 \simeq 6.2$ for $k < k_1$, while for $k > k_1$ it obeyed, over one or two octaves, the minus-three law with a constant $C_\omega \simeq 4$. This estimate of C_ω differs from Kraichnan's theoretical one, which was mentioned earlier. However, in experiments to investigate the decay of two-dimensional turbulence (when F was put equal to zero after 50 time steps) Lilly (1971, 1972a, b) estimated $C_\omega \simeq 2$, which was close to the theory. He also confirmed Batchelor's laws for the decay of turbulence, $\Omega = \frac{1}{2}At^{-2}$ and $\epsilon_\omega = At^{-3}$. These, however, do not agree with the minus-three law, but with (13.3′). The validity of the minus-three law was also confirmed in numerical experiments by Gavrilin *et al.* (1972), Gavrilin and Mirabel (1972), and Mirabel (1974). Mirabel (1974) presented experimental data describing the decay of turbulence in which the initial field ψ was a random Gaussian field with a spectrum equal to $(k_0^2/\pi)(k^2 + k_0^2)^{-2}$ ($C_\omega \simeq 1.54$). He also presented data concerning the evolution of two-dimensional fields of passive contaminants, which is described by the equation

$$\frac{\partial \vartheta'}{\partial t} + \frac{\partial(\psi, \vartheta')}{\partial(x, y)} - \chi \, \Delta\vartheta' = \epsilon_{ij} \frac{\partial \bar{\vartheta}}{\partial x_i} \frac{\partial \psi}{\partial x_j}. \tag{13.21}$$

In this case, the average contaminant concentration gradient, $\partial\bar{\vartheta}/\partial x_j$, was set constant for the first 500 time steps, and zero thereafter. The initial field $\vartheta'(\mathbf{x}, 0)$ was put equal to zero. After 1500–2000 steps, the $\vartheta'(\mathbf{x}, t)$-spectrum (13.5), with a coefficient $B_\omega \simeq 1.56$, was obtained.

All this evidence for the laws of spectral enstrophy transfer was doubted by Herring *et al.* (1974), was performed the most detailed quantitative experiments on the decay of two-dimensional turbulence. They took the initial field as being periodic in squares of size 2π, as being a random Gaussian field with a spectrum equal to $v_0^2(kk_0^{-1})\, e^{-k/k_0}$, and as having a viscosity equal to 5×10^{-3}, 2.5×10^{-3} or 1×10^{-3}. Calculations were carried out on a 64×64-point or 128×128-point spatial grid; the Reynolds number $\mathrm{Re} = E_\eta^{-1/3} \nu^{-1}$ varied from 50 to 100. The vorticity equation (13.20) was integrated with the second-order Arakawa finite-difference method and Orszag's spectral method (1971; see also Fox and Orszag, 1973). These calculations also incorporated a two-dimensional mathematical simulation of three-dimensional turbulence, developed by Orszag and Patterson (1972) and Herring *et al.* (1973). The calculations led to the conclusion that a correct simulation of the inertial range of two-dimensional turbulence, i.e., the resolution of all scales that contribute appreciably to the enstrophy dissipation rate ϵ_ω, requires 512×512 grid points for $\nu = 1 \times 10^{-3}$. From this viewpoint, the quantitative experiments mentioned earlier did not have sufficient resolution (in Lilly's experiments Re = 315, 411, 537, while in the experiments carried out by Deem and Zabusky Re $\simeq$ 2200).

On the other hand, according to the calculations, the large-scale turbulence components (those with $k \leqslant 10$) are, in fact, independent of Re for Re = 150–1100. Hence, they can be reliably simulated at small values of Re.

We also call attention to the numerical experiments investigating the absolute statistical equilibrium with either positive or negative temperatures that were carried out by some investigators, the most typical being those (Seyler *et al.*, 1975) that investigated the numerical integration of the truncated two-dimensional spectral vorticity equation in an inviscid fluid with smooth initial data. It should be noted, in particular, that, in states with negative temperatures ($\alpha < 0$), a configuration with two large vortices of opposite sign is quite typical. It is interesting to mention in this context also a number of papers treating the numerical integration of the equations for a system of point vortices. These include calculations of the motion of 4008 vortices in a rectangle, which formed two large vortices of opposite sign at negative temperatures (Montgomery *et al.*, 1972), and investigations into the evolution of a system of similar vortices ($N = 100$) with an initially uniform distribution around a circumference (Sedov, 1976; Morikawa and Swenson, 1971; Murty and Sancara Rao, 1970). These configurations rotate with a constant angular velocity around the center. For $N > 7$, the motion becomes unstable (Havelock, 1971); the spectrum (13.16) then develops energy and enstrophy cascading ranges. Thereafter, the system reaches a statistical equilibrium, with secondary instabilities that result in large compact vortex groups. Finally, we note the derivation and numerical integration of model equations that describe the irreversible approach to a stationary self-similar state of two- and three-dimensional turbulence and passive contaminant fields.

Let us now explain why the properties of large-scale turbulence in the atmosphere and in the ocean (with scales that exceed the effective height of the atmosphere and depth of the ocean) can be expected to reveal some features of two-dimensional turbulence, despite the fact that the large-scale flow in the atmosphere or the ocean is quasi-two-dimensional rather than two-dimensional. In reality, the hydrodynamic fields change significantly with height, and vertical motions that connect the horizontal motion at various heights are of paramount importance in the dynamics. Hence, in adiabatic processes the vorticity of fluid particles is not conserved; on the contrary, the vortex lines are stretched. These facts can be accounted for as follows. First, in adiabatic processes not only the entropy η of fluid particles is conserved, but also the so-called *potential vorticity* (Monin, 1969; Obukhov, 1962a; Ertel, 1942). If we assume that large-scale three-dimensional motions in the atmosphere and in the ocean are non-divergent, then we obtain for this vorticity: $\Omega_* = \boldsymbol{\Omega}_a \cdot \nabla\eta$, where $\boldsymbol{\Omega}_a = \text{rot}\,\mathbf{u} + 2\boldsymbol{\omega}$ is the absolute vorticity and $\boldsymbol{\omega}$ is the angular velocity vector of the Earth. In the ocean, η is the so-called *pseudo-entropy*, i.e., the entropy reduced to constant salinity both isopycnically and isobarically (Monin, 1973b). Adiabatic processes are characterized by the equations

$$\frac{\mathrm{d_h}\eta}{\mathrm{d}t} + w\,\frac{\partial\eta}{\partial z} = 0; \qquad \frac{\mathrm{d_h}\Omega_*}{\mathrm{d}t} + w\,\frac{\partial\Omega_*}{\partial z} = 0, \tag{13.22}$$

where $\mathrm{d_h}/\mathrm{d}t$ represents the material derivative with respect to horizontal motion, z is the height, and w is the vertical velocity. Removing w from (13.22), we obtain

$$\frac{\mathrm{d_h}\Omega_*}{\mathrm{d}t} - \frac{\partial\Omega_*/\partial z}{\partial\eta/\partial z}\,\frac{\mathrm{d_h}\eta}{\mathrm{d}t} = 0. \tag{13.23}$$

Second, in the case of large-scale processes, (13.23) can be used to derive an approximate conservation law for horizontal motion that is quite similar to that of vorticity conservation in two-dimensional hydrodynamics. In large-scale processes, the field $\eta(\mathbf{x})$ is quasi-horizontal, so that the vector $\nabla\eta$ is approximately vertical. Consequently, $\Omega_* \simeq \Omega_{az}\, \partial\eta/\partial z$ and (13.23) reduces to

$$\frac{d_h \ln \Omega_{az}}{dt} + \left(\frac{\partial\eta}{\partial z}\right)^{-1} \frac{d_h}{dt}\frac{\partial\eta'}{\partial z} - $$
$$- \left(\frac{\partial\eta}{\partial z}\right)^{2} \left(\frac{\partial^2\eta}{\partial z^2} + \frac{\partial\eta}{\partial z}\frac{\partial \ln \Omega_{az}}{\partial z}\right) \frac{d_h\eta'}{dt} = 0, \qquad (13.23')$$

where the prime designates the component $\eta' = \eta - \eta_0$ of the field η that depends on the horizontal coordinates and on time. It is small compared to η_0. Hence, in the coefficients of $d_h/dt\ \partial\eta'/\partial z$ and $d_h\eta'/dt$ in (13.23) one can substitute η_0 for η and neglect the term $\partial\eta_0/\partial z\ \partial \ln \Omega_{az}/\partial z$ compared with $\partial^2\eta_0/\partial z^2$. As a result, (13.23') takes the form

$$\frac{d_h}{dt}\left(\ln \Omega_{az} + \frac{\partial}{\partial z}\frac{\eta'}{\partial\eta_0/\partial z}\right) = 0. \qquad (13.24)$$

A similar relation for the atmosphere was derived by Gavrilin and Monin (1969). Assuming the large-scale processes to be quasi-hydrostatic, we have

$$\frac{\partial\eta_0}{\partial z} = \left(\frac{\partial\eta}{\partial p}\right)_0 \left(\frac{\partial p_0}{\partial z} - c_0^2\frac{\partial\rho_0}{\partial z}\right) = \left(\frac{\partial\eta}{\partial p}\right)_0 \frac{\rho_0 c_0^2 N^2}{g},$$

where c_0 is the speed of sound and N is the Brunt–Väisälä frequency. In a similar way,

$$\eta' = \left(\frac{\partial\eta}{\partial p}\right)_0 (p' - c_0^2\rho') \simeq -\left(\frac{\partial\eta}{\partial p}\right)_0 c_0^2\rho' \simeq \left(\frac{\partial\eta}{\partial p}\right)_0 \frac{c_0^2}{g}\frac{\partial p'}{\partial z}.$$

Therefore,

$$\left(\frac{\partial\eta_0}{\partial z}\right)^{-1} \eta' \simeq \frac{1}{\rho_0 N^2}\frac{\partial p'}{\partial z}.$$

Moreover, if the fact that large-scale processes are quasi-geostrophic is taken into account, i.e., $|\mathrm{rot}_z\, \mathbf{u}| \leqslant |2\omega_z| = f$, (13.24) can be written approximately as

$$\frac{d_h\omega_*}{dt} = 0; \qquad \omega_* = \Omega_{az} + f\frac{\partial}{\partial z}\frac{1}{\rho_0 N^2}\frac{\partial p'}{\partial z}. \qquad (13.24')$$

The quantity ω_* can be also called a potential vorticity, so that $\frac{1}{2}\overline{\omega_*^2}$ is a *potential enstrophy*. Hence, with enstrophy replaced by potential enstrophy, the statistical properties of two-dimensional turbulence discussed above prove to be valid also for large-scale turbulence in the atmosphere and in the ocean (see Gavrilin *et al.*, 1972; Charney, 1971). Charney integrated the simplified equation for potential vorticity, which reads

$$\frac{\partial\, \Delta\psi}{\partial t} + \frac{\partial(\psi, \Delta\psi)}{\partial(x, y)} - \nu\, \Delta\Delta\psi + \frac{\Delta\psi}{\tau} = F, \qquad (13.25)$$

where

$$F = (\Delta\psi + f)\Delta\varphi - \nabla\varphi \cdot \nabla\Delta\psi - w^* \frac{\partial\,\Delta\psi}{\partial p} -$$

$$- \frac{\partial\left(w^*, \frac{\partial\varphi}{\partial p}\right)}{\partial(x, y)} - \nabla w^* \cdot \nabla \frac{\partial\psi}{\partial p}.$$

Here φ and w^* are the horizontal velocity potential and the isobaric equivalent of the vertical velocity component, related by $\Delta\varphi + \partial w^*/\partial p = 0$. Equation (13.25) was integrated in terms of a two-parametric model, with $\psi = \psi_1(x, y, t) + \Psi(p)\psi_2(x, y, t)$. The integral of $\Psi(p)$ over all p was equal to zero and that of $\Psi^2(p)$ was equal to one. As a result, (13.25) yielded two expressions for ψ_1 and ψ_2. These were integrated over the range $0 \leqslant (x, y) \leqslant L$, with impermeable boundaries at $y = 0$ and $y = L$, and periodic boundary conditions at $x = 0$ and L. The vertical velocity of excitation was set as a stationary sinusoidal wave, with wave numbers $k_x = k_y = 3$. The initial conditions for vorticity were set at zero. The energy level of the barotropic component of the motion, ψ_1, became stationary after 1200 time steps, and that of the baroclinic component, ψ_2, after 1400 time steps. After 1800 time steps the energy spectrum showed a pronounced inertial range for $k \geqslant 10$, which obeyed a minus-three law.

Note that Manabe *et al.* (1970) discovered the minus-three law in the range of zonal wavenumbers $k = 2\pi a \cos\theta/l = 8$–$20$ (a is the Earth's radius, θ is the co-latitude, l is the wavelength) in kinetic energy spectra obtained by numeric integration of the equations of a model for the general circulation of the atmosphere. For $l \simeq 1200$–4000 km, the minus-three law is also observed in the empirical spectra of large-scale meteorological fields (Gavrilin *et al.*, 1972). As an example, Figure 13.1 exhibits one-dimensional kinetic energy spectra of large-scale meteorological fields obtained from empirical data by Saltzman and Fleicher (1962), Horn and Bryson (1963), Wiin–Nielsen (1967), Kao and Wendell (1970), and Julian *et al.* (1970). These spectra graphically demonstrate the range obeying the minus-three law.

The theorem of the adiabatic invariance of potential vorticity, e.g., in the form of (13.24′), thus makes it possible to find an analogy between quasi-two-dimensional large-scale turbulence in the atmosphere and the ocean and two-dimensional turbulence in an incompressible fluid. However, there is a difference between the two, which was the subject of a special study by Rhines (1973, 1975, 1977). Putting $\Omega_{az} = \Delta\psi + f$ in (13.24) and using the quasi-geostrophic approximation $p' \simeq \rho_0 f\psi$, we can present the potential vorticity equation in the form

$$\frac{\partial L\psi}{\partial t} + \frac{\partial(\psi, L\psi)}{\partial(x, y)} + \beta\frac{\partial\psi}{\partial x} - \nu\,\Delta\Delta\psi + R\,\Delta\psi = F. \tag{13.26}$$

Here,

$$L = \Delta + \frac{\partial}{\partial z}\frac{f^2}{N^2}\frac{\partial}{\partial z}$$

is the three-dimensional elliptic operator, $\beta = \mathrm{d}f/\mathrm{d}y$ is the meridional derivative of the Coriolis parameter, and the term $R\,\Delta\psi$ describes bottom friction. The basic difference between this equation and the two-dimensional vorticity equation (13.20) lies in the

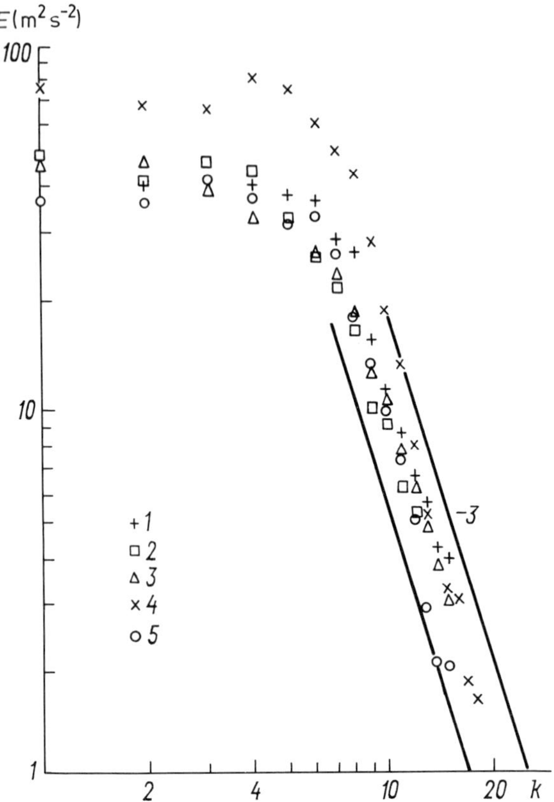

Fig. 13.1. One-dimensional kinetic energy spectra of large-scale meteorological fields, by Saltzman and Fleicher (1962) (1), Rossby (1969) (2), Wiin–Nielsen (1967) (3), Kao and Wendell (1970) (4), and Julian *et al.* (1970) (5).

the term $\beta\, \partial\psi/\partial x$. This leads to the appearance of Rossby waves among the solutions of (13.26), which is not the case with (13.20). Substitution of the three-dimensional operator L for the operator Δ, which results in baroclinic next to the barotropic modes, is of less importance, since the basic properties of (13.26) hold also for barotropic situations, i.e., at $L = \Delta$, which is the subject of our discussion. If the initial field ψ in (13.26) is presented as densely packed vortices, with a narrow-band spectrum with a peak at wavenumber k_0, then the relative importance of turbulence and Rossby waves can be expressed as the ratio of the nonlinear term in the left-hand side of (13.26) to the term $\beta\, \partial\psi/\partial x$. This ratio is $\delta = U/c_\varphi = 2k_0^2 U/\beta$, where U is the root-mean-square velocity of the motion ($\frac{1}{2}U^2 = E$) and $C_\varphi = \beta/(2k_0^2)$ is the phase velocity of Rossby waves with wavenumber k_0. For $\delta > 1$, the dynamics are determined mainly by vortices, for $\delta < 1$ they are determined mainly by Rossby waves. The boundary $\delta = 1$ corresponds to the wavenumber

$$k_\beta = \left(\frac{\beta}{2U}\right)^{1/2}. \qquad (13.27)$$

For typical atmospheric conditions, k_β^{-1} = 1000 km, which corresponds to a vortex diameter $2\pi k_\beta^{-1} \cos\theta \simeq 3100$ km at 30° latitude; the corresponding zonal wavenumber is 3 when $\theta = 45°$ and $U = 15$ m s^{-1}, and 5 when $\theta = 60°$ and $U = 10$ m s^{-1}. For typical conditions in the ocean, k_β^{-1} = 70 km when $\theta = 60°$ and $U = 5$ m s^{-1}. This corresponds to a vortex diameter of about 70 km, which fits synoptic vortex observations in the open ocean reasonably well. Defining a typical fluid velocity $\hat{U}(k)$ at wavenumber k by the relation $\frac{1}{2}\hat{U}^2(k) = kE(k)$, we can express the boundary between turbulence and waves at wavenumber k by means of the expression $\hat{U}(k) = c_\varphi(k)$ or $[2kE(k)]^{1/2} = \beta/(2k^2)$. This yields the boundary spectrum

$$E_\beta(k) = c_1 \beta^2 k^{-5}, \tag{13.28}$$

where $c_1 = 1/8$. The condition $E(k) > E_\beta(k)$ corresponds to turbulence, $E(k) < E_\beta(k)$ to waves. The boundary spectrum has a very steep slope, since the Rossby frequency $\omega = -\beta k_x k^{-2}$ rapidly increases with wavelength $2\pi/k$. The possibility of spectral saturation reaching (13.28) for $k \ll k_2$ with a continuous flux of energy into the vicinity of the wavenumber k_1 is still open to discussion, since the saturated spectrum can depend not only on β and k, but also on the location of the central peak $\hat{k}$: the velocity difference that affects vortices of scale k^{-1} is chiefly determined by vortices with scales near $\hat{k}^{-1}$.

Let $k_0 \gg k_\beta$, so that the field ψ is initially nothing else but turbulence. The initially narrow spectral peak will obviously broaden, i.e.,

$$\frac{\partial}{\partial t} \int (k - \bar{k})^2 E(k)\, \mathrm{d}k > 0,$$

where $\bar{k}$ is the mean wavenumber with respect to energy, used before in (13.19). Since the total enstrophy cannot increase,

$$\frac{\partial}{\partial t} \int k^2 E(k)\, \mathrm{d}k \leqslant 0,$$

and, since the total energy remains constant, we have $\partial\bar{k}/\partial t < 0$. More exactly, the use of the self-similar Batchelor spectrum,

$$E(k, t) = E^{3/2} t f(E^{1/2} k t),$$

yields

$$\frac{\partial}{\partial t} \frac{1}{\bar{k}} = TU, \tag{13.29}$$

where $T = [\int f(\xi)\, \mathrm{d}\xi] / [\int \xi f(\xi)\, \mathrm{d}\xi]$ is a numerical constant.

The wavenumber $\bar{k}$ of the dominant vortices and the typical frequency $\bar{\omega} = U\bar{k}$ decrease. Over a time period of order

$$(TUk_\beta^{-1}) \simeq [T(\beta U)^{1/2}]^{-1},$$

which is independent of k_0, the wavenumber $\bar{k}$ then reaches the value k_β, where the vortices change into Rossby waves. Thereafter, $\bar{k}$ goes on decreasing due to the weak interactions between the Rossby waves. This decrease is much slower, since the wave interactions demand not only the superposition of sets of three waves in space, but also resonance between their wavenumbers and frequencies. A decrease in the frequency due to such interactions is accounted for by the fact that, in a resonant triad, the wave with the highest frequency appears to be unsteady with respect to the increasing energy of the other two waves (Hasselmann, 1967; Rhines, 1975). The dispersion relation $\omega = -\beta k_x k^{-2}$ results in a shift of the ψ-disturbances eastward, while a decrease in the frequency leads to anisotropy of these inhomogeneities, i.e., to a prevalent increase in their meridional wavenumbers k_y. Therefore, this leads to the development of zonal flows (or, at the limit, steady-state zonal flows in narrow latitudinal bands with directions that vary from band to band either towards the East or the West). However, owing to the dissipation of wave energy in space and the resulting decay of the waves, this limit is not always attainable. A tendency of wave interactions to transfer energy to other directions in space rather than to other wavelengths results in a decrease of the T defined in (13.29) when turbulence is replaced by waves. This decrease in T at $\bar{k} \simeq k_\beta$ means that the vortices would not reach global sizes if the Rossby waves did not radiate energy and if the turbulence developed by following its own laws. A decrease in T due to the β-effect, i.e., slowing down the energy cascade towards large scales, decelerates the enstrophy cascade towards small scales. As a result, when enstrophy is dissipated, the spectrum must lose its small-scale structure and become narrower and steeper.

Numerical integrations of the barotropic equation for potential vorticity, i.e., (13.26) with $L = \Delta$, carried out by Rhines (1975) by a technique close to that reported by Herring *et al.* (1974), confirmed these predictions. This resulted in $T \simeq 3 \times 10^{-2}$ in turbulent conditions. As $\delta^{-1} = \beta/(2k^2 U)$ increased, T decreased rapidly; in wave conditions (i.e., even at $\delta = 1$), it proved to be five times smaller (i.e. $T \simeq 6 \times 10^{-3}$). Note that these results fit the theoretical estimates of resonant wave interaction effects fairly well, as reported by Kenyon (1967), Lorenz (1972), Longuet–Higgins and Gill (1967) and Gill (1974). These authors gave values of T from 3×10^{-3} to 6×10^{-3}. In numerical experiments carried out by Rhines, a typical width of the developed spectrum, $\langle k^2 \rangle \langle k \rangle^{-2} - 1$, and its shape were 0.34 and $k^{-4.3}$, respectively, with the β-effect neglected, and 0.1 and $k^{-5.4}$ with the β-effect taken into account. Note also the results of quantitative experiments carried out by Holloway and Hendershot (1977) and their good agreement with theoretical calculations based on the Markovian quasi-normal closure of the equations of two-dimensional turbulence with the β-effect.

Let us now return to the baroclinic equation for potential vorticity, (13.26), in which

$$L = \Delta + \frac{\partial}{\partial z} \frac{f^2}{N^2} \frac{\partial}{\partial z}.$$

In this case, waves arise with lengths shorter than the Rossby deformation radius $L_R = (N/f) \cdot H$ (where H is the effective height of the atmosphere or ocean). For these waves, $L \simeq \Delta$; i.e., the vertical interaction between different layers of fluid is negligible and the layers develop independently. These waves behave in a different way than those with $L > L_R$. In the latter case, different layers interact strongly in the vertical direction. In other words, they behave as a single layer, so that the fluid is effectively barotropic. Therefore, if the mean wavenumber $\bar{k}(t)$ decreases to $k_R = 2\pi/L_R$, the vortices must tend to become barotropic. The difference between motions at various levels then must decrease. Quantitative experiments with Rhines' two-level model (1977) confirmed these predictions and demonstrated that this process runs its course rapidly in the absence of perturbing factors (discussed below) and for a not very loose packing of vortices.

So, when expanding, the vortices tend to become barotropic, anisotropic (stretched in the east–west direction), and to move westwards. Opposing tendencies can result, first of all, from small vortices that are excited by external forces or internal baroclinic instabilities of large vortices. Quantitative experiments by Rhines demonstrated that extremely long baroclinic Rossby waves ($L \gg L_R$) are unstable (meridional motions are particularly unstable). They rapidly decompose into vortices of size L_R, which subsequently develop as described above. If baroclinic zonal flows evolve at a sufficient rate, then the vertical velocity differences ΔU remain stable for $\beta L_R^2/\Delta U > 1$, so that the vortices do not necessarily become barotropic.

Fig. 13.2. The Gaussian vortex transformation over a sinusoidal seabed topography, by Rhines (1977).

The second exception occurs when large-scale motions in Rossby waves turn into small-scale ones with increased enstrophy as they are reflected from the western shore of the ocean. At the eastern shore the enstrophy decreases. Since the motion of the small-scale vortices is slower, both energy and enstrophy accumulate at the western shore and dissipate at the eastern shore. Finally, seabed inhomogeneities result in Rossby wave dissipation and in enstrophy generation in the form of inhomogeneous spatial distributions of small-scale topographical vortices. They also make the motion baroclinic (Rhines, 1977; Rhines and Bretherton. 1973). A typical spectrum of the seabed topography is rather flat: $E_h(k) \simeq k^{-3/2}$ or k^{-2}, the latter corresponding to white noise in the spectrum of seabed slopes. The topography tends to smooth out the kinetic energy spectrum. Seabed inhomogeneities of a scale exceeding that of the initial vortices add to the β-effect. The isolines f/h then assume the role of latitude circles, the wavenumber

$$k_h = [h|\nabla(f/h)|/2U]^{1/2}$$

replaces k_β, and the anisotropy manifests itself in the tendency of the fluid to move along the isolines f/h. Figure 13.2 presents a drawing that resembles an Aztec mask. It corresponds to a Gaussian vortex over a sinusoidal topography $h \simeq \sin x \sin y$ (Rhines, 1977).

The interplay of the excitation, dissipation, and the three opposing tendencies must bring about different quasi-stationary statistical conditions of synoptic vortices and Rossby waves in various regions of the World Ocean.

14. HORIZONTAL TURBULENCE SPECTRA

Ozmidov and Yampolsky (1965) were the first to calculate spectra of horizontal large-scale turbulence in the ocean, from data of two-month observations of flow velocities in the North Atlantic Ocean. The primary series were smoothed by a cosine filter with a smoothing parameter of 24 h. The spectrum, averaged over all observation series, distinctly showed a section that was well approximated by a power law with the exponent $-5/3$. This was used to prove the applicability of the theory of locally isotropic turbulence to large-scale processes in the ocean. Although it seemed paradoxical at that time, this result had been obtained somewhat earlier by Ozmidov (1964), by the statistical processing of a series of data on flow velocities obtained with a current propeller meter.

The idea of considering a series of long-term observations of hydrophysical characteristics in the ocean to be realizations of random functions was first put forward by V. B. Shtokman (1941), who calculated the correlation functions of horizontal velocity components in the Caspian Sea as far back as 1941. In 1956–59 correlation functions for temperature and flow velocities were calculated by Japanese investigators (Hikosaka and Higano, 1959; Nan'niti, 1956, 1957; Nan'niti and Yasui, 1957). The correlation functions showed variations of different periods, which were induced by surface and internal waves. The approximation of the correlation functions by power laws resulted in exponents ranging from 0 to 1.9. Sometimes the exponent was close to 2/3, which could be interpreted as evidence for the applicability of the theory of locally isotropic turbulence. However, no clear conclusion was drawn by the authors.

Haurwitz *et al.* (1959) calculated temperature correlation functions using data from several months of observations. The records were obtained with thermistors located at depths of 50 and 500 m near the Bermuda Islands. The correlation functions revealed variations that resulted from internal waves. However, the form of the functions was not compared with any theoretical expressions. This operation was performed by Bortkovsky (1962), who plotted a structure function from data on the water temperature inside a 10° square of the Atlantic Ocean. Bortkovsky approximated the plots of the structure function by a linear relation of the argument over the range between 20 and 500 km. For smaller arguments the temperature structure function was approximated by power laws with an exponent of 2/3 by Piskunov (1957) and Nemchenko and Tishunina (1963). In the latter publication, it was pointed out that this approximation of the structure function is reliable up to a certain value of the argument only. For large arguments the exponent of the approximating curve became smaller. The argument at which the exponent changed ranged from 5 to 48 miles in different regions of the Atlantic Ocean.

Correlations and structure functions for velocity components were plotted by Ozmidov (1962) from data obtained by long-term measurements carried out at buoy stations. The first part of the structure function was fairly well approximated by power laws with an exponent of 2/3. However, the curves were affected by pronounced fluctuations at the inertial period. To interpret the shape of these structure functions, Ozmidov (1962) proposed a superposition model of the velocity field in the sea. This model consists of random large-scale turbulent motions, obeying the laws of locally isotropic turbulence, and ordered periodic motions. In terms of this model the structure function is given by

$$D(\tau) = 2A^2 \sin^2 \frac{\omega\tau}{2} + c\epsilon^{2/3}\bar{u}^{2/3}\tau^{2/3}. \tag{14.1}$$

The averaging scale must be divisible by the period of the orderly fluctuations. Here, A and ω are the amplitude and frequency of the orderly fluctuations, $\bar{u}$ is the mean velocity, which appears in the structure function as a result of substituting time for distance with the frozen-turbulence hypothesis, and c is a universal constant.

Figure 14.1 shows a plot of (14.1) and the experimental structure function obtained by observations at a level of 100 m depth in the Black Sea. The agreement between theory and experiment is seen to be fairly good. When estimated by curve-fitting, the dissipation rate turns out to be $\epsilon = 1.4 \times 10^{-4}$ cm^2 s^{-3}.

During the years which followed, propeller records of velocity were employed by a number of investigators to plot correlation functions of large-scale turbulence. Glinsky (1965) analyzed the measurements taken at three buoy stations in the Atlantic Ocean. Mizinov (1965) and Volzhenkov and Istoshin (1965) used measurements of currents in the Atlantic Ocean and in other regions. A number of correlation functions were plotted by Ovsiannikova (1965) from observations of velocities in the Black Sea. All the correlation functions were affected by inertial fluctuations.

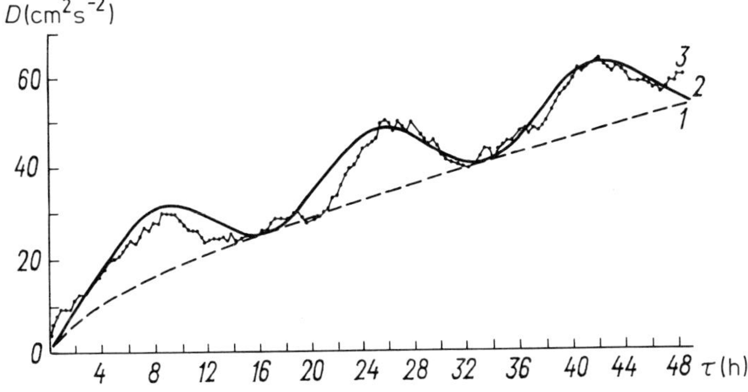

Fig. 14.1. Structure function $D(\tau)$ for the velocity measured at 100 m depth in the Black Sea (Ozmidov, 1962). (1) 2/3-power law, (2) function (14.1), (3) experimental curve.

Haurwitz *et al.* (1959) seem to have been the first to calculate spectral functions for large-scale temperature fluctuations; somewhat later this was also done by Roden (1963). The spectra revealed peaks at the frequencies of tidal and inertial fluctuations and of internal waves. Ozmidov (1965b) obtained a pronounced tidal maximum in the spectrum of large-scale velocity fluctuations based on data for the Atlantic Ocean. The spectral functions were calculated by the method of analytical filtration, with cosine filter centers of 3, 6, 12, 24, 48, and 96 h. The experimental points were approximated by a 5/3-power law, with the coefficient chosen such that the theoretical curve passed through points in the high-frequency part of the spectrum. An attractive fit between theory and experiment was observed up to periods of 8–10 h. At larger periods the experimental points went above the 5/3-power law. After reaching a maximum at $T = 25$ h, the spectral function $E(\omega)$ dropped sharply. This shape of $E(\omega)$ could be attributed to the superposition of decaying periodic motions and random turbulent components that obey the laws of locally isotropic turbulence (Ozmidov, 1965b). Decaying fluctuations in the ocean can be treated as results of pulse effects produced by, e.g., gales. Gale-excited inertial fluctuations exist for a certain time period even after the gale has ceased. This phenomenon was observed, in particular, by Day and Webster (1965) who calculated the energy of inertial fluctuations as a function of time. The velocity spectrum, when expressed as the sum of an exponentially decaying fluctuation of frequency ω_0 and random turbulent noise, is given by (Ozmidov, 1965b):

$$E(\omega) = c\left\{\frac{\lambda\cos\varphi - (\omega - \omega_0)\sin\varphi}{\lambda^2 + (\omega_0 - \omega)^2} + \frac{\lambda\cos\varphi - (\omega + \omega_0)\sin\varphi}{\lambda^2 + (\omega_0 + \omega)^2}\right\} + E_0(\omega). \tag{14.2}$$

Here c is a constant, λ and φ are the decay factor and the initial phase shift of the fluctuations, and $E_0(\omega)$ is the turbulent noise spectrum.

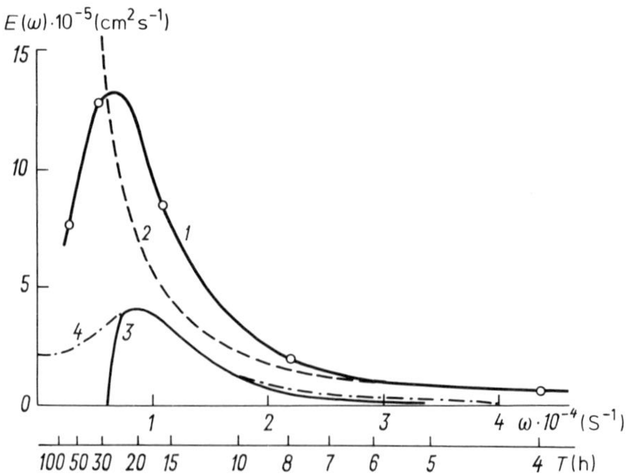

Fig. 14.2. A spectrum plotted from thirty-day measurements of the flow at a buoy station in the Atlantic Ocean at 200 m depth (Ozmidov, 1965). (1) Experimental curve, (2) 5/3-power law, (3) the difference between curves 1 and 2, (4), the energy spectrum of decaying periodic motion.

The approximation of the experimental spectrum by (14.2), under the assumption that $E_0(\omega) \simeq \omega^{-5/3}$, proves to be fairly good (Figure 14.2).

The spectra of large-scale turbulence calculated from data collected over two months of observations of flows in the Atlantic Ocean (Ozmidov and Yampolsky, 1965) demonstrated that maxima at the tidal and inertial frequencies, and sections that obey universal relations, are typical of the ocean. Although the levels and some other parameters of the spectra can vary, depending on the geographical location of the observation point and on the hydrometeorological conditions, the basic qualitative features of $E(k)$ are sufficiently universal and reflect the general laws of energy partition in the ocean over the range of scales considered.

The existence of universal segments in the spectra of large-scale turbulence, which first seemed improbable, was later interpreted as follows (Ozmidov, 1965c, d). The existence of universal conditions demands, first of all, that the Reynolds number of the flow under study is large. When calculating the Reynolds number, one relates the characteristic scale to the scale of the energy supply. In the case of tidal and inertial fluctuations, this time scale is about a day (at temperate latitudes). Although the motions excited by these forces generally are strongly anisotropic and not universal, the resulting smaller-scale motion (with characteristic time scales of several hours down to tens of minutes) can become horizontally isotropic, with the spectral fluxes of energy ϵ and enstrophy ϵ_ω serving as determining parameters. The feasibility of the 5/3-power law and the minus-three law then is determined by the relation between ϵ and ϵ_ω and by the possibility of neglecting either of the two parameters.

If, in the ocean, there exist several discrete zones of energy supply, which are separated by ranges in which the input of energy is small, then the spectral distribution of kinetic energy must have a specific character. This is the main difference between the ocean and flows with an energy supply scale that coincides with the scale of the flow, as observed, e.g., in gravity currents (rivers, laboratory flumes) or in pressure flows (water tubes and wind tunnels). In those flows, the Reynolds number is calculated from the typical scale of the flow and the possibility of universal components in the spectrum is determined by this Re. If a flow has several energy supply zones and hence several characteristic scales, the Reynolds number also has several values, Re_i.

The energy transferred from the outside to the ocean water at the scales L_i will then be redistributed among other scales of motion. Since the mechanism of turbulent vortex stretching is rather chaotic, it is natural to assume that vortices can lose the prevailing orientation which is characteristic of the formations supplied with energy when the process starts at a scale that differs significantly from L_i. If the range of scales between neighboring values of L_i is sufficiently large, it can have regions that obey universal laws of turbulence. The parameters ϵ and ϵ_ω for each area must differ by the additional energy and enstrophy in the energy-supply zones of the ocean (Beliayev *et al.*, 1973a).

A generalized energy spectrum of ocean water motion based on this discrete energy-supply scheme was proposed by Ozmidov (1965b, d) on the basis of the spectral processing of data collected from the long-term measurement of flows in the Atlantic Ocean and measurements of velocity fluctuations by low-inertia devices. The generalized spectrum proposed shows three energy-supply zones: a zone of wind waves with a characteristic scale of 10 m, a zone of inertial and tidal fluctuations with a characteristic scale of 10 km, and a zone of global energy supply with a characteristic scale of 1000 km (Figure

14.3). Between these zones, and at scales smaller than the wind-wave scale, there are areas that obey universal laws of turbulence. The scheme proposed is, of course, nothing but a substantially simplified model of real phenomena in the ocean. In specific regions, in different seasons, under different climatic conditions, and at different depths, the function $E(k)$ can vary significantly: maxima and areas obeying universal laws of turbulence can either appear or disappear. However, though schematic, this spectral model reflects the general features of energy supply in the ocean and has served as a basis for a better understanding of energy spectra in the ocean.

The applicability of the Kolmogorov laws to temperature inhomogeneities in the upper ocean was shown by Ivanov *et al.* (1968) for the scale range between 1 and 10 km. Structure functions calculated from long-term temperature records obtained with towed devices proved to obey the –2/3-power law. On the other hand, the corresponding spectra obtained by Saunders (1972) for the range between 3 and 100 km were better approximated by a power law with exponent –2.2. This can either be attributed to turbulence properties that deviate from the Kolmogorov laws, or to energy sources inside the range considered, i.e., to non-universal spectra.

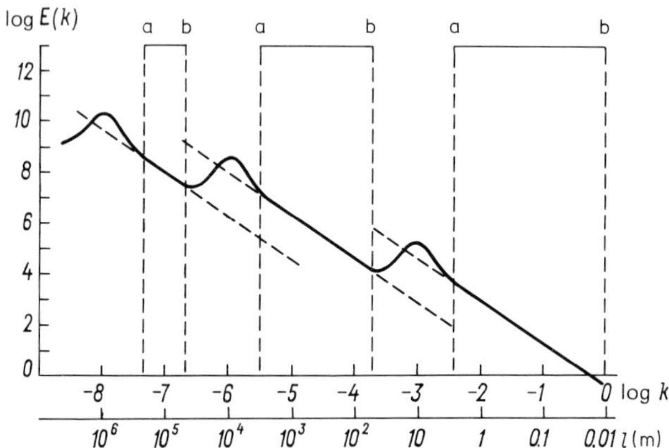

Fig. 14.3. Generalized scheme of the energy distribution over different-scale motions of ocean waters (Ozmidov, 1965). (a, b) are zones of universal turbulence laws.

The applicability of the laws of locally isotropic turbulence to processes that occur at various scales was confirmed by diffusion experiments in the ocean. The first experiments with discrete particles, performed by Richardson and Stommel (1948) and Stommel (1949), confirmed the applicability of the 4/3-power law for the horizontal diffusivity at scales ranging between several centimeters and several meters. More extensive studies of the diffusion of discrete particles (sheets of paper and buoys with radar reflectors) carried out by Ozmidov (1957, 1959) in artificial reservoirs and in the Pacific Ocean showed this result to be valid also for processes up to scales of 2 to 3 km. The generalization of a large number of diffusion experiments over a wide range of scales enabled Okubo and Ozmidov (1970) to plot the horizontal turbulent diffusivity K_l versus the scale l, with l ranging approximately from 10 m to several thousands of kilometers. Figure 14.4 shows that the experimental values of $K_l(l)$ (on a logarithmic scale) agree

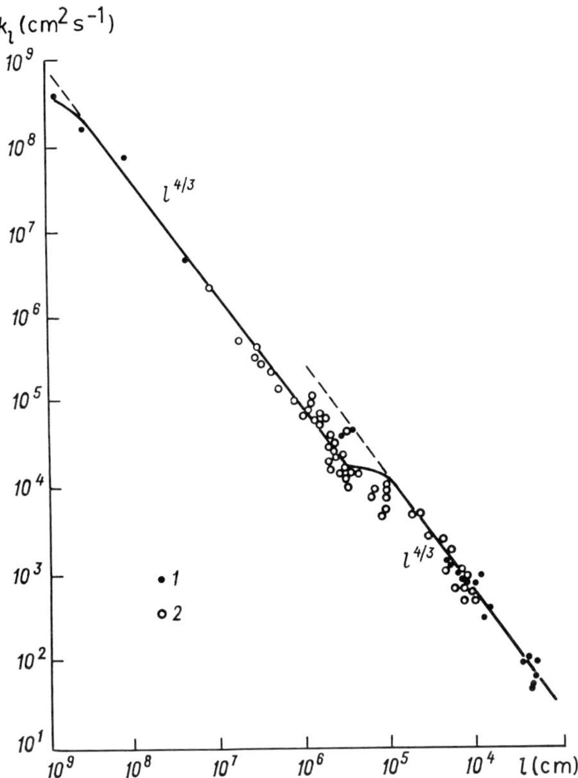

Fig. 14.4. Plots of the horizontal turbulent diffusion coefficient K_l vs the scale l, from diffusion experiments in the ocean (Okubo and Ozmidov, 1970). (1) data obtained before 1962, (2) those collected between 1962 and 1968.

fairly well with local straight lines with a 4/3-power law. The scale of 1–10 km corresponds to a transient region where experimental points leave one straight line for another with a lower coefficient, i.e., with a lower turbulent energy dissipation rate. When estimated for the large- and small-scale regions, ϵ is 10^{-4} and 10^{-5} cm^2 s^{-3}, respectively.

To obtain detailed spectra with pronounced neighboring maxima, one needs to perform a long-term series of measurements. For this purpose, as well as for investigations of the spatial structure of hydrophysical variables, special field experiments (mentioned above) were initiated. Spectra of large-scale turbulence were calculated by Vasilenko and Krivelevich (1974) and Vasilenko *et al.* (1976) from Polygon-70 data. The spectra corresponding to measurement levels of 25, 100, 500, and 1500 m reveal pronounced maxima at inertial (some 42 h) and half-day tidal periods, and a pronounced minimum at a four-day period. A characteristic feature of the spectra is a decrease in their level with depth, i.e., a decrease in the large-scale energy observed experimentally. It is interesting to note that the minimum that corresponds to the four-day period broadens with depth.

Spectral minima at periods of several days were also observed by Thompson (1971) and Rhines (1971), who plotted spectra for levels of 100, 500, 1000, and 2000 m at frequencies from 10 h to a month (Figure 14.5). To this end, they used data obtained during several years of measurements taken at permanent buoy station D (39°20′ N, 70° W) by the Woods Hole Oceanographic Institute. A similar minimum, located somewhat to the left of the inertial peak, was observed on the 1500 m spectrum calculated from data obtained in the MODE program (Gould *et al.*, 1974). The presence of this 'synoptic minimum' therefore should be regarded as a regular feature of the spectra of large-scale turbulence in the ocean. At the same time, the rise in $E(\omega)$ at long periods demonstrates the presence of large-scale (synoptic) eddies. The discovery of these eddies in the Polygon-70 data is of essential importance for oceanography.[5]

Synoptic eddies were also found in the data obtained by American and English investigators in the Sargasso Sea during the MODE-1-program which ran from March to July 1973. The horizontal sizes of synoptic eddies during the MODE-1 experiment were about 90–95 km, which was somewhat lower than the estimates obtained in Polygon-70. This discrepancy can be accounted for by a smaller Rossby radius in the MODE-1 region compared to that in the Polygon-70 region. The anticyclonic synoptic vortex measured in MODE-1 moved eastwards with a mean velocity of about 2 cm s^{-1}, i.e., much slower than the vortices observed in Polygon-70. The rotational velocities in the former case, however, were somewhat higher than those in the latter. Temperature measurements made it possible to conclude that the main anticyclonic vortex penetrated down to the bottom. If the vortices observed are interpreted as Rossby waves, then a vortex can be simulated by the superposition of several barotropic and baroclinic waves.

Polygon-70 measurements demonstrated that synoptic vortices occupied the entire observation area during the experiment, with alternating cyclonic and anticyclonic rotation in the vortices. The mean vortex size (the distance between vortex centers) was 110–120 km at 300 m depth and decreased to 100 km at a depth of 1000 m. The water velocity in the vortices was about 10 cm s^{-1}, while the translational velocity of the vortex was about 5–6 cm s^{-1}. Hence, the kinetic energy of the large-scale eddies is much higher than the mean kinetic energy of the water.

One more type of large-scale vortex in the ocean is the so-called frontal synoptic eddy or 'ring'. These eddies result from the baroclinic instability of strong boundary flows, such as the Gulf Steam and the Japan Current. In the Northern Hemisphere, the meanders created by this instability leave the core of the flow and form cyclonic eddies, with a lower temperature (compared with the environment) to the right and a higher temperature to the left. In the Gulf Stream, between Cape Hatteras and the Grand Banks, there develop, on average, five pairs of frontal eddies per year. The eddy diameters are, as a rule, approximately 200 km, and the temperature difference between the eddies and the surrounding water can reach 10–12°C. The velocity in the rings is, in most cases, several centimeters per second. The translational velocities of cold rings in the Gulf Stream are approximately the same. Their lifetime can be 2–3 years. Warm anticyclonic rings are usually of a somewhat smaller size, and their lifetimes are approximately 6 months.

Next to synoptic and frontal eddies, there can exist large-scale vortex formations of other types. For instance, the upper ocean is often characterized by eddies with horizontal sizes of 5–50 km. In the Atlantic Ocean, to the south of the North Atlantic

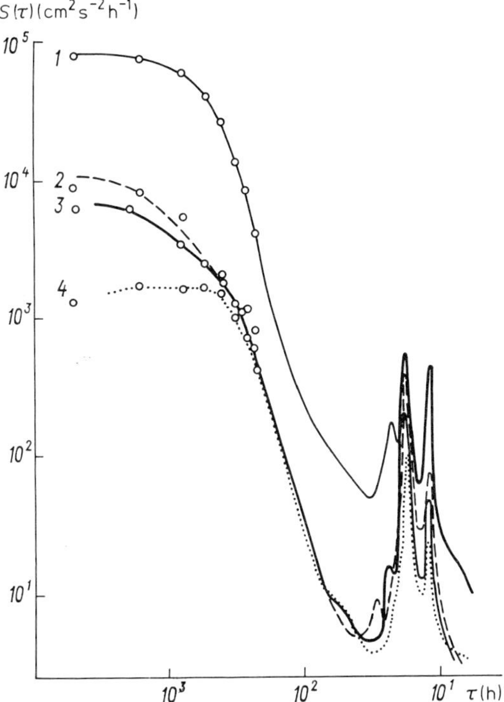

Fig. 14.5. Kinetic energy spectra of the horizontal flow during six-moneths of observations at station D (Rhines, 1971). Curves 1–4 correspond to 100, 500, 1000, and 2000 m depths, respectively.

Drift, weak cyclonic eddies develop, which most likely stem from meanders in the flow. In American publications, these are called 'big babies'.

The large-scale turbulence spectrum was shown above to obey universal laws between different energy-supply zones. However, it is so far not clear from experimental data which conditions are predominant, the inertial range of three-dimensional turbulence, with a 5/3-power law, or two-dimensional turbulence with a minus-three law. For example, next to the examples mentioned above that attest to a 5/3-power law in large-scale turbulence, one can refer to Webster (1966), who calculated large-scale velocity spectra from data obtained at a long-term buoy station. The spectra showed an external input of energy at inertial and tidal scales and a spectrum following the 5/3-power law at smaller scales. The applicability of the 5/3-power law to large-scale turbulence in the ocean, with scales from 40 to 1000 km, was pointed out by Wyrtki (1967), who used hydrological survey data for the North Pacific Ocean in his calculations. However, the spectra of large-scale velocity fluctuations calculated by American scientists from measurements taken at the station D produced more diversified results. For instance, some spectra between energy-supply zones obeyed the 5/3-power law fairly well (Webster, 1969), while other spectra, corresponding to scales from 2 days to 2 weeks, were better approximated by the minus-three law (Thompson, 1971). Figure 14.6 is an example of a spectrum with a slope of −3.

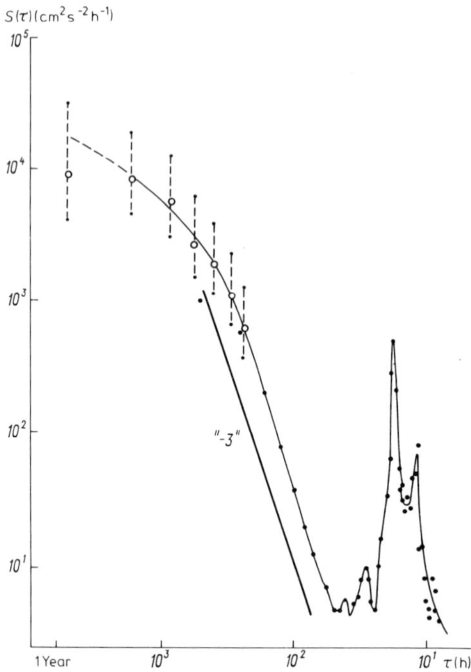

Fig. 14.6. Specimen of a spectrum with a −3 slope, from measurements at station D (Thompson, 1971).

In most cases, however, large-scale turbulence in the ocean seems to be not purely three- or two-dimensional, but intermediate between the two. To verify this statement, we calculated the slopes of 47 spectra of large-scale velocity fluctuations reported by the Woods Hole Oceanographic Institute (1965, 1966, 1967, 1970, 1971, 1974, 1975). At frequencies from about 5 to 0.005 cycle h^{-1} the mean slope of the spectra proved to be −2.11. The slope ranged from −1.87 to 2.70. As usual, the spectra analyzed have pronounced peaks at the inertial and tidal periods. When estimated separately, the slopes of the spectra at frequencies above and below the inertial and tidal ones had slightly differing values: −2.34 in the former case and −1.92 in the latter. Thus, as would be expected, the large-scale turbulence appeared to fit the two-dimensional theory with a better approximation than the theory of chaotic motions at smaller scales. To summarize the results of the calculations, it should be noted once more that universal two- and three-dimensional turbulence, over the scale range considered, are rarely observed. Intermediate cases are more typical of conditions in the ocean. Let us discuss in more detail the conditions that correspond to this phenomenon.

Let energy and enstrophy be generated in the vicinity of a certain wavenumber k_*. According to two-dimensional turbulence theory, the energy then propagates toward $k \to 0$, while the enstrophy propagates toward $k \to \infty$. However, real currents may demonstrate deviations from this ideal situation. A part of the energy can transferred toward large eddies by the vorticity cascade, and a part of the enstrophy toward small scales. This is the case, e.g., for a limited range of wavenumbers in a vortex flow. Panchev (1976) proposed the following approximation for the energy and enstrophy spectra:

$$E(k) = \gamma_1 \epsilon_{\omega 1}^{2/3} k^{-3} \left(1 - \frac{k^2}{k_1^2}\right)^{2/3}; \tag{14.3}$$

$$\Omega(k) = \gamma_2 |\epsilon_2|^{2/3} k^{-5/3} \left(1 - \frac{k_2^2}{k^2}\right)^{2/3}. \tag{14.4}$$

Here $k_1 = (\epsilon_{\omega 1}/\epsilon_1)^{1/2}$, $k_2 = (\epsilon_{\omega 2}/\epsilon_2)^{1/2}$, ϵ_2 and $\epsilon_{\omega 1}$ are the energy and enstrophy fluxes in the directions typical of two-dimensional turbulence, ϵ_1 and $\epsilon_{\omega 2}$ are those directed the other way, with $\epsilon_1 \ll |\epsilon_2|$, $|\epsilon_{\omega 2}| \ll \epsilon_{\omega 1}$, and $k_1 \gg k_* \gg k_2$. The coefficients γ_1 and γ_2 are dimensional constants.

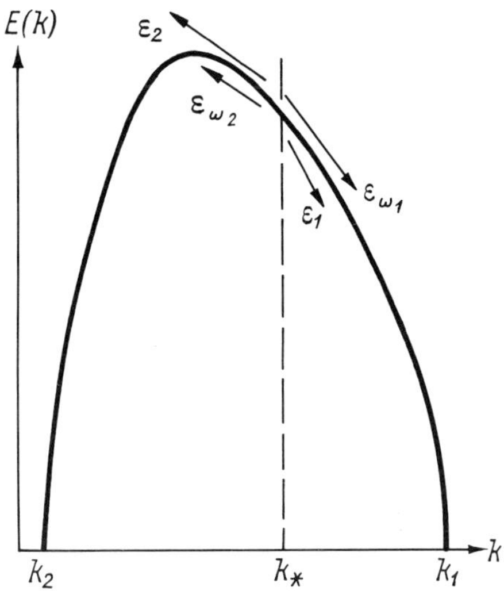

Fig. 14.7. The energy distribution over wavenumber: directions of energy and enstrophy fluxes in two-dimensional turbulence over a limited range of scales (Panchev, 1976).

Figure 14.7 shows a plot of the spectrum (14.3). It is seen that the slopes -3 and $-5/3$ are possible only near k_*, in the immediate vicinity of the site of energy and enstrophy generation. At other wavenumbers the slopes deviate toward either lower or higher values. Thus, it seems quite natural that experimentally-determined slopes can differ from -3 and $-5/3$ at certain wavenumbers. The slopes that correspond to different values (ranges) of k_* in cases with a noticeable generation (input) of energy and enstrophy differ still further. For example, when one of the sources is located in the baroclinic instability zone of the large-scale circulation (generating of synoptic-scale eddies) and the other sources occur at inertial and tidal motion scales, the energy and enstrophy fluxes will overlap in the range of wavenumbers between the sources, but also to their left and right. Of course, the slopes of the spectra then depend heavily on wavenumber and on the relative intensity of the sources. In the long run, the spectral slopes then depend on the mean large-scale hydrological and weather conditions in the region considered. As a result, these conditions must be taken into account in field experiments. Undoubtedly, this will contribute to a deeper insight into the generation mechanisms and properties of large-scale turbulence. These problems can be tackled with the help of extensive studies of regions in which the mean hydrometeorological conditions are typical of the ocean. The Polygon-70 program and the recently completed Soviet–American POLYMODE program are convincing examples of this type of investigation.

Notes

[1] A similar principle is employed in flow meters currently being developed at some institutions in the U.S.S.R. and abroad on the basis of the Doppler effect in scattered laser-generated light beams (see, e.g., Kuznetsov, 1974).

[2] The arrangement of the sensors and the program of measurements in the polygons varied, naturally, depending on the conditions. Therefore, the situation mentioned above should be regarded as an example. Other arrays of instruments and other techniques were used, e.g., in the expeditions of the Institute of Marine Hydrophysics of the Ukranian Academy of Sciences (Kolesnikov and Panteleyev, 1975).

[3] When analyzing data concerning fluctuation measurements in the ocean, remember that the fluctuation estimates obtained may not be precise enough, due to the severe difficulties of the absolute calibration of the small-inertia sensors employed.

[4] These examples illustrate the difficulties encountered in the absolute calibration of fluctuation instruments. This results in some inaccuracy in fluctuation measurements.

[5] This discovery was registered at the U.S.S.R. Committee for Inventions and Discoveries under the authorship of L. M. Brekhovskih, V. G. Kort, M. N. Koshliakovi, and L. M. Fomin in November 1978, under No. 207.

References

Agafonova, E. G. and Monin, A. S., 1972: 'Statistics of temperature and salinity in the upper Atlantic Ocean', *Doklady Ak. Nauk SSSR* **207**, 586–588.

Ahlers, G., 1974: 'Low-temperature studies of the Rayleigh–Benard instability and turbulence', *Phys. Rev. Lett.* **33, 20,** 1185–1188.

Baller, D. J., 1972: 'Viscous overturning – a mechanism for oceanic microstructure', *Trans. Amer. Geophys. Union* **53**, 415.

Bartlett, M. S., 1955: *An introduction to stochastic processes, with special reference to methods and application*, Cambridge Univ. Press, England.

Barenblatt, G. I., 1978: 'Dynamics of, turbulent spots and intrusions in a stably stratified fluid', *Atm. Ocean. Phys.* **14**, 139–145.

Batchelor, G. K., 1959: 'Small-scale variation of convected quantities like temperature in a turbulent fluid. Part 1. General discussion and the case of small conductivity', *J. Fluid Mech.* **5**, 113–133.

Batchelor, G. K., 1969: 'Computation of the energy spectrum in homogeneous two-dimensional turbulence', *Phys. Fluids. Suppl. 2* **12**, 233–238.

Bell, T. H., 1974: 'Internal wave-turbulence interpretation of ocean fine structure', *Geophys. Res. Lett.* **1**, 253–255.

Beliayev, V. S., 1973: 'A technique for the detection and processing of information on small-scale oceanic turbulence'. In: *Studies on ocean turbulence*, Moscow, Nauka, pp. 44–48.

Beliayev, V. S. and Gezentsvei, A. N., 1977: 'Fine structure of the temperature field in the region of the Kuroshio'. *Atm. Ocean. Phys.* **13**, 663–668.

Beliayev, V. S. *et al.*, 1973a: 'Some experimental data on hydrophysical field fluctuations in the upper ocean'. In: *Studies on ocean turbulence*, Moscow, Nauka, pp. 20–31.

Beliayev, V. S. *et al.*, 1974a: Spectra of small-scale velocity fluctuations and of kinetic energy dissipation in the ocean'. In: *Studies on the variability of hydrophysical fields in the ocean*, Moscow, Nauka, pp. 31–41.

Beliayev, V. S. *et al.*, 1975a: 'Relationships between small-scale turbulence parameters and local stratification conditions in the ocean', *Atm. Ocean. Phys.* **11**, 448–452.

Beliayev, V. S. *et al.*, 1975b: 'A study of the relation between the conductivity fluctuation characteristics of water and features of vertical temperature profiles in the ocean', *Atm. Ocean. Phys.* **11**, 680–683.

Beliayev, V. S. and Liubimtsev, M. M., 1977: 'Running structure functions as an indicator of turbulence intermittency', *Atm. Ocean. Phys.* **13**, 918–924.

Beliayev, V. S. *et al.*, 1973b: 'Turbulent energy dissipation rate and the rate of equalization of temperature inhomogeneities in the ocean', *Atm. Ocean Phys.* **9**, 669–672.

Beliayev, V. S. *et al.*, 1974b: 'A thousand spectra of oceanic turbulence', *Doklady Ak. Nauk* **217**, 1053–1056.

Beliayev, V. S. *et al.*, 1974c: 'An experiment in the measurement of deep-ocean turbulence', *Atm. Ocean. Phys.* **10**, 319–324.

Beliayev, V. S. and Ozmidov, R. V., 1979: 'Turbulence spectra in a stratified ocean', Proc. Symposium on Stratified Flows, Novosibirsk, *Sibirskoe Otdelenie Ak. Nauk SSSR*, pp. 6–7.

Beliayev, V. S. and Ozmidov, R. V., 1971: 'Distributions of velocity vector components in the ocean', *Atm. Ocean. Phys.* **7**, 344–347.

Beliayev, V. S. *et al.*, 1974d: 'Empirical distributions of one-dimensional spectral densities of velocity and electrical-conductivity fluctuations in the ocean', *Oceanology* **14**, 651–655.

Beliayev, V. S. *et al.*, 1976: 'Diversity of small-scale turbulence regimes in the ocean', *Oceanology* **16**, 128–130.

Beliayev, V. S. *et al.*, 1975c: 'Comparative measurements of small-scale fluctuations of hydrophysical fields in the ocean with different types of instrument', *Oceanology* **15**, 368–371.

Benilov, A. Yu., 1969: 'Experimental data on small-scale ocean turbulence', *Atm. Ocean. Phys.* **5**, 288–299.

Benilov, A. Yu., 1973: 'Generation of ocean turbulence by surface waves', *Atm. Ocean. Phys.* **9**, 160–164.

Benilov, A. Yu. and Lozovatsky, I. D., 1974: 'On the inertial range in turbulence spectra of stratified liquids'. In: *Studies on the variability of hydrological fields in the ocean*, Moscow, Nauka, pp. 83–91.

Benilov, A. Yu. and Filiushkin, B. N., 1970: 'Application of methods of linear filtration to an analysis of fluctuations in the surface layer of the sea', *Atm. Ocean. Phys.* **6** 477–482.

Bernstein, R. L. and White, W. B., 1974: 'Time and length scales of baroclinic eddies in the central North Pacific Ocean', *J. Phys. Oceanogr.* **4**, 613–624.

Blackman, R. B. and Tukey, J. W., 1958: *The measurement of power spectra from the point of view of communication engineering*, New York, Dover.

Borkovsky, M. M., 1972: 'The digital hydrophysical probe 'Aïst' '. In: *Automation of scientific research of seas and oceans*, Sebastopol, MGI AN USSR, pp. 41–46.

Bortkovsky, R. S., 1962: 'Methods of calculating heat and salt flows in the ocean examplified by certain areas in the Atlantic Ocean', *Trudy Instituta Okeanologii Akademii Nauk* **56**, 130–228.

Bolgiano, R., 1959: 'Turbulent spectra in a stably stratified atmosphere', *J. Geophys. Res.* **64**, 2226–2229.

Bolgiano, R., 1962: 'Structure of turbulence in stratified media', *J. Geophys. Res.* **67**, 3015–3023.

Bowden, K. F., 1962: 'Measurements of turbulence near the seabed in a tidal current', *J. Geophys. Res.* **67**, 3181–3186.

Bowden, K. F., 1964: 'Turbulence', *Annual Rev. Oceanogr. Mar. Biol.* **2**, 11–31.

Bowden, K. F. and Fairbairn, L. A., 1952: 'Further observations of turbulence fluctuations in a tidal current', *Phil. Trans.* **A244**, 335–356.

Bowden, K. F. and Fairbairn, L. A.: 'Measurements of tubulence fluctuations and Reynolds stresses in a tidal current', *Proc. Roy. Soc.* **A237**, 422–438.

Bowden, K. F. and Howe, M. R., 1963: 'Observations of turbulence in a tidal current', *J. Fluid Mech.* **17**, 271–284.

Bowden, K. F., 1965: 'Turbulence'. In: *The Sea*, Leningrad, Gidrometeoizdat, pp. 434–461.

Brekhovskih, L. M. (ed.), 1974: *Oceanic Acoustics*, Moscow, Nauka.

Brekhovskih, L. M., 1972: 'Resonant excitation of internal waves by nonlinear interactions of surface waves', *Atm. Ocean. Phys.* 8, 112–117.

Brekhovskih, L. M., 1971: 'Some results of the hydrophysical experiment on a test range established in the tropical Atlantic Ocean', *Atm. Ocean. Phys.* 7, 332–343.

Brekhovskih, L. M., 1976: 'Physical oceanographic problems discussed at the XVIth General Assembly, of the International Union of Geodesy and Geophysics', *Oceanology* **16**, 105–107.

Bulgakov, N. P., 1975: *Oceanic convection*, Moscow, Nauka.

Busse, F. H., 1967: 'The stability of finite amplitude cellular convection and its relation to an extremum principle', *J. Fluid Mech.* **30**, 625–649.

Busse, F. H., 1972: 'The oscillatory instability of convection rolls in a low Prandtl number fluid', *J. Fluid Mech.* **52**, 97–112.

Charney, J. G., 1971: 'Geostrophic turbulence', *J. Atmos. Sci.* **28**, 1087–1095.

Chistiakov, A. I., 1973a: 'Intermediate ascent–stabilizing devices used in towed complexes for oceanologic studies'. In: *Studies on oceanic turbulence*, Moscow, Nauka, pp. 139–142.

Chistiakov, A. I., 1973b: 'Auto-control of towed devices for oceanologic measurements'. In: *Studies on oceanic turbulence*, Moscow, Nauka, pp. 143–148.

Cook, I. and Taylor, J. B., 1972: 'Stationary states of two-dimensional turbulence', *Phys. Rev. Lett.* **28**, 82–84.

Csanady, G. T., 1973: *Turbulent diffusion in the environment*, Dordrecht, D. Reidel.

Davis, S. J. and White, C. M., 1928: 'An experimental study of the flow of water in pipes of rectangular section', *Proc. Roy. Soc.* **A119**, 92–107.

Day, C. G. and Webster, F., 1965: 'Some current measurements in the Sargasso Sea', *Deep-Sea Res.* **12**, 805–814.

Deem, G. S. and Zabusky, N. J., 1971: 'Ergodic boundaries in numerical simulations of two-dimensional turbulence', *Phys. Rev. Lett.* **27**, 396–399.

Defant, A., 1921a: 'Die Bestimmung der Turbulenzgrössen der atmosphärischen Zirkulation süssertropischen Breiten', *Sitz. Akad. Wien. Mathem-naturw. Klasse*, No. 130(2A).

Defant, A., 1921b: 'Die Zirkulation der Atmosphäre in den gemässigten Breiten der Erde', *Geograf. Ann.* **3**, 209–266.

Defant, A., 1954: 'Turbulenz und Vermischung im Meer', *Dtsch. Hydrogr. Z.* **7**, 2–14.

Delisi, D. P. and Orlanski, J., 1975: 'On the role of density jumps in the reflection and breaking of internal gravity waves', *J. Fluid Mech.* **69**, 445–464.

Desniansky, V. N. and Novikov, E. A., 1974a: 'Evolution of turbulence spectra towards similarity', *Atm. Ocean. Phys.* **10**, 76–81.

Desniansky, V. N. and Novikov, E. A., 1974b: 'Simulation of cascade processes in turbulent flows', *Prikl. Mat. i Mekh.* **38**, 507–513.

Dolzhansky, F. V. *et al.*, 1974: *Non-linear systems of the hydrodynamic type*, Moscow, Nauka.

Edwaras, S. F. and Taylor, J. B., 1974: 'Negative temperature states of two-dimensional plasmas and vortex fluids', *Proc. Roy. Soc. London* **A36**, 257–271.

Ellison, T. H., 1967: 'Kolmogoroff's similarity theory of small-scale motion and the effect of fluctuations in dissipation', unpublished manuscript. In: Monin, A. S. and Yaglom, A. M. (1975), *Statistical Fluid Mechanics*, Vol. 2, MIT Press, p. 647.

English, W. W., 1953: 'Thermal microstructure in the ocean', *Bull. Amer. Met. Soc.* **34**.

Ertel, H., 1942: 'Ein neuer hydrodynamischer Wirbelsatz', *Meteor. Zeitschr.* **59**, 277–281.

Fedorov, K. N., 1976: *Fine thermohalinic structure of the ocean*, Leningrad, Gidrometeoizdat.

Fjørtoft, R., 1953: 'On the changes in the spectral distribution of kinetic energy for two-dimensional non-divergent flow', *Tellus* **5**, 225–230.

Fofonoff, N. P., 1962: 'Physical properties of sea water'. In: *The Sea, Physical Oceanography*, Vol. 1, New York, Wiley, pp. 3–30.

Fox, D. G. and Orszag, S. A., 1973a: 'Inviscid dynamics of two-dimensional turbulence', *Phys. Fluids* **16**, 169–171.

Fox, D. G. and Orszag, S. A., 1973b, 'Pseudospectral approximation of two-dimensional turbulence', *J. Comp. Phys.* **11**, 612–619.

Gargett, A. E., 1967: 'An investigation of the occurrence of oceanic turbulence with respect to fine structure', *J. Phys. Oceanogr.* **6**, 139–156.

Garret, C. J. and Munk, W. H., 1971: 'Internal wave spectra in the presence of fine structure', *J. Phys. Oceanogr.* **1**, 196–202.

Garret, C. J. and Munk, W. H., 1972a: 'Space–time scales of internal waves', *Geophys. Fluid Dynam.* **2**, 225–264.

Garret, C. J. and Munk, W. H., 1972b: 'Oceanic mixing by breaking internal waves', *Deep-Sea Res.* **19**, 823–832.

Garnitch, N. G. and Miropolsky, Yu. Z., 1974: 'On some properties of the fine thermal structure of the ocean', *Oceanology* **14**, 474–480.

Gavrilin, B. L. and Mirabel, A. P., 1972: 'Digital simulation of two-dimensional turbulence decay', *Trudy Gidrometeor. Nauchno issledov. tsentr. SSSR* **100**, 117–121.

Gavrilin, B. L. *et al.*, 1972: 'The energy spectrum of synoptic processes', *Atm. Ocean. Phys.* **8**, 275–280.

Gavrilin, B. L. and Monin, A. S., 1969: 'On the primitive and balanced hydrodynamic equations of meteorological fields', *Atm. Ocean. Phys.* **5**, 435–441.

Geisenberg, W., 1948: 'On the statistical theory of turbulence', *Z. Phys.* **124**, 628–657.

Gibson, C. H. and Schwartz, W. H., 1963: 'The universal equilibrium spectra of turbulent velocity and scalar fields', *J. Fluid Mech.* **16**, 365–384.

Gibson, C. H. *et al.*, 1974: 'Turbulent diffusion of heat and momentum in the ocean'. In: *Turb. Diffus. Environ. Pollut.*, Advances in Geophys., Vol. 18A, New York, pp. 357–370.

Gibson, C. *et al.*, 1975: 'Joint Soviet–American measurements of ocean turbulence during the eleventh cruise of the 'Dmitriy Mendeleyev' ', *Oceanology* **15**, 135–137.

Gill, A. E., 1974: 'The stability of planetary waves on an infinite beta plane', *Geophys. Fluid. Dyn.* **6**, 29–47.

Gill, A. E. *et al.*, 1974: 'Energy partition in the large-scale ocean circulation and the production of mid-ocean eddies', *Deep-Sea Res.* **21**, 499–528.

Gisina, F. A., 1966: 'The effect of mean velocity and temperature gradients on the spectral characteristics of turbulence', *Atm. Ocean. Phys.* **2**, 487–491.

Gisina, F. A., 1969: 'Spectral characteristics of turbulence in a thermally stratified atmosphere', *Atm. Ocean. Phys.* **5**, 137–141.

Glinsky, N. T., 1965: 'Spectral studies of internal waves. Investigations by the MGG program', *Okeanologicheskiye Issledovaniya*, pp. 24–36.

Gould, W. J. *et al.*, 1974: 'Preliminary field results for the Mid-Ocean Dynamics Experiment (MODE-0)', *Deep-Sea Res.* **21**, 911–931.

Grant, H. L. *et al.*, 1968a: 'The spectrum of temperature fluctuations in turbulent flow', *J. Fluid Mech.* **34**, 423–442.

Grant, H. L. and Moilliet, A., 1962: 'The spectrum of the cross-stream component of turbulence in a tidal stream', *J. Fluid. Mech.* **13**, 237–240.

Grant, H. L. *et al.*, 1959: 'Spectrum of turbulence at very high Reynolds number', *Nature* **184**, 808–810.

Grant, H. L. *et al.*, 1963: 'Turbulent mixing in the thermocline', *XIII Gen. Assembly IUGG*, Vol. 6, Berkeley, p. 137.

Grant, H. L. *et al.*, 1968b: 'Some observations of the occurrence of turbulence in and above thermocline', *J. Fluid Mech.* **34**, 443–448.

Grant, H. L. *et al.*, 1960: 'Turbulent measurements in inshore waters', *Proc. Gen. Assembly IUGG*, Helsinki, Intern. Ass. Phys. Oceanogr. pp. 107–108.

Grant, H. L. *et al.*, 1962: 'Turbulence spectra from a tidal channel', *J. Fluid Mech.* **12**, 241–268.

Gregg, M. C., 1975: 'Microstructure and intrusions in the California Current', *J. Phys. Oceanogr.* **5**, 253–278.

Gregg, M. C. and Cox, C. S., 1972: 'The vertical microstructure of temperature and salinity', *Deep-Sea Res.* **19**, 355–376.

Gregg, M. C. *et al.*, 1973: 'Vertical microstructure measurements in the central North Pacific', *J. Phys. Oceanogr.* **3**, 458–469.

Gruza, G. V., 1961: *Microturbulence in the general circulation of the atmosphere*, Leningrad, Gidrometeoizdat.

Gurvich, A. S., 1966: 'Probability distribution of the square of the velocity differential in a turbulent flow', *Atm. Ocean. Phys.* **2**, 659–661.

Gurvich, A. S., 1967: 'On the probability distribution of the squared temperature difference at two points of a turbulent current', *Doklady Ak. Nauk. SSSR* **172**, 554–557.

Gurvich, A. S. and Zubkovsky, S. L., 1963: 'Experimental estimation of turbulent energy dissipation fluctuations', *Izv. AN SSSR*, ser. geofiz., 1856–1858.

Gurvich, A. S. and Zubkovsky, S. L., 1965: 'Measurement of the fourth and sixth correlation moments of a velocity gradient', *Atm. Ocean. Phys.* **1**, 460–463.

Gurvich, A. S. and Yaglom, A. M., 1967: 'Breakdown of eddies and probability distributions for small-scale turbulence', *Phys. Fluids* **10**, No. 9 (pt. 11), 59–65.

Hasselmann, K., 1967: 'A criterion for non-linear wave stability', *J. Fluid Mech.* **30**, 737–739.

Haurwitz, B. *et al.*, 1959, 'On the thermal unrest in the ocean'. In: *The Atmosphere and sea in motion*, Rossby Mem. Volume 4, 74–94.

Havelock, T. H.: 1931, 'The stability of motion of rectilinear vortices in rings', *London, Edin. and Dublin Philos. Mag.*, 1st ser., **11**, 617–633.

Hayes, S. P. *et al.*, 1975: 'Measurements of vertical finestructure in the Sargasso Sea', *J. Geophys. Res.* **80**, 314–319.

Heisenberg, W., 1948: 'Zur statistischen Theorie der Turbulenz', *Zs. Phys.* **124**, 628–657.

Herring, I. R. *et al.*, 1974: 'Decay of two-dimensional homogeneous turbulence', *J. Fluid. Mech.* **66**, 417–444.

Herring, I. R. *et al.*, 1973: 'Growth of uncertainty in decaying isotropic turbulence', *J. Atm. Sci.* **30**, 997–1006.

Hikosaka, S. and Higano, R., 1959: 'Water fluctuations off Muroto', *Shikoku-Records of Oceanogr.*, Works on Japan, No. 3.

Holloway, G. and Hendershot, M. C., 1977: ' Stochastic closure for non-linear Rossby waves', *J. Fluid Mech.* **82**, 747–765.

Hopf, F., 1942: 'Abweichung einer periodischen Lösung von einer stationären Lösung einer Differential-System', *Ber.-Sächs. Akad. Wiss.* **94**, 3–22.

Hopf, E., 1948: ' A mathematical example displaying features of turbulence', *Commun. Pure Appl. Math* **1**, 303–322.

Horn, L. and Bryson, R., 1963: 'An analysis of the geostrophic kinetic energy spectrum of large-scale atmospheric turbulence', *J. Geophys. Res.* **68**, 1059–1064.

Howard, L. N., 1961: 'Note on a paper of John W. Miles', *J. Fluid. Mech.* **10**, 509–512.

Howells, J., 1060: 'An approximate equation for the spectrum of a conserved scalar quantity in a turbulent fluid', *J. Fluid Mech.* **9**, 104–106.

Ichiye, T., 1951: 'Theory of oceanic turbulence', *Oceanogr. Mag.* **3**, 83–89.

Inoue, E., 1952: 'Turbulent fluctuations in temperature in the atmosphere and oceans', *J. Meteorol. Soc. Japan* **30**, 289–295.

Inoue, E., 1960: 'An interpretation of turbulence measurements in tidal currents near the seabed', *Proc. 10th Japan Nat. Congress Appl. Mech.*, pp. 247–250.

Ivanov, V. N. *et al.*, 1968: 'On the structure of the temperature field in the upper ocean', *Doklady Ak. Nauk SSSR* **183**, 1304–1307.

Ivanov, Yu. A. *et al.*, 1974: 'Internal gravitation waves in the ocean'. In: *Studies on Variability of Hydrophysical Fields in the Ocean*, Moscow, Nauka, pp. 91–98.

Joyce, G. and Montgomery, D., 1972: 'Simulation of "negative temperature" in the stability of line vortices', *Phys. Lett.* **39**, 371–372.

Joyce, G. and Montgomery, D., 1973: 'Negative temperature states for a two-dimensional guiding-center plasma', *J. Plasma Phys.* **10**, 107–121.

Julian, P. *et al.*, 1970: 'On the spectral distribution of large-scale atmospheric kinetic energy', *J. Atmos. Sci.* **27**, 376–387.

Kamenkovich, V. M., 1967: 'On the coefficient of eddy diffusion and eddy viscosity in large-scale oceanic and atmospheric motions', *Atm. Ocean. Phys.* **3**, 777–781.

Kamenkovich, V. M., 1973: *Principles of ocean dynamics*, Leningrad, Gidrometeoizdat.

Kamenkovich, V. M. and Monin, A. S. (eds.), 1978: *Oceanology*, Vol. 1, *Hydrophysics in the ocean*, Moscow, Nauka.

Kao, S. K. and Wendell L., 1970: 'The kinetic energy of the large-scale atmospheric motion in wave-number-frequency space', *J. Atmos. Sci.* **27**, 359–375.

Kao, T. W., 1976: 'Principal stage of wake collapse in a stratified fluid. Two-dimensional theory', *Phys. Fluids* **19**, 1071–1074.

Karabasheva, E. I. *et al.*, 1976: 'Small-scale turbulence in the Norwegian Sea and in the Gulf Stream', *Oceanology* **15**, 27–30.

Kazansky, A. B. and Monin, A. S., 1960: 'On turbulence conditions above the near-surface air layer', *Izv. Ak. Nauk SSSR*, ser. geofiz., 165–168.

Kazansky, A. B. and Monin, A. S., 1961: 'On dynamic interactions between the atmosphere and the Earth's surface', *Izv. Ak. Nauk SSSR*, ser. geofiz., 786–788.

Keller, B. S. and Yaglom, A. M., 1970: 'Effects of energy dissipation on the shape of the turbulence spectrum in the short-wave range', *Izv. Ak. Nauk SSSR, Liquid and Gas Mechanics*, 70–79.

Kenyon, K., 1967: 'Discussion', *Proc. Roy. Soc.* **A299**, No. 1457, pp. 141–144.

Kitaigorodsky, S. A., 1960: 'Calculations of the depth of the wind-mixed layer in the ocean, *Izv. Ak. Nauk SSSR*, ser. geofiz., 341–425.

Kolesnikov, A. G., 1959: 'Some results of direct measurements of vertical turbulent exchange in a sea'. In: *Some problems and data on oceanological studies*, pp. 20–28.

Kolesnikov, A. G., 1960: 'Vertical turbulent exchange in a stratified sea', *Izv. Ak. Nauk SSSR*, ser. geofiz., 1614–1623.

Kolesnikov, A. G. (ed.), 1970: *Modern facilities for oceanic studies*, Sebastopol, MGI AN USSR.

Kolesnikov, A. G. *et al.*, 1969: 'Progress in facilities for hydrophysical studies'. In: *Techniques and equipment for hydrophysical studies*, Kiev, Naukova Dumka.
Kolesnikov, A. G. and Panteleyev, N. A., 1975: 'Studies of small-scale turbulence structure in the ocean'. In: *Studies on the structure of turbulence in the ocean*, Sebastopol.
Kolesnikov, A. G. *et al.*, 1972: 'An automated system for the study of turbulent processes in the ocean'. In: *Automation of scientific research of seas and oceans*, Sebastopol. MGI, AN USSR, pp. 24–34.
Kolesnikov, A. G. *et al.*, 1963: 'A deep-water autonomous turbulence meter as a device for detecting turbulent fluctuations of velocity and temperature in the ocean', *Oceanology* **3**, 911–923.
Kolesnikov, A. G. *et al.*, 1958: 'Instruments and techniques for detecting small-scale turbulent fluctuations of temperature and velocity in the sea', *Izv. Ak. Nauk SSSR*, ser. geofiz., 405–413.
Kolesnikov, A. G. *et al.*, 1966: 'An automatic shipboard system for measuring basic hydrophysical variables in the ocean'. In: *Papers of the 2nd Congress on Oceanology*, Kiev, Naukova Dumka, pp. 126–131.
Kolesnikov, A. G. *et al.*, 1959: 'Instruments for turbulence studies in natural waters'. In: *New methods and facilities for the investigation of fluvial processes*, Moscow, Ak. Nauk SSSR, pp. 180–194.
Kolmogorov, A. N. 1941: 'On the logarithmically normal particle size distribution caused by particle crushing', *Doklady Ak. Nauk SSSR* **31**, 99–102.
Kolmogorov, A. N., 1941b: 'Local turbulent structure in incompressible fluids at very high Reynolds number', *Doklady Ak. Nauk SSSR* **30**, 299–303.
Kolmogorov, A. N., 1962a: 'Mécanique de la turbulence', *Coll. Intern. du CNRS à Marseille*, Paris, CNRS, pp. 447–458.
Kolmogorov, A. N., 1962b: 'A refinement of previous hypotheses concerning the local structure of turbulence in a viscous incompressible fluid at high Reynolds number', *J. Fluid Mech.* **13**, 82–85.
Kolmogorov, A. N. *et al.*, 1971: 'Experimental studies of turbulence in the open ocean'. In: *The world ocean*, Proc. Joint Oceanographic Assembly, Tokyo, p. 197.
Kontoboitseva, N. V., 1958: 'Temperature fluctuation measurements in the upper sea', *Izv. AK. Nauk SSSR*, ser. geofiz., 86–92.
Kontoboitseva, N. V., 1962: 'On investigations of some turbulence characteristics in disturbances', *Vestnik Moskovskogo Univ.*, ser. fiz., astronom., 35–41.
Korchashkin, N. N., 1970: 'The effect of fluctuations of energy dissipation and temperature dissipation on locally isotropic turbulent fields', *Atm. Ocean. Phys.* **6**, 560–561.
Korchashkin, N. N., 1976: 'Statistical characteristics of the vertical fine structure of hydrophysical fields in the ocean', *Oceanology* **16**, 343–346.
Korchashkin, N. N., 1977: 'Distribution laws of fine structure characteristics of oceanic hydrophysical fields'. COMECON Coordination Centre for Studies of Chemical, Physical, Biological and other Processes in the most Important Ocean Areas, Inform. Bull. No. 5, Moscow, pp. 57–63.
Korchashkin, N. N. and Ozmidov, R. V., 1973: 'Some problems of investigating the fine structure of hydrophysical fields in the ocean with probing devices'. In: *Studies of Oceanic Turbulence*, Moscow, Nauka, pp. 154–160.
Koshliakov, M. N. and Sorokhtin, O. G., 1972: 'The recording and workup of turbulence measurements in the ocean', *Atm. Ocean. Phys.* **8**, 36–41.
Koshliakov, M. N. and Monin, A. S., 1978: 'Synoptic eddies in the ocean', *Ann. Rev. Earth Planet. Sci.* **6**, 495–523.
Kraichnan, R. H., 1967: 'Inertial ranges in two-dimensional turbulence', *Phys. Fluids* **10**, 1417–1428.
Kraichnan, R. H., 1971: 'Inertial-range transfer in two- and three-dimensional turbulence', *J. Fluid Mech.* **47**, 525–535.
Kraichnan, R. H., 1974a: 'Convection of passive scalar by a quasi-uniform random straining field', *J. Fluid Mech.* **64**, 737–762.
Kraichnan, R. H., 1974b: 'On Kolmogorov's inertial range theories', *J. Flud Mech.* **62**, 305–330.
Kraichnan, R. H., 1975: 'Statistical dynamics of two-dimensional flow', *J. Fluid Mech.* **67**, 155–175.
Krishnamurti, R., 1973: 'Some further studies on the transition to turbulent convection', *J. Fluid Mech.* **60**, 285–303.
Kuznetsov, M. I., 1974: 'Laser doppler meter of liquid flow velocity', *Trudy Vsesoyuznogo. nauchno-issledov. Inst. Fiz. Tekhnich i Radiotekhnich. Izmerenii* **14**(44), pp. 47–49.

Kushnir, V. M. and Paramonov, A. N., 1972: 'A hydrophysical adaptive measuring system'. In: *Automation of scientific research of seas and oceans*, Sebastopol, MGI Ak. Nauk USSR, pp. 213–219.

Landau, L. D., 1944: 'On the problem of turbulence, *Doklady Ak. Nauk SSSR* **44**, 339–342.

Landau, L. D. and Lifshitz, M. E., 1953: *The mechanics of continuous media*, Moscow, Gostehizdat.

Lee, T. D., 1952: 'On some statistical properties of hydrodynamical and magneto-hydrodynamical fields', *Quart. Appl. Math.* **10**, 69–74.

Leith, C. E., 1968: 'Diffusion approximation for two-dimensional turbulence', *Phys. Fluids* **11**, 671–673.

Lemmin, U. *et al.*, 1974: 'The development from two-dimensional to three-dimensional turbulence generated by breaking waves', *J. Geophys. Res.* 79, 3442–3448.

Leroy, C. C., 1969, 'Development of simple equations for accurate and more realistic calculation of the speed of sound in seawater', *J. Acoust. Soc.* **46**, 216–226.

Liberman, L., 1951: 'The effect of temperature inhomogeneities in the ocean on the propagation of sound', *J. Acoust. Soc.* **23**.

Lilly, D. K., 1969, 'Numerical simulation of two-dimensional turbulence', *J. Phys. Fluids* **12**, 240–249.

Lilly, D. K., 1971: 'Numerical simulation of developing and decaying two-dimensional turbulence', *J. Fluid Mech.* **45**, 395–415.

Lilly, D. K., 1972a: 'Numerical simulation studies of two-dimensional turbulence. Part I', *Geophys. Fluid Dyn.* **3**, 289–319.

Lilly, D. K., 1972b: 'Numerical simulation studies of two-dimensional turbulence. Part II', *Geophys. Fluid Dyn.* **4**, 1–28.

Lin, C. C., 1943: *On the motion of vortices in two dimensions*. Toronto, Univ. Toronto Press.

Long, R. R., 1969: 'A theory of turbulence in stratified fluids', *J. Fluid Mech.* **42**, 349–365.

Longuet-Higgins, M. S., 1969: 'On wave breaking and the equilibrium spectrum of wind-generated waves', *Proc. Roy. Soc.* **A310**, No. 1501, pp. 151–159.

Longuet-Higgins, M. S. and Gill, A. E., 1967: 'Resonant interactions between planetary waves', *Proc. Roy. Soc.* **A299**, No. 1456, pp. 120–140.

Lorenz, E. N., 1953: 'The interaction between a mean flow and random disturbances', *Tellus* **5**, 238–250.

Lorenz, E. N., 1960: 'Maximum simplification of the dynamic equations', *Tellus* **12**, 243–254.

Lorenz, E. N., 1963: 'The mechanics of vacillation', *J. Atmosph. Sci.* **20**, 448–464.

Lorenz, E. N., 1972: 'Barotropic instability of Rossby wave motion', *J. Atmosph. Sci.* **29**, 258–264.

Lozovatsky, I. D., 1974: 'On the budget of turbulent energy in the ocean'. In: *Studies on the variability of hydrological fields in the ocean*, Moscow, Nauka, pp. 71–75.

Lozovatsky, I. D. *et al.*, 1977: 'Bottom turbulence in stratified enclosed seas'. In: *Bottom turbulence*, C. J. Nihoul (ed.), Amsterdam, Elsevier, pp. 49–57.

Lozovatsky, I. D., 1977: 'Investigation of small-scale temperature inhomogeneities in the southern part of the Baltic Sea', *Oceanology* **17**, 135–138.

Lozovatsky, I. D., 1978: 'On the effect of the fine structure of the temperature field on the internal wave spectra in the sea', *Atm. Ocean. Phys.* **14**, 86–87.

Liubimtsev, M. M., 1976: 'Structure of kinetic energy dissipation in the ocean', *Atm. Ocean. Phys.* **12**, 763–764.

Ludlam, F. A., 1967: 'Characteristics of billow clouds and their relation to clear-air turbulence', *Quart. J. Roy. Meteor. Soc.* **93**, 419–435.

Lumley, J. L., 1964: 'The spectrum of nearly inertial turbulence in a stably stratified fluid', *J. Atmosph. Sci.* **21**, 99–102.

Mamayev, O. I., 1964: 'Simplified dependence between density, temperature, and salinity of sea water', *Izv. Ak. Nauk SSSR*, ser. geofiz., 309–311.

Manabe, S. *et al.*, 1970: 'Simulated climatology of a general circulation model with a hydrological cycle. Part III. Effect of increased horizontal computational resolution', *Mon. Wea. Rev.* **96**, 175–212.

McIntyre, M. E., 1969: 'Diffusive destabilization of the baroclinic circular vortex', *Geophys. Fluid Dyn.* **1**, 19–58.

McLanghlin, J. B. and Martin, P. C.,, 1975: 'Transition to turbulence in a statistically forced fluid system', *Phys. Rev.*, Ser. A, **12**, 186–203.

Miles, J. M., 1961: 'On the stability of a heterogeneous shear flow', *J. Fluid. Mech.* **10**, 496–508.

Mizinov, P. I., 1965: 'On statistical characteristics of sea flows under various synoptic conditions', *Trudy gosudarstvennogo okeanogr. Inst.*, pp. 76–83.

Mirabel, A. V., 1974: 'A numerical model of the evolution of the energy spectra and concentration spectra of a passive additive in two-dimensional turbulence', *Atm. Ocean. Phys.* **10**, 128–131.

Miropolsky, Yu. Z. and Filiushkin, B. N., 1971: 'Temperature fluctuations in the upper ocean at scales comparable to internal gravity waves', *Atm. Ocean. Phys.* 7, 523–535.

Moller, A. B. and Riste, T., 1975: 'Neutron-scattering study of transitions to convection and turbulence in nematic para-azoxyanisole', *Phys. Rev. Lett.* **34**, 996–999.

Monin, A. S., 1950: 'Dynamic turbulence in the atmosphere', *Izv. Ak. Nauk SSSR*, ser. geogr. geofiz. **14**, 232–254.

Monin, A. S., 1956a: 'On large-scale turbulence exchange in the Earth's atmosphere', *Izv. Ak. Nauk SSSR*, ser. geofiz. 4, 452–463.

Monin, A. S., 1956b: 'On turbulent diffusion in the atmospheric near-surface layer', *Izv. Ak. Nauk SSSR*, ser. geofiz. **12**, 1461–1473.

Monin, A. S., 1962: 'On turbulence spectra in a temperature-inhomogeneous atmosphere', *Izv. Ak. Nauk SSSR*, ser. geofiz. 3, 397–407.

Monin, A. S., 1965a: 'On the symmetry properties of turbulence in the near-surface layer of air', *Atm. Ocean. Phys.* **1**, 25–30.

Monin, A. S., 1965b: 'Structure of an atmospheric boundary layer', *Atm. Ocean. Phys.* **1**, 153–157.

Monin, A. S., 1965c: 'On the atmospheric boundary layer with inhomogeneous temperature', *Atm. Ocean. Phys.* **1**, 287–292.

Monin, A. S., 1967a: 'Equations of turbulent motion', *Prikl. Mat. Mekh.* **31**, 1057–1068.

Monin, A. S., 1967b: 'Equations for finite-dimensional probability distributions of turbulence', *Doklady Ak. Nauk SSSR* **177**, 1036–1038.

Monin, A. S., 1969a: 'Oceanic turbulence' (Material from the Intern. Symp., Vancouver, Canada, 11–14 June, 1968), *Atm. Ocean. Phys.* **5**, 120–125.

Monin, A. S., 1969b: *Weather Forecasting as a Problem in Physics*, Moscow, Nauka. Also: Cambridge, MIT Press, 1975.

Monin, A. S., 1970a: 'Main features of turbulence in the sea', *Oceanology* **10**, 184–189.

Monin, A. S., 1970b: 'On turbulent flows in the ocean', *Doklady Ak. Nauk. SSSR* **193**, 1038–1040.

Monin, A. S., 1973a: 'Turbulence and microstructure in the ocean', *Uspekhi Fiz. Nauk* **109**, 333–354.

Monin, A. S., 1973b: 'The hydrothermodynamics of the ocean', *Atm. Ocean. Phys.* **9**, 602–604.

Monin, A. S., 1977: 'On the generation of oceanic turbulence', *Atm. Ocean. Phys.* **13**, 798–803.

Monin, A. S., 1980: 'Small-scale three-dimensional turbulence'. In: *Turbulence in the ocean*, Heidelberg, Springer-Verlag.

Monin, A. S. *et al.*, 1974: *Variability of the world ocean*, Leningrad, Gidrometeoizdat.

Monin, A. S. *et al.*, 1966: 'Problems of physical oceanography discussed at the 2nd International Oceanographic Congress', *Atm. Ocean. Phys.* **2**, 749–753.

Monin, A. S. *et al.*, 1970: 'On density stratification in the ocean', *Doklady Ak. Nauk SSSR* **191**, 1277–1279.

Monin, A. S. and Obukhov, A. M., 1953: 'Dimensionless characteristics of turbulence in the atmospheric near-surface layer', *Doklady Ak. Nauk SSSR* **93**, 257–260.

Monin, A. S. and Obukhov, A. M., 1954: 'Basic turbulent mixing laws in the atmospheric near-surface layer', *Trudy Geofiz. Instit. Ak. Nauk SSSR* **24**(151), 163–187.

Monin, A. S. *et al.*, 1973: 'On the vertical meso- and microstructure of flows in the ocean', *Doklady Ak. Nauk SSSR* **208**, 833–836.

Monin, A. S. and Yaglom, A. M., 1965: *Statistical Hydromechanics*, Part 1, Moscow, Nauka.

Monin, A. S. and Yaglom, A. M., 1967: *Statistical Hydromechanics*, Part 2, Moscow, Nauka.

Monin, A. S. and Yaglom, A. M., 1971: *Statistical Fluid Mechanics*, Cambridge, MIT Press, vol. 1.

Monin, A. S. and Yaglom, A. M., 1975: *Statistical Fluid Mechanics*, Cambridge, MIT Press, vol. 2.

Montgomery, D., 1972: 'Two-dimensional vortex motion and "negative temperature"', *Phys. Lett.* **A39**, 7–8.
Montgomery, D. and Joyce, G., 1974: 'Statistical mechanics of "negative temperature" states', *Phys. Fluids* **17**, 1139–1145.
Montgomery, D. *et al.*, 1972: 'Three-dimensional plasma diffusion in a very strong magnetic field', *Phys. Fluids* **15**, 815–819.
Montgomery, D. and Tappert, F., 1972: 'Conductivity of a two-dimensional guiding center plasma', *Phys. Fluids* **15**, 683–687.
Montgomery, R. B., 1939: 'Ein Versuch der verticalen und seitlichen Austausch in der Tiefe der Sprung-schicht im äquatorialer Atlantischen Ozean zu bestimmen', *Ann. Hydrogr. Marit. Meteorol.* 67, 242–246.
Morikawa, G. K. and Swenson, E. V., 1971: 'Interacting motion of rectilinear geostrophic vortices', *Phys. Fluids* **14**, 1058–1073.
Morozov, E. G., 1974, 'Experimental study on the breaking of internal waves', *Oceanology* **14**, 17–20.
Moroshkin, K. V., 1948: 'On energy dissipation in the Baltic Sea', *Meteorologiya i Gidrologiya*, No. 3, 41–51.
Moroshkin, K. V., 1951: 'On the variability of energy dissipation in stationary wind flows in the Baltic Sea', *Trudy Inst. Okean. Ak. Nauk* **6**, 49–58.
Munk, W. H. *et al.*, 1949: 'Diffusion in Bikini Atoll', *Trans. Amer. Geophys. Union* **30**, 59–66.
Munk, W. H. and Wimbush, M., 1971: 'The benthic boundary layer'. In: *The Sea*, New York, Interscience, Vol. 4(1), pp. 731–758.
Murty, G. S. and Sancara Rao, K., 1970: 'Numerical study of the behavior of a system of parallel line vortices', *J. Fluid Mech.* **40**, 595–602.
Nabatov, V. N. *et al.*, 1974: 'On the application of hot-wire anemometers to a flow with temperature fluctuations'. In: *Studies on the variability of hydrophysical fields in the ocean*, Moscow, Nauka, pp. 162–174.
Nan'niti, T., 1956: 'On the structure of ocean currents', *Papers in Meteorol. and Geoph.* 7, 161–170.
Nan'niti, T., 1957: 'Eulerian correlation of temperature in the atmosphere and oceans', *Papers in Meteor. and Geoph.* 8, 236–240.
Nan'niti, T., 1962: 'A study of the temperature of the surface layer in the Pacific Ocean near Japan', *Papers in Meteor. and Geoph.* **13**, 196–205.
Nan'niti, T., 1964a: 'Some observed results of oceanic turbulence', *Studies in Oceanography*, Japan, pp. 211–215.
Nan'niti, T., 1964b: 'Oceanic turbulence', *Oceanogr. Mag.* **16**, 35–46.
Nan'niti, T. and Yasui, M., 1957: 'Fluctuations of water temperature off Turumi, Yakohama', *J. Oceanogr. Soc. Japan* **13**, (1).
Nasmyth, P. W., 1972: 'Some observations of turbulence in the upper layer of the ocean', *Rapports et proces-verbaux des reunions* **162**, 19–24.
Nasmyth, P. W., 1973: 'Turbulence and microstructure in the upper ocean', *Mem. Soc. Roy. Sci. Liege*, Series 6, **4**, 47–56.
Neal, V. T., Neshyba, S. J., and Denner, W. W., 1969: 'Thermal stratification in the Arctic Ocean', *Science* **166**, 373–374.
Nemchenko, V. I., 1962a: 'Apparatus for temperature mesurements at various levels in the upper layer of the sea'. *Trudy Morskogo Gigrofiz. Instit.* **26**, 78–81.
Nemchenko, V. I., 1962b: 'A device for detecting temperature distribution by depth in a sea', *Oceanology* **2**, 727–730.
Nemchenko, V. I. and Tishunina, V. I., 1963: 'Studies on the thermal structure of the upper North Atlantic Ocean. Experiments in the IGY Program', *Okeanologicheskiye Issledovaniya*, No. 8, 97–103.
Nihoul, J. C. J. (ed.), 1977: *Bottom Turbulence*, Amsterdam, Elsevier.
Novikov, E. A., 1961: 'On the energy spectrum of a turbulent flow of incompressible fluids', *Doklady Ak. Nauk SSSR* **139**, 331–334.
Novikov, E. A., 1965: 'High-order correlations in turbulent flow', *Atm. Ocean. Phys.* **1**, 455–459.

Novikov, E. A., 1966: 'Mathematical simulation of the intermittency of a turbulent flow', *Doklady Ak. Nauk SSSR* **168**, 1279–1282.

Novikov, E. A., 1969: 'Scale similarity for random fields', *Doklady Ak. Nauk SSSR* **184**, 1072–1075.

Novikov, E. A., 1971: 'Intermittency and scale similarity in the structure of turbulent flow', *Prikl. Matem. Mekhan.* **35**, 266–277.

Novikov, E. A., 1974: 'Statistical irreversibility of turbulence and spectral energy transfer'. In: *Turbulent flows*, Moscow, Nauka, pp. 85–94.

Novikov, E. A., 1975: 'Dynamics and statistics of a system of vortices', *Zhurnal Eksp. Teor. Fiz.* **68**, 1868–1882.

Novikov, E. A. and Stewart, R. W., 1964: 'Turbulence intermittency and spectra of energy dissipation fluctuations', *Izv. Ak. Nauk SSSR*, ser. geofiz., No. 3, 409–413.

Nozdrin, Yu. V., 1974: 'Influence of buoyant forces on the spectra of turbulent processes in the ocean', *Oceanology* **14**, 647–651.

Nozdrin, Yu. V., 1975: 'Fine structure of small-scale ocean turbulence', Cand. Thesis, Inst. of Oceanology, USSR Academy of Sciences, Moscow.

Noss, J., 1957: 'Turbulence effects on the speed of sound in the ocean'. In: *Problems of modern physics*, No. 11, pp. 82–88.

Obukhov, A. M., 1949: 'Structure of temperature field in a turbulent flow', *Izv. Ak. Nauk SSSR*, ser. geofiz., **13**(1), 58–69.

Obukhov, A. M., 1959: 'Buoyancy force effects on the temperature structure in a turbulent flow', *Doklady Ak. Nauk SSSR* **125**, 1246–1248.

Obukhov, A. M., 1960: 'On integral invariants in systems of the hydrodynamic type', *Doklady Ak. Nauk SSSR* **184**, 309–317.

Obukhov, A. M., 1962a: 'On stratified fluid dynamics', *Doklady Ak. Nauk SSSR* **145**, 1239–1242.

Obukhov, A. M., 1962b: 'Some specific features of atmospheric turbulence', *J. Geoph. Res.* **67**, 3011–3014.

Obukhov, A. M., 1962c: 'Some specific features of atmospheric turbulence', *J. Fluid Mech.* **13**, 77–81.

Ogura, Y., 1952: 'The theory of turbulent diffusion in the atmosphere', *J. Meteor. Soc. Japan* **30**, 53–58.

Ogura, Y., 1958: 'On the isotropy of large-scale disturbances in the upper troposphere', *J. Meteor.* **15**, 376–382.

Ogura, Y., 1962: 'Energy transfer in a normally distributed and isotropic turbulent velocity field in two-dimensions', *Phys. Fluids* **5**, 395–401.

Okubo, A. and Ozmidov, R. V., 1970: 'Empirical dependence of the coefficient of horizontal turbulent diffusion in the ocean on the scale of the phenomenon in question', *Oceanology* **6**, 308–309.

Onsager, L., 1949: 'Statistical hydrodynamics', *Nuovo Cimento Sippl.* **6**, 279–287.

Orlansky, J. and Bryan, K., 1969: 'Formation of a thermocline step structure by large-amplitude internal waves', *J. Geoph. Res.* **74**, 6975–6983.

Orszag, S. A., 1970: 'Indeterminacy of the moment problem for intermittent turbulence', *Phys. Fluids* **13**, 2211–2212.

Orszag, S. A., 1971: 'Numerical simulation of incompressible flows with simple boundaries: accuracy', *J. Fluid Mech.* **49**, 75–112.

Orszag, S. A. and Patterson, G. S., 1972: 'Numerical simulation of three-dimensional homogeneous isotropic turbulence', *Phys. Rev. Lett.* **28**, 76–79.

Osborn, T. R., 1974: 'Vertical profiling of velocity microstructure', *J. Phys. Oceanogr.* **4**, 109–115.

Osborn, T. R. and Cox, C. S., 1972: 'Oceanic finestructure', *Geoph. Fluid Dynam.* **3**, 321–345.

Ovsiannikova, O. A., 1965: 'Forecast of non-periodic flows in the sea based on the method of linear extrapolation of random processes', *Trudy Gosud. Okeanogr. Inst.*, No. 85, pp. 173–184.

Ozmidov, R. V., 1957: 'Experiments on horizontal turbulent diffusion in a sea and in an artificial basic of moderate depth', *Izv. Ak. Nauk SSSR*, ser. geofiz., **6**, 756–764.

Ozmidov, R. V., 1958: 'On calculating the horizontal turbulent diffusion of contaminant spots in a sea', *Dokl. Ak. Nauk SSSR* **120**, 761–763.

Ozmidov, R. V., 1959: 'Studies on medium-scale horizontal turbulent exchange in the ocean using the method of radar observations of floating buoys', *Dokl. Ak. Nauk SSSR* **126**, 63–65.

Ozmidov, R. V., 1960: 'On the turbulent energy dissipation rate in the sea and the dimensionless universal constant in the 4/3 power law', *Izv. Ak. Nauk SSSR*, ser. geofiz. 8, 1234–1237.

Ozmidov, R. V., 1961: 'Theory of turbulence and turbulence in the sea', *Trudy Inst. Okeanol. Ak. Nauk* **52**, 97–114.

Ozmidov, R. V., 1962: 'Statistical characteristics of horizontal large-scale turbulence in the Black Sea', *Trudy Inst. Okeanol. Ak. Nauk* **60**, 114–129.

Ozmidov, R. V., 1964: 'Some data the large-scale characteristics of the horizontal velocity component in the ocean', *Izv. Ak. Nauk SSSR*, ser. geofiz. **11**, 1708–1719.

Ozmidov, R. V., 1965a: 'Turbulence and mixing in the sea'. In: *Progress in Sciences, Geophysics, 1963*, Moscow, VINITI AN SSSR, pp. 242–258.

Ozmidov, R. V., 1965b: 'Energy distribution between oceanic motions of different scales', *Atm. Ocean. Phys.* **1**, 257–261.

Ozmidov, R. V., 1965c: 'On the turbulent exchange in a stably stratified ocean', *Atm. Ocean. Phys.* **1**, 493–497.

Ozmidov, R. V., 1965d: 'On some features of energy spectra of oceanic turbulence', *Doklady Ak. Nauk SSSR* **161**, 828–831.

Ozmidov, R. V., 1966: 'The scales of oceanic turbulence', *Oceanology* **6**, 325–328.

Ozmidov, R. V., 1967: 'Turbulence and turbulent mixing in the ocean', *Oceanology* **7**, 665–670.

Ozmidov, R. V., 1968: *Horizontal turbulence and turbulent exchange in the ocean*, Moscow, Nauka.

Ozmidov, R. V., 1969: 'Turbulence and mixing in the ocean'. In: *Progress in sciences, Geophysics (1968)* Moscow, VINITI *Ak. Nauk SSSR*, pp. 125–156.

Ozmidov, R. V., 1971a: 'The second expedition of the scientific research ship 'Dmitriy Mendeleyev' ', *Oceanology* **II**, 173–177.

Ozmidov, R. V., 1971b: 'Studies of small-scale ocean turbulence. The ninth expedition of the 'Akademicik Kurchatov' ', *Vestnik Ak. Nauk SSSR* **9**, 60–66.

Ozmidov, R. V., 1973a: 'Experimental studies on small-scale ocean turbulence carried out in the P. P. Shirshov Institute of Oceanology, USSR Academy of Sciences'. In: *Studies of ocean turbulence*, Moscow, Nauka, pp. 3–19.

Ozmidov, R. V. (ed.), 1973b: *Studies on oceanic turbulence*, Moscow, Nauka.

Ozmidov, R. V. (ed.), 1974a: *Studies on the variability of hydrophysical fields in the ocean*, Moscow, Nauka.

Ozmidov, R. V., 1974b: 'The eleventh cruise of the R/V 'Dmitriy Mendeleyev' ', *Okeanologiya* **14**, 779–783.

Ozmidov, R. V., 1975: 'The first All-Union Symposium on the study of Small-Scale Ocean Turbulence', *Oceanology* **15**, 385–388.

Ozmidov, R. V., 1977: 'Turbulence and mixing in the ocean'. In: *Progress in Sciences: Oceanology*, Vol. 4, Moscow, VINITY Ak. Nauk SSSR, pp. 88–111.

Ozmidov, R. V., 1978a: 'The nineteenth cruise of the 'Dmitriy Mendeleyev' ', *Oceanology* **18**, 370–373.

Ozmidov, R. V., 1978b: *Turbulence in the Upper Ocean*, VINITI SSSR, Progress in Sciences, Fluid and Gas Mechanics Series, Vol. 12, pp. 52–143.

Ozmidov, R. V., 1979: 'The twenty-second cruise of the 'Dmitriy Mendeleyev' ', *Oceanology* **19**, 627–630.

Ozmidov, R. V. and Beliayev, V. S., 1973: 'Some peculiarities of turbulence in stratified flows'. *Int. Symp. on Stratified Flows, Novosibirsk*, 1972, New York, Amer. Soc. of Civil Engineers, pp. 593–599.

Ozmidov, R. V. *et al.*, 1980: 'Measurement of the fine structure of hydrophysical fields and turbulence from the manned underwater vehicle Pisces', *Oceanology* **20**, 153–156.

Ozmidov, R. V. *et al.*, 1970: 'Some features of the turbulent energy transfer and the transformation of turbulent energy in the ocean', *Atm. Ocean. Phys.* **6**, 160–163.

Ozmidov, R. V. *et al.*, 1973: 'Experimental studies on the diffusion of contaminants introduced into the sea'. In: *Studies on oceanic turbulence*, Moscow, Nauka, pp. 64–78.

Ozmidov, R. V. and Fedorov, K. N., 'Turbulence in the ocean', (11th Int. Liege Colloq. on Ocean Hydrodyn. and 2nd Symp. on Turbulence in the Ocean, Liege, 1979), *Atm. Ocean. Phys.* **15**, 935–937.

Ozmidov, R. V. and Yampolsky, A. D., 1965: 'Some statistical characteristics of velocity and density fluctuations in the ocean', *Atm. Ocean. Phys.* **1**, 357–360.

Paka, V. T., 1972: 'Complex measurements of physical fields under towing conditions in the ocean'. In: *Automation of scientific research of seas and oceans*, Sebastopol, MGI Ak. Nauk USSR, pp. 82–86.

Paka, V. T., 1974: 'Technique of measurements with towed sensors'. In: *Hydrophysical and hydrostatic investigations in the Atlantic and Pacific Oceans*, Moscow, Nauka, pp. 9–14.

Palevich, L. G. *et al.*, 1973: 'Metrological preparation of turbulence meters for field experiments'. In: *Studies on oceanic turbulence*, Moscow, Nauka, pp. 161–168.

Palm, E. *et al.*, 1967: 'On the occurrence of cellular motion in Bénard convection', *J. Fluid Mech.* **30**, 651–661.

Panchev, S., 1975: 'On the existence of power-law relationships in oceanic turbulence spectra', *Atm. Ocean. Phys.* **11**, 381–383.

Panchev, S., 1976: 'Spectra of large-scale geophysical turbulence in inertial ranges', *Blgarsko Geofizichno Spisanie* **2**, 3–11.

Panteleyev, N. A., 1960: 'Turbulence studies in the upper layer of the Antarctic, Indian and Pacific Oceans', *Trudy Okeanogr. Komiss.* **10**, 133–140.

Pekeris, C. L. and Shkoller, B., 1967: 'Stability of plane Poiseuille flow to periodic disturbances of finite amplitude in the vicinity of the neutral point'. *J. Fluid Mech.* **29**, 31–38.

Phillips, O. M., 1967: *The Dynamics of the Upper Ocean*, London, Cambridge Univ. Press.

Phillips, O. M., 1971: 'On spectra measured in an undulating layered medium', *J. Phys. Oceanogr.* **1**, 1–6.

Piskunov, A. K., 1957: 'Some data on temperature fluctuations in the sea', *Trudy Taganrogsk. Radiotekhnich. Inst.* **3**(2), 27–31.

Plakhin, E. A. and Fedorov, K. N., 1972: 'Certain aspects of the variability of the temperature field in the mixing layer of water in the tropical Atlantic Ocean', *Atm. Ocean. Phys.* **8**, 369–375.

Pochapsky, T. E., 1963: 'Measurement of small-scale motions in the ocean with neutrally-buoyant floats', *Tellus* **15**, 352–362.

Pochapsky, T. E., 1966: 'Measurement of deep water movements with instrumented neutrally buoyant floats'. *J. Geoph. Res.* **71**, 2491–2505.

Pozdynin, V. D., 1974: 'Statistical estimates of the parameters of small-scale turbulence in the ocean'. In: *Studies on variability of hydrophysical fields in the ocean*, Moscow, Nauka, pp. 50–61.

Pozdynin, V. D., 1976: 'Some statistical patterns of small-scale ocean turbulence', *Oceanology* **16**, 453–456.

Reynolds, W. C. and Potter, M. C., 1967: 'Finite-amplitude instability of parallel shear flows', *J. Fluid Mech.* **27**, 465–482.

Rhines, P. B., 1971: 'A note on long-period motion at site D', *Deep-Sea Res.* **18**, 21–26.

Rhines, P. B., 1973: 'Observations of the energy-containing ocean eddies and theoretical models of waves and turbulence', *Boundary Layer Meteorol.* **4**, 345–360.

Rhines, P. B., 1975: 'Waves and turbulence on a β-plane', *J. Fluid Mech.* **69**, 417–443.

Rhines, P. B., 1977: 'The dynamics of unsteady currents'. In: *The Sea*, Vol. 6, pp. 189–318.

Rhines, P. B. and Bretherton, F. B., 1973: 'Topographic Rossby waves in a rough-bottomed ocean', *J. Fluid Mech.* **61**, 583–607.

Richardson, L. F., 1926: 'Atmospheric diffusion shown on a distance-neighbour graph', *Proc. Roy. Soc.* **A110**, No. 756, pp. 709–737.

Richardson, L. F. and Stommel, H., 1948: 'Note on eddy diffusion in the sea', *J. Meteorol.* **5**, 238–240.

Robinson, A. R., 1975: 'The variability of ocean currents', *Rev. Geophys. Space Phys.* **13**, 598–601.

Roden, G. L., 1963: 'On sea level, temperature, and salinity variations in the tropical Pacific Ocean and on Pacific Ocean Islands', *J. Geoph. Res.* **68**, 455–472.

Rossby, C. G., 1936: 'Dynamics of steady ocean currents in the light of experimental fluid mechanics', *Papers in Phys. Oceanogr. and Meteorol.* **5**, 43.

Rossby, C. G., 1939: 'Note on the shear stress caused by large-scale lateral mixing', *Proc. 5th Int. Congr. Appl. Mech.*, Cambridge.

Rossby, C. G., 1947: 'On the distribution of angular velocity in gaseous envelopes under the influence of large-scale horizontal mixing process', *Bull. Am. Met. Soc.* **28**, 53–68.

Rossby, A. T., 1969: 'A study on Bénard convection with and without rotation', *J. Fluid Mech.* **36**, 309–335.

Ruelle, D., 1975: 'Strange attractors as a mathematical explanation of turbulence', *Lect. Notes Physics.* **12**, 292–299.

Ruelle, D., 1976: 'The Lorenz attractor and the problem of turbulence'. In: *Turbulence and Navier-Stokes equations*, Springer–Verlag. Also: *Lect. Notes in Math.*, No. 565, pp. 146–158.

Ruelle, D. and Takens, F., 1971: 'On the nature of turbulence', *Communs. Math. Phys.* **20**, 167–192.

Saffman, P. G., 1971: 'On the spectrum and decay of random two-dimensional vorticity distributions at large Reynolds numbers', *Stud. Appl. Math.* **50**, 377–383.

Saint-Guily, B., 1956: 'Sur la théorie des courants marins induits par le vent', *Ann. Ins. Oceanogr.* **33**, 64.

Saltzman, B., 1962: 'Finite amplitude free convection as an initial value problem', *J. Atmosph. Sci.* **19**, 329–341.

Saltzman, B. and Fleicher, A., 1962: 'Spectral statistics of the wind at 500 mb', *J. Atmosph. Sci.* **19**, 195–204.

Sanford, T. B., 1975: 'Observations of the vertical structure of internal waves', *J. Geophys. Res.* **80**, 3861–3871.

Saunders, P. M., 1971: 'Anticyclonic eddies formed from shoreward meanders of the Gulf Stream', *Deep-Sea Res.* **18**, 1207–1219.

Saunders, P. M., 1972a: 'Comment on "wavenumber frequency spectra of temperature in the free atmosphere"', *J. Atmosph. Sci.* **29**, 197–199.

Saunders, P. M., 1972b: 'Space and time variability of temperature in the upper ocean', *Deep-Sea Res.* **19**, 467–480.

Schülter, A., Lortz, D., and Busse, F. A., 1965: 'On the stability of steady finite amplitude convection', *J. Fluid Mech.* **23**, 129–144.

Sedov, Yu. B., 1976: 'Evolution of the energy spectra of a system of eddies', *Izv. Ak. Nauk SSSR, Mekhanika Zhidkosti i Gaza*, No. 6, 43–48.

Seidler, G., 1974: 'The fine-structure contamination of vertical velocity spectra in the deep ocean', *Deep-Sea Res.* **21**, 37–46.

Seyler, C. E., 1974: 'Partition function for a two-dimensional plasma in the random-phase approximation', *Phys. Rev. Lett.* **32**, 515–517.

Seyler, C. E. *et al.*, 1975: 'Two-dimensional turbulence in inviscid fluids or guiding center plasures', *Phys. Fluids* **18**, 803–813.

Shevtsov, V. P. and Volkov, P. O., 1973: 'A method for measuring the vertical profiles of ocean currents from a ship', *Oceanology* **13**, 916–920.

Shishkov, Yu. A., 1974: 'Meteorological auxiliary observations during experimental turbulence studies in the ocean'. In: *Studies of the variability of hydrophysical fields in the ocean*, Moscow, Nauka, pp. 66–71.

Shtokman, V. B., 1940: 'On turbulent exchange in the south and middle Caspian Sea', *Izv. Ak. Nauk SSSR*, ser, geofiz., No. 4, 569–592.

Shtokman, V. B., 1941: 'On the fluctuations of horizontal velocity components in the sea due to large-scale turbulence', *Izv. Ak. Nauk SSSR*, ser. geogr. geofiz., No. 4–5, 476–486.

Shtokman, V. B., 1946: 'On the energy dissipation of stationary flows in the sea', *Trudy Inst. Okeanol. Ak. Nauk* **1**, 47–63.

Shtokman, V. B., 1947: 'On the energy dissipation of stationary flows excited by unsteady winds in a closed homogeneous basin', *Meteorol. i Gidrometeorol.*, No. 6, 53–64.

Shtokman, V. B. and Ivanovsky, I. I., 1937: 'Stationary studies of flows off the west shore of the middle Caspian Sea', *Meteorol. i Gidrolog.*, No. 4–5, 154–159.

Shtokman, V. B., *et al.*, 1969: 'Long-term measurements of the variability of physical parameters at ocean polygons as a new step in ocean investigations', *Dokl. Ak. Nauk SSSR* **186**, 1070–1073.

Shur, G. H., 1962: 'Experimental studies on the energy spectra of atmospheric turbulence', *Trudy Tsentralnaya Aerologicheskaya Obs.*, No. 43, pp. 79–90.

Simpson, J. H. and Woods, J. D., 1970: 'Temperature microstructure in a freshwater thermocline', *Nature* **226**, 832–834.

Speranskaya, A. A., 1960: 'Turbulent exchange in the layer below the ice in a water basin', *Trudy Okeanogr. Komiss.* **10**(1), 147–151.

Speranskaya, A. A., 1964: *Experimental studies on the turbulent exchange in fresh-water basins*, Cand. Thesis, MGU, Moscow.

Starr, V., 1953: 'Note concerning the nature of the large-scale eddies in the atmosphere', *Tellus* **5**, 494–498.

Starr, V., 1968: *Physics of negative viscosity phenomena*, McGraw Hill.

Stepanov, V. N., 1965: 'Basic types of water structure in the world ocean', *Oceanology* **5**, 777–786.

Stern, M., 1960: 'The "salt-fountain" and thermohaline convection', *Tellus* **12**, 172–175.

Stern, M. E. and Turner, J. S., 1969: 'Salt fingers and convective layers', *Deep-Sea Res.* **16**, 497–511.

Stevenson, R. E., 1974: 'Observation from Skylab of mesoscale turbulence in ocean currents', *Nature* **250**, 638–640.

Stevenson, R. E., 1975: 'Real-time observation of spectral ocean features from manned spacelab; related to pollution', *Proc. Interdisciplinary Symp. IUGG 16th Gen. Assemb.*, Grenoble, p. 171.

Stewart, R. W. and Grant, H. L., 1962: 'Determination of the rate of dissipation of turbulent energy near the sea surface in the presence of waves', *J. Geoph. Res.* **67**, 3177–3180.

Stewart, R. W. *et al.*, 1970: 'Some statistical properties of small-scale turbulence in an atmospheric boundary layer', *J. Fluid. Mech.* **41**, 141–152.

Stommel, H., 1949: 'Horizontal diffusion due to oceanic turbulence', *J. Mar. Res.* **8**, 199–225.

Stommel, H., 1962: 'Examples of mixing and self-exciting convection on the T–S-diagram', *Oceanology* **2**, 205–209.

Stommel, H. *et al.*, 1956: 'The oceanographic curiosity of the perpetual salt fountain', *Deep-Sea Res.* **3**, 152–153.

Stommel, H. and K. N. Fedorov, 1967: 'Small-scale structure in temperature and salinity near Timor and Mindanao', *Tellus* **19**, 306–325.

Stuart, J. T., 1971: 'Nonlinear stability theory', *Ann. Rev. Fluid Mech.* **3**, 347–370.

Suda, K., 1936: 'On the dissipation of energy in density currents', *Geophys. Mag.* **10**, 131–243.

Sverdrup, H. U., 1939: 'Lateral mixing in the deep water of the South Atlantic Ocean', *J. Mar. Res.* **2**, 195–207.

Swallow, J., 1969: 'A deep eddy off Cape St. Vincent, *Deep-Sea Res.* **16**, 285–296.

Swallow, J. and Worthington, L. V., 1961: 'An observation of deep countercurrents in the western North Atlantic Ocean', *Deep-Sea Res.* **8**, 1–19.

Tait, R. J. and Howe, M. R., 1971: 'Thermohaline staircase', *Nature* **231**, 178–179.

Taylor, G. I., 1918: 'Tidal friction in the Irish Sea', *Phil. Trans. Roy. Soc.* **A220**, 1–92.

Taylor, J. B., 1972: 'Negative temperatures in two-dimensional vortex motion', *Phys. Lett.* **A40**, 1–2.

Taylor, J. B. and McNamara, B., 1971: 'Plasma diffusion in two dimensionals', *Phys. Fluids* **14**, 1492–1499.

Tchen, C. M., 1953: 'On the spectrum of energy in turbulent shear flow', *J. Res. Nat. Bur. Standards* **50**, 51–62.

Tchen, C. M., 1954: 'Transport processes as foundation of the Heisenberg and Obukhov theories of turbulence', *Phys. Res.* **93**, 4–14.

Tchen, C. M., 1975: 'Cascade theory of turbulence in a stratified medium', *Tellus* **27**, 1–14.

Thompson, R., 1971: 'Topographic Rossby waves at a site north of the Gulf Stream', *Deep-Sea Res.* **18**, 1–19.

Thorpe, S. A., 1973: 'Turbulence in stably stratified fluids: a review of laboratory experiments', *Boundary Layer Meteor.* **5**, 95–119.

Trokhan, A. M. and Stefanov, S. P., 1974: 'Optical-acoustic methods of measuring turbulence characteristics'. In: *Meteorologiya v Gidrofizicheskikh Ismereniyakh*, Trudy Vsesyuznyi Nauchno-Issledovatelski Inst. Fiz. – Tech. i Radiotech. Izm., **14**(44), 40–46.

Turner, J. S., 1965: 'The coupled turbulent transport of salt and heat across a sharp density interface', *Int. J. Heat and Mass Transfer* **8**, 759–767.

Turner, J. S., 1967: 'Salt fingers across a density interface', *Deep-Sea Res.* **14**, 599–611.

Turner, J. S. and Stommel, H., 1964: 'A new case of convection in the presence of combined vertical salinity and temperature gradients', *Proc. U.S. Nat. Acad. Sci.*, Vol. 52, pp. 49–53.

Vahala, G., 1972: 'Transverse diffusion and conductivity in a magnetized equilibrium plasma', *Phys. Rev. Lett.* **29**, 93–95.

Vahala, G. and Montgomery, D., 1971: 'Kinetic theory of a two-dimensional magnetized plasma', *J. Plasma Phys.* **6**, 425–439.

Van Atta, C. W., 1971: 'Influence of fluctuations in local dissipation rates on turbulent scalar characteristics in an inertial subrange', *Phys. Fluids* **14**, 1803–1804.

Van Atta, C. W., 1973: 'On the moments of turbulent velocity derivatives for large Reynolds numbers, Technical Note F-49, Charles Kolling Res. Lab., Dept. Mech. Eng., University of Sydney, Australia.

Van Atta, C. W. and Chen, W. Y., 1970: 'Structure functions of turbulence in the atmospheric boundary layer over the ocean', *J. Fluid. Mech.* **44**, 145–159.

Van Atta, C. W. and Parc, J., 1972: 'Statistical self-similarity and inertial subrange turbulence'. In: *Statistical models and turbulence, Lecture Notes in Physics*, Springer Verlag, No. 12, pp. 402–426.

Van Atta, C. W., 1974: 'Influence of fluctuations in dissipation rates on some statistical properties of turbulent scalar fields', *Atm. Ocean Phys.* **10**, 443–446.

Van Atta, C. W. and Yeh, T. T., 1975: 'Evidence for similarity of internal intermittency in turbulent flows at large Reynolds numbers', *J. Fluid Mech.* **71**, 417–440.

Vasilenko, V. M. and Krivelevich, L. M., 1974: 'Statistic characteristics of flows from experimental data obtained from a polygon in the Atlantic Ocean'. In: *Studies on the variability of hydrodynamic fields in the ocean*, Moscow, Nauka, pp. 76–82.

Vasilenko, V. M. *et al.*, 1975: 'Fluctuations of the turbulent energy dissipation rate and of the high-order structure functions of the velocity field in the ocean', *Atm. Ocean Phys.* **11**, 580–584.

Vasilenko, V. M. *et al.*, 1976: 'Current velocity spectra and horizontal turbulent eddy viscosity in the Atlantic Ocean', *Oceanology* **16**, 27–29.

Volzhenkov, V. A. and Istoshin, Yu. V., 1965: 'Applications of spectral functions to studies on variability of oceanic characteristics', *Trudy Tsentralnyi Inst. Prognozov*, pp. 103–107.

Volkov, A. P. *et al.*, 1974: 'Two-coordinate Doppler flow-velocity meter'. In: *Hydrophysical and hydro-optical studies in the Atlantic and Pacific Oceans*, Moscow, Nauka, pp. 18–20.

Volochkov, A. G. and Korchashkin, N. N., 1977: 'Some results of a statistical analysis of the fine structure of the thermodynamic fields in the ocean', *Atm. Ocean Phys.* **13**, 146–147.

Vorobiov, V. P. *et al.*, 1973: 'Thermodynamic velocity fluctuations meter'. In: *Studies on Oceanic Turbulence*, Moscow, Nauka, pp. 149–153.

Vorobiov, V. P. and Palevich, L. G., 1974: 'On a detector for fine structure in the ocean'. In: *Studies on the variability of hydrophysical fields in the ocean*, Moscow, Nauka, pp. 155–161.

Webster, F., 1961: 'The effect of meanders on the kinetic energy balance of the Gulf Stream', *Tellus* **13**, 392–401.

Webster, F., 1965: 'Measurements of eddy fluxes of momentum in the surface layer of the Gulf Stream, *Tellus* **17**, 239–245.

Webster, F., 1966: 'Some measurements of the spectrum of oceanic fluctuations', *Proc. 2nd Int. Cong. Oceanology*, Moscow, Nauka.

Webster, F., 1969: 'Turbulence spectra in the ocean', *Deep-Sea Res.* **16**, 357–368.

Whitehead, J. A. 1971: 'Cellular convection', *Am. Sci.* **59**, 444–451.

Wiin–Nielsen, A., 1967: 'On the annual variation and spectral distribution of atmospheric energy', *Tellus* **19**, 540–559.

Williams, A. J., 1974: 'Salt fingers observed in the Strait of Gibraltar, *Science* **185**, 941–943.

Williams, R. W. and Gibson, C. H., 1974: 'Direct measurement of turbulence in the Pacific Counter Current', *J. Phys. Oceanogr.* **4**, 104–108.

Willis, G. E. and Deardorff, J. W., 1970: 'The oscillatory motions of Rayleigh convection', *J. Fluid Mech.* **44**, 661–672.

Wilson, W. D., 1960: 'Speed of sound in seawater as function of temperature, pressure, and salinity, *J. Acoust. Soc. Amer.* **32**, 641–644.

Woods Hole Oceanographic Institute: A compilation of moored current meter observations, Tech.

Report (unpublished MS), Woods Hole, U.S.A., 1965, Vol. 1; 1966, Vol. 2; 1967, Vol. 3; 1970, Vol. 4; 1971, Vol. 5; 1974, Vol. 6; 1974, Vol. 7; 1975, Vol. 8.

Woods, J. D., 1968a: 'An investigation of some physical processes associated with the vertical flow of heat through the upper ocean', *Meteor. Mag.* 97, 65–72.

Woods, J. D., 1968b: 'Diurnal behaviour of the summer thermocline off Malta', *Dtsch. Hydrogr. Z.* **21**, 106–108.

Woods, J. D., 1968c: 'Wave-induced shear instability in the summer thermocline', *J. Fluid Mech.* **32**, 791–800.

Woods, J. D., 1974: 'Diffusion due to fronts in the rotation subrange of turbulence in the seasonal thermocline', *Houille Blanche* **29**, 589–597.

Wu, J., 1969: 'Mixed region collapse with inertial wave generation in a density-stratified medium', *J. Fluid Mech.* **35**, 531–544.

Wunsh, C., 1972: 'The spectrum, from two years to two minutes, of temperature in the main thermocline offshore Bermuda, *Deep-Sea Res.* **19**, 577–593.

Wyrtki, K., 1967: 'The spectrum of ocean turbulence over distances between 40 and 1000 km', *Dtsch. Hydrogr. Z.* **20**, 176–186.

Yaglom, A. M., 1966: 'On the influence of fluctuations in energy dissipation on the turbulence characteristics in the inertial interval', *Doklady Ak. Nauk SSSR* **166**, 49–52.

Name Index

Subject Index